Health Informatics

For further volumes:
http://www.springer.com/series/1114

Alain Venot • Anita Burgun • Catherine Quantin
Editors

Medical Informatics, e-Health

Fundamentals and Applications

 Springer

Editors
Alain Venot
Université Paris 13
Paris
France

Anita Burgun
Université Paris 5
Paris
France

Catherine Quantin
Université de Bourgogne
Dijon
France

ISSN 1431-1917
ISBN 978-2-8178-0477-4   ISBN 978-2-8178-0478-1 (eBook)
DOI 10.1007/978-2-8178-0478-1
Springer Paris Heidelberg New York Dordrecht London

Library of Congress Control Number: 2013948360

Translation from the French language edition 'Champs Informatique Médicale, e-Santé – Fondements et applications' sous la direction de Alain Venot, avec la collaboration de Anita Burgun et Catherine Quantin, © Springer-Verlag France, Paris, 2013; ISBN: 978-2-8178-0337-1

Printed on acid-free paper

Springer is part of Springer Science+Business Media (www.springer.com)

Dedicated to François Grémy, successively Professor of Medical Informatics at the universities of Paris and Montpellier. As the first chairman and moderator of IFIP-TC4, François Grémy is widely held to be the first president of its renamed and refocused successor, the International Medical Informatics Association (IMIA). He played a major role in the development of medical informatics, particularly in France, and he received the first IMIA Medical Informatics Award of Excellence in 2004.

Preface

Medical informatics has matured into a true scientific discipline, with its own international conferences, journals, exhibitions, research laboratories, calls for proposals for national and international research funding, and master's courses for the training of students.

Both basic and applied aspects are taught, not only in all areas of health (medicine, dentistry, pharmacy, health and social sciences, nursing, public health) but also in many other domains, including life sciences, engineering and economics.

Medical informatics is a complex and rapidly changing discipline. Few books are published in this domain, and they rapidly become obsolete. This book is the fruit of collaboration between several authors, all of whom teach medical informatics in France and perform research in this field. It contains 18 chapters, all of which include learning objectives, recommendations for further reading and information retrieval, exercises and bibliographic references.

Chapter 1 presents the main areas of medical informatics and links to other scientific domains, such as computer sciences, biostatistics, biomedical engineering and public health. Key scientific societies, journals, conferences and exhibitions are listed.

Chapter 2 focuses on the broad terminological resources developed for health. These resources are an essential tool for the normalisation of concept expression and for information coding in health, both these operations being prerequisites to the use of health information.

Chapter 3 presents the various types of knowledge resources available in the domain of health, the principles underlying their indexing, the means developed for accessing these documents and quality criteria for health information.

Chapter 4 deals with the representation of patient data in health information systems. The problems posed by the computerisation of patient records are analysed, and ways of structuring and standardising data are presented.

In Chapters 5 and 6, the reader is introduced to the principles of medical image processing and analysis. These chapters provide an understanding of the ways in which these methods and techniques can improve interventions, providing surgeons with the possibility of using robots and achieving higher-quality care.

Chapters 7, 8, 9 and 10 concern individual and collective decision support. They provide the reader with an understanding of the theoretical foundations underlying computerised decision-making methods for diagnosis and treatment. Medical economic aspects of decision-making are presented to illustrate the way in which information systems have led to sophisticated methods for pricing activities in hospitals. The various methods and tools facilitating decision-making in the field of public health are explained to the reader.

Chapter 11 deals with the respect of ethical and legal aspects during the collection, archiving and processing of health data, particularly in Europe. Methods for securing records and information exchanges are presented.

Chapters 12, 13, 14, 15 and 16 deal with the principal applications of medical informatics, such as the development of hospital information systems, shared medical records and the computerisation of medical and dental offices. The concept of e-health is presented, together with telemedicine and telehealth, including teleservices for everyday life and social welfare.

Chapter 17 explains how medical informatics can facilitate research, whether basic or clinical, in the domains of epidemiology and public health.

Finally, Chapter 18 highlights the role of human factors and ergonomics in the applications of medical informatics.

This book is designed to be a study tool. It contains many international comparisons of developments and actions in the field.

The writing of this book was supported by the French College of Biostatistics and Medical Informatics Teachers, presided by **Prof. Pascal Roy** (Lyon 1 University).

Paris 13 University, Bobigny, France   Prof. Alain Venot
Paris 5 University, Paris, France   Prof. Anita Burgun
Burgundy University, Dijon, France   Prof. Catherine Quantin

Contents

Chapter 1
Medical Informatics as a Scientific Discipline

A. Venot, A. Burgun, S. Després, and P. Degoulet

Abstract This chapter will provide a brief history of the birth and development of medical informatics, followed by a description of its principal domains. These domains include bioinformatics, which relates to molecular and cellular aspects of medicine, clinical informatics, which deals with patient data and medical knowledge relating to the care of individual patients, and public health informatics, which brings together the tools, techniques and applications for reasoning at the population level. Links with other disciplines, including subdisciplines of computer sciences, biostatistics and biomedical engineering, have also been developed. There are many scientific societies for medical informatics, operating at the national, continental and international levels. These societies are presented here, together with the principal journals, scientific conferences and exhibitions in this field.

Keywords Medical informatics history • Biomedical informatics • Bioinformatics

A. Venot (✉) • S. Després
LIM&BIO EA 3969, UFR SMBH Université Paris 13, 74 rue Marcel Cachin,
Bobigny Cedex 93017, France
e-mail: alain.venot@univ-paris13.fr

A. Burgun • P. Degoulet
CRC INSERM U872 eq 22, René Descartes University, 15 rue de l'Ecole de Médecine,
75006 Paris, France
e-mail: patrice.degoulet@egp.aphp.fr

A. Venot et al. (eds.), *Medical Informatics, e-Health*, Health Informatics,
DOI 10.1007/978-2-8178-0478-1_1, © Springer-Verlag France 2014

After reading this chapter you should:

- Be able to explain what medical informatics is
- Be aware of the principal links between medical informatics and various other areas, such as scientific computing, biostatistics and biomedical engineering
- Be able to cite the major associations and national and international scientific societies in the field of medical informatics
- Be aware of the principal journals and international conferences for researchers to communicate their results and know how to submit articles and communications
- Be aware of some of the research structures and sources of funding for research in medical informatics.

1.1 Development of Medical Informatics

Medical informatics has gradually established itself as a scientific discipline (Shortliffe and Blois 2006; Geissbuhler et al. 2011; McCray et al. 2011). In this chapter, we present a brief history of this discipline, its key areas and links with other disciplines. We describe the national and international societies for medical informatics and discuss the major journals and congresses in this field, together with sources of research funding.

Interest in the potential value of computers for use in medicine first emerged in the 1960s and has rapidly grown ever since. From the outset, the goals were ambitious, including the replacement of paper medical records with electronic records and assisting or replacing doctors in diagnostic or therapeutic procedures. Early in the development of medical informatics, a number of key issues, such as terminology systems in health, medical knowledge modelling and reasoning, rapidly emerged.

A new scientific community developed. In 1968, a working group of researchers working specifically in the field of medical informatics was established within the International Federation for Information Processing (IFIP). One of its notable participants was Professor François Grémy (Degoulet et al. 2005), a French researcher. This committee became increasingly powerful, leading to the creation in 1974 of an international scientific society, the International Medical Informatics Association (IMIA). In the same year, the IMIA organised the first world congress in this field, Medinfo.

This medical informatics community has organised itself into a network of national and international associations. It has set up its own conventions, journals and standards bodies. In academia, research laboratories have been set up, together with specific Masters and Ph.D. programmes.

This research has generated a number of commercial products. Software vendors and database editors in the domain of health have set up companies developing products for the computerisation of hospitals, medical and dental offices, imaging departments and pharmacies.

Public partners, firmly convinced of the potential value of information technology (IT) to improve health systems and reduce costs, have been established in many countries to provide research funding at the national or supranational level.

IT management rapidly developed to reach a certain level of maturity, but the computerisation of healthcare processes proved complex. After 50 years of development, IT has spread throughout the health sector, but further progress and active research are required (Haux 2010). For example:

– Electronic medical records are widespread, but their routine use in everyday practice remains difficult and time-consuming for health professionals. Physicians, for example, continue to input many patient data in natural language, whereas the use of terminology systems to code these data would make more elaborate functions possible.
– Many clinical studies have shown that diagnostic errors are often made in the various medical specialities. Much effort has been made over the last 30 years to develop software to assist doctors with diagnosis, but these programs do not yet seem to have come of age.
– Many physicians appreciate computer assistance in the drafting of prescriptions, because it reduces the risk of unsafe prescriptions, but handwritten prescriptions are still very abundant.

1.2 The Main Areas of Medical Informatics

Since the 2000s, the term "medical informatics", which was originally coined in Europe, has spread to the United States and around the world, to describe this discipline. However, the term "biomedical informatics" has recently gained in popularity, as it ensures that rapid developments in bioinformatics are not obscured (Shortliffe and Blois 2006). This term encompasses bioinformatics as a subdomain.

In this book, we have deliberately avoided including a chapter dealing explicitly with bioinformatics, although some concepts specific to this area are presented in Chap. 17. Our goal was to focus on the field of health, without including purely biological applications.

The diagram below outlines the major areas of biomedical informatics and its various microscopic and macroscopic levels (Fig. 1.1).

Bioinformatics deals with the molecular and cellular levels of medicine. Very sophisticated methods have been developed for the analysis of gene sequences. Bioinformatics gained prominence with the advent of functional genomics, proteomics, transcriptomics and high-throughput sequencing. Research is progressive and continually adding to our knowledge, and there is now a need for a rational

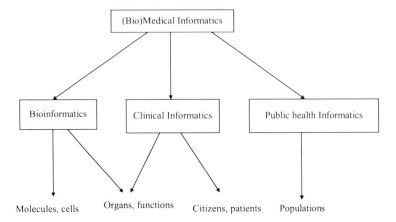

Fig. 1.1 (Bio)Medical informatics and its various subdomains

framework to explain large imbalances and their translation to the cellular or molecular level.

The principal applications in personalised medicine and health concern diagnosis and treatment. Functional genomics data from microarrays are useful for studies of the relationships between genes and for understanding the determinants and molecular mechanisms underlying the severity of a given disease in a given individual. They can also be used clinically, to select a drug treatment suitable for use in an individual with a particular genetic background (Kulikowski and Kulikowski 2009).

Clinical and bioclinical informatics deal with patient data and medical knowledge relating to the care of individual patients. Clinical informatics aims to provide methodological and technical solutions for data and knowledge representation, organisation, capture, storage, interrogation, interpretation, communication and use in practice (Mitchell et al. 2011). Computer support can be used in the presence of patients or remotely, through the use of telemedicine tools.

Specific methods are applied to populations, in a domain known as public health informatics. Public Health aims to develop educational, preventive, curative and social activities to improve the overall health of populations. If such actions are to be effective, they require the support of information systems designed with medical informatics methods. The WHO is now insisting that all states should consider the quality of their health information systems and seek to improve them.

Public health informatics brings together the tools, techniques and applications for reasoning not at an individual level, but at the population level. Tools for following cohorts, disease registries or vigilance systems have been derived from this approach. This is the case, for example, for pharmacovigilance, in which suspected adverse event reports for particular drugs are collected from various sites and efforts must then be made to group them on the basis of their similarities.

There is thus an important need for the aggregation of information, data mining and automatic alerts, to support decision-making with a potentially major impact on the health of the population.

1.2.1 Medical and Computer Sciences

Developments in medical informatics are based on methods and tools developed by computer scientists, although some of the problems encountered in the domain of health are original and lead to advances in computational methods.

The major areas of IT providing techniques widely used in medical informatics include:

Databases: the computerization of medical records is based on the use of database management systems (DBMS). One particular problem is posed by the need to develop and use specific terminology systems to represent and code the symptoms, clinical signs, disease and treatment of patients.

Artificial intelligence: the components of artificial intelligence are widely used in medical informatics. Decision-making is one of the core activities of physicians. It is difficult because the information on which decisions are based is plentiful but scattered and heterogeneous, and may be ambiguous or implicit, and because the mechanisms of medical reasoning may be complex. Knowledge engineering methods are used to model medical knowledge, to design inference engines for decision support systems and to develop health ontologies. Machine learning methods are used for medical data mining and language processing can be used for automatic structuring of the information (diseases, treatments) contained in the hospital and surgical reports.

Networks: computer networks support data exchange, providing the user with easy access to large volumes of data increasing the availability of data for patient management. However, the use of networks generates specific problems in medicine relating to the standardisation of content in terms of the semantics of messages, data security and privacy and the need to respect the rights of patients.

Automatic processing and analysis of medical images: medical imaging devices generate digital images and methods of contrast enhancement, 2D or 3D reconstruction, segmentation and the automatic registration of images are widely used.

The methods used in geographic information systems are useful for the development of epidemiological surveillance systems (simulation and visualisation of epidemics of influenza or measles, for example).

1.2.2 Medical Informatics and Biostatistics

Biostatistics provides a methodological framework for clinical research, which involves humans and is designed to generate new information about diseases or to assess and compare new treatments or new diagnostic procedures. Biostatistics can be used to make inferences and to draw conclusions about populations from observations or measurements on samples of individuals. Medical informatics and biostatistics differ principally in that many of the developments in the domain of medical informatics involve no probabilistic modelling. However, these two disciplines are closely linked and complementary in many areas.

Some of the methods developed by biostatisticians are used in the development of decision support systems in medical informatics. A particular example of this is provided by the use of Bayes' theorem for computer-aided diagnosis. Indeed, purely statistical techniques make little use of medical knowledge and are far from the reasoning of physicians. Conversely, purely logical methods do not take into account a very important type of information: the prevalence of disease and the frequency of signs in disease. This probabilistic dimension, represented by Bayes' theorem, significantly enriches the reasoning process and much work in medical informatics is based on this approach. In addition, purely statistical data mining approaches can be used for the analysis of large collections of data, and computer-based methods derived from machine learning techniques may also be applied.

1.2.3 Medical Informatics and Biomedical Engineering

Biomedical engineering involves the application of engineering methods and techniques to the medical domain, with the aim of developing devices useful for disease diagnosis and non-pharmacological treatments. This field combines physics, computer science, medicine and biology. Digital images are processed to compensate for their imperfections and merged images are produced by several methods, for automatic comparisons of the same object at successive time points, to highlight certain features of the image and to calculate areas or quantitative parameters of clinical interest. Such treatments may be applied to static images or to sequences of images. Some devices developed especially for surgical robotics and interventional radiology are largely based on medical image processing methods. Biomedical engineering covers the area of signal processing (analysis of physiological signals or applications to help patients to continue living at home (e.g. fall detection)).

1.3 Scientific Societies in the Domain of Medical Informatics

Many scientific societies have been set up in the domain of medical informatics over the last 40 years. These associations are federated at the continental level and these continental organisations themselves belong to the IMIA (International Medical Informatics Association).

1.3.1 The IMIA and Its Federations

Six continental federations of national associations have been formed:

- APAMI: Asia Pacific Association for Medical Informatics http://www.apami. org/;
- EFMI: European Federation For Medical Informatics (http://www.helmholtz-muenchen.de/ibmi/efmi/);
- Helina: African Region (http://www.helina-online.org/);
- IMIA LAC: Regional Federation of Health Informatics for Latin America and the Caribbean (http://www.imia-medinfo.org/new2/node/159);
- IMIA North America;
- MEAHI: Middle East Association for Health Informatics http://www.imia-medinfo.org/new2/node/160.

More than two dozen working groups bring together representatives from particular domains (e.g. computing and genomic medicine, dental computing).

The IMIA (http://www.imia.org/) has several types of activities and products:

- The World Congress of Medical Informatics, MEDINFO, which is organised under the auspices of the IMIA, every 3 years;
- A "Yearbook of Medical Informatics", reproducing a series of articles of particular importance published in various journals during the course of the year.

1.3.2 The National Societies

Many countries have national medical informatics associations. In the US, the American Medical Informatics Association (AMIA) is particularly active and organises two annual conferences. (https://www.amia.org/).

In France, the AIM (originally known as the Association for Computing Applications in Medicine) has been in existence since 1968 (http://france-aim. org/).

Table 1.1 The main international journals in medical informatics

Name of the journal	Free access Y/N	Editor	Scientific domain
Artificial Intelligence in Medicine http://www.aiimjournal.com/	N	Elsevier	Theory and practice of artificial intelligence in medicine
BMC Medical Informatics and Decision Making www.biomedcentral.com/bmcmedinformdecismak	Y	BioMed Central	Information technologies and decision-making
International Journal of Medical Informatics www.elsevier.com/locate/ijmedinf/	N	Elsevier	Clinical informatics
Journal of Biomedical Informatics www.elsevier.com/locate/yjbin	N	Elsevier	New methodologies and techniques
Journal of the American Medical Informatics Association http://jamia.bmj.com/	Y (after 12 months)	BMJ group	Clinical informatics
Journal of Medical Internet Research http://www.jmir.org/	Y	Eysenbach	e-health, clinical applications of the Internet
Methods of Information in Medicine http://www.schattauer.de/en/magazine/subject-areas/journals-a-z/methods	N	Schattauer	Methodologies for the processing of health information

1.4 The Main Journals and Scientific Conferences in Medical Informatics

1.4.1 The Scientific Journals

Medical informatics researchers submit articles reporting the results of their studies to various international scientific journals (Fu et al. 2011; Spreckelsen et al. 2011). Table 1.1 provides the details of several of these journals (name, publisher, type of access and fields covered).

The websites of these journals provide recommendations and guidelines for the writing and submission of articles. The submitted articles are then sent to at least three "reviewers" competent in the domain, who analyse the article and provide their opinions and comments. The editor summarises the recommendations made by the reviewers and transmits his or her decision to the authors. The decision may be the rejection of the article, a request for modifications and improvements (the authors are asked to provide a covering letter with their resubmission, explaining how they have taken into account the reviewers' criticisms and suggestions), or direct acceptance with no modifications requested (very unusual). The review and resubmission process for an article may take between a few weeks and a few months.

1.4.2   The Conferences

Conferences are organised by national or supranational scientific societies, at various intervals. Some of these conventions are particularly interesting because the resulting articles (but not the posters) are indexed in Medline and can be found by searching the PubMed search engine developed by the National Library of Medicine.

These conferences include:

- MEDINFO: World Congress of Medical Informatics, held every 3 years under the auspices of the International Medical Informatics Association (IMIA). Articles are published in the series "Studies in Health Technology and Informatics";
- The annual conference of the AMIA (American Medical Informatics Association), held annually (in Washington or in another American city), bringing together researchers from the US and elsewhere.
- The MIE (Medical Informatics Europe) Congress, organised annually under the auspices of the European Federation for Medical Informatics (EFMI). Articles are published in the series "Studies in Health Technology and Informatics".

The submission process is similar to that for journal articles, with the conference website providing instructions for authors. Medical software is often presented and exhibited at these conferences. The exhibitions organised by the HIMSS (Healthcare Information and Management Systems Society http://www.himss.org) are particularly useful in this respect. HIMSS is represented in North America, Europe and Asia.

1.5   Structuring and Funding of Research in Medical Informatics

Most medical informatics research is carried out in university laboratories, but the medical software industry also plays a role in research and development. Medical informatics research is funded by national and supranational organisations.

- For the USA, the various funding bodies are listed on the site of the AMIA (https://www.amia.org/informatics/research/agencies.asp);
- The European Union annually publishes tenders for projects including research linking academic and industrial partners (http://ec.europa.eu/information_society/activities/health/research/index_en.htm);

1.6   For More Information

Internet search engine queries with the keywords "medical informatics history" return a series of links to articles tracing the history of the development of medical informatics.

The websites of the major associations and medical informatics journals and conferences provide tips for authors and information about submission procedures.

Exercises

Q1 Go to the European Union website and look for tenders in medical informatics for this year and next year.
What type of file should be created to respond to such tenders?

Q2 What are the principal medical informatics conferences scheduled for the next year? You should visit the websites of the major societies in this field to find the answer.

References

Degoulet P, Haux R, Kulikowski C, Lun KC (2005) François Grémy and the birth of IMIA. 1st IMIA/UMIT Medical Informatics Award of Excellence given to Professor Grémy. Methods Inf Med 44:349–351

Fu LD, Aphinyanaphongs Y, Wang L, Aliferis CF (2011) A comparison of evaluation metrics for biomedical journals, articles, and websites in terms of sensitivity to topic. J Biomed Inform 44:587–594

Geissbuhler A, Kimura M, Kulikowski CA et al (2011) Confluence of disciplines in health informatics: an international perspective. Methods Inf Med 50:545–555

Haux R (2010) Medical informatics: past, present, future. Int J Med Inform 79:599–610

Kulikowski CA, Kulikowski CW (2009) Biomedical and health informatics in translational medicine. Methods Inf Med 48:4–10

McCray AT, Gefeller O, Aronsky D et al (2011) The birth and evolution of a discipline devoted to information in biomedicine and health care, as reflected in its longest running journal. Methods Inf Med 50:491–507

Mitchell JA, Gerdin U, Lindberg DA et al (2011) 50 years of informatics research on decision support: what's next? Methods Inf Med 50:525–535

Shortliffe EH, Blois MS (2006) The computer meets medicine and biology: emergence of a discipline. Biomedical informatics. In: Shortliffe EH, Cimino JJ (eds) Computer applications in health care and biomedicine, 3rd edn. Springer, New York, pp 3–45

Spreckelsen C, Deserno TM, Spitzer K (2011) Visibility of medical informatics regarding bibliometric indices and databases. BMC Med Inform Decis Mak 11:24

Chapter 2
Medical Vocabulary, Terminological Resources and Information Coding in the Health Domain

C. Duclos, A. Burgun, J.B. Lamy, P. Landais, J.M. Rodrigues, L. Soualmia, and P. Zweigenbaum

Abstract This chapter explains why it is hard to use medical language in computer applications and why the computer must adopt the human interpretation of medical words to avoid misunderstandings linked to ambiguity, homonymy and synonymy. Terminological resources are specific representations of medical language for dedicated use in particular health domains. We describe here the components of terminology (terms, concepts, relationships between concepts, definitions, constraints). The various artefacts of terminological resources (e.g. thesaurus, classification, nomenclature) are defined. We also provide examples of the dedicated use of terminological resources, such as disease coding, the indexing of biomedical publications, reasoning in decision support systems and data entry into electronic medical records. ICD 10, SNOMED CT, and MeSH are among the terminologies used in the examples. Alignment methods are described, making it possible to identify equivalent terms in different terminologies and to bridge

C. Duclos (✉)
LIM&BIO EA 3969, UFR SMBH Université Paris 13, 74 rue Marcel Cachin, 93017 Bobigny Cedex, France
e-mail: catherine.duclos@avc.aphp.fr

A. Burgun
Centre de recherche des cordeliers, 15 rue de l'école de Médecine, 75006 Paris, France

J.B. Lamy
UFR SMBU Université Paris 13, 74 rue Marcel Cachin, 93017 Bobigny Cedex, France

P. Landais
Université de Montpellier 1, 641 avenue du Doyen Gaston Giraud, 34093 Montpellier Cedex 5, France

J.M. Rodrigues
Université Jean Monnet, 10 rue Tréfilerie, 42023 Saint Etienne Cedex 2, France

L. Soualmia
Université de Rouen, Place Emile Blondel, 76821 Mont Saint Aignan Cedex, France

P. Zweigenbaum
LIMSI-CNRS, BP 133, 91403 Orsay, France

A. Venot et al. (eds.), *Medical Informatics, e-Health*, Health Informatics,
DOI 10.1007/978-2-8178-0478-1_2, © Springer-Verlag France 2014

different domains in health. We also present plans for multi-terminological servers, such as the UMLS (Unified Medical Language Systems), which provide a key vocabulary linking heterogeneous health terminologies in different languages.

Keywords Terminology • Classification • Semantics • Language • Controlled vocabulary • Coding scheme

After reading this chapter you should:

- Know the characteristics of medical language and the notions of synonymy, homonymy and ambiguity,
- Understand the requirement for the formalisation of medical language for the computerisation of health activities,
- Understand the notion of a "concept",
- Be aware of the various components used for the development of terminological systems,
- Know the definition of the different terminological systems and the issues they address,
- Be aware of the various uses of terminological systems and be able to provide examples suitable for a dedicated use,
- Be able to use terminological servers,
- Be aware of the major dedicated terminological resources in the domain of health.

2.1   Introduction

Healthcare professionals use specific health-related terminologies to express entities as diverse as diagnoses, findings, procedures, laboratory tests, drugs, anatomy, biological findings or genetics. Health terminology is complex and multifaceted.

The computerisation of health systems requires the recording and storage of large amounts of information about the health of patients and populations and expectations are high for the "intelligent" use of such information.

Humans can understand and reason from words, based on an understanding of their meaning, but computers can only compare text strings. For a computer to be able to understand medical language, resources that convey meaning are required. Terminological resources provide lists of organised concepts for specific health domains. These representations, when used to model information, can convey, contextually, the meaning of health information, enabling the computer to use this information correctly.

In this chapter, we discuss the characteristics of medical language, the need to normalise the expression of medical concepts for their use by computers and how such concepts can be represented through various artefacts (e.g. thesauri, classifications, nomenclatures), the components of which we describe here.

We then illustrate the use of these terminological resources by particular cases and present tools for viewing terminological resources (multi-terminological servers).

2.2 The Medical Vocabulary and Its Properties

2.2.1 Medical Vocabulary

Medical vocabulary has evolved with the historical development of medicine and surgery. Medical terms have been translated from Greek to Arabic, Arabic to Latin and Latin to modern languages. The coexistence of these various languages, Latin, Greek and Arabic, and of various schools of thought, such as the Aristotelian or Platonic schools, has made it difficult to develop an unequivocal single medical vocabulary. The simultaneous use of several linguistic systems has led to multiple synonyms.

2.2.2 Establishment of Medical Terms

Most medical terms are borrowed from Greek and Latin. They consist of a radical, possibly associated with a prefix or a suffix. The radical is the root of the word (for example, the radical "pharmac", from the Greek *pharmakon* refers to drugs in *pharmac*y or *pharmac*ology). A prefix is an element in front of a word, which modifies the meaning (e.g. the prefix *a-* indicates absence, as in *a*mnesia). A suffix is placed at the end of the word and also modifies its meaning (e.g. the suffix *-itis* indicates inflammation, of the larynx in laryng*itis*, of a node in aden*itis*, or a joint in arthr*itis*).

Radicals

The *kine* radical refers to movement, as in *kine*tics or a*kine*sia; here the "a" indicates an absence of movement, one of the characteristics of Parkinson's disease.

The association of the radical *cyt*(o)- (cell) and the suffix *–logy* (study), results in "cytology", the study of the cell. Similarly, the association of *histo-* with the suffix *-logy* gives histology, the study of tissues.

Prefixes

Absence or deprivation:	a- (*a*mnesia), *an*- (*an*aemia), ab-(*ab*stinence), *in*-(*in*somnia), *im*-(*im*maturity).
Number:	0, *nulli*- : a *nulli*para is a woman who has never given birth; 1, *primi*-: a *primi*para has given birth once; n, *multi*-: a *multi*para has already given birth several times.
Quantity:	much, *poly*-: *poly*uria = much urine; little, *olig*-: *olig*uria, little urine, pauci- *pauci*symptomatic.
Frequency:	fast, *tachy*-: *tachy*cardia = fast heart rate; slow, *brady*-:*brady*cardia = slow heart rate; often, *pollaki*- : *pollaki*uria = needing to urinate frequently; rare, *spanio*-: *spanio*menorrhea = a decrease in the frequency of periods.
Site:	in the middle, *mid*renal, *meso*colon; in front of, *prerenal*; behind, *retro*caval; above, *supra*tentorial; below, *hypo*gastrium; next to, *para*umbilical; around, *peri*carditis; at the base, *rhiz*arthrosis; at the end, *acro*megaly.
Resemblance:	self; *auto*graft; the same, *homo*zygotic twins; different, *hetero*geneous, *hetero*zygotic twins.
Function:	normal, *eu*thyroidism; abnormal, *dys*thyroidism; high, *hyper*thyroidism; low, *hypo*thyroidism

Suffixes

- *algia* means pain, arthr*algia*, joint pain
- *osis* refers to degeneration, as in adenomat*osis*

Suffixes (continued)
- *lysis* indicates destruction: auto*lysis* for self-destruction, osteo*lysis* for bone destruction
- *ectasia* or – *cele* for dilation: bronchi*ectasis*, varico*cele*
- *sten* refers to the narrowing of the lumen of a conduit: coronary *sten*osis
- *stasis* means stagnation, chole*stasis* is bile accumulation
- *rrhoea* refers to a flow, as in rhino*rrhoea* for runny nose, or ameno*rrhoea*, the cessation of menstrual periods
- *oma* denotes a malignant tumor, carcin*oma*
- *tomy*, indicates an incision or opening: phlebo*tomy*, opening a vein, gastro*tomy* opening the stomach to insert a feeding tube, for example
- *stomy* is used to indicate surgical procedures in which stomata are created: colo*stomy*, creation of a stoma from the skin to the colon
- *pexy* indicates attachment: cysto*pexy*, for example, is the attachment of the bladder to the abdominal wall
- *ectomy* means removal, excision: nephr*ectomy*
- *plasty* means repair, rhino*plasty* for nose reconstruction

Some medical terms may be eponymous. In other words, they may include the name of a person (e.g. Dupuytren's disease, Hodgkin's lymphoma).

Some medical terms are acronyms, formed from the first letters of a group of words and generally pronounced letter by letter, although some create collections of letters than can be pronounced like words in their own right (BBS for Besnier, Boeck and Schaumann sarcoidosis, NSAIDs for non-steroidal anti-inflammatory drugs, and MI for myocardial infarction).

2.2.3 Properties of Medical Language: Synonymy, Polysemy, Vagueness, Ambiguity

Medical language, like any language, can be difficult to understand because of ambiguities, leading to various possible interpretations of individual words. These ambiguities may result from the definition of a given word or acronym not being universal (for example, the acronym VIP is interpreted as vasoactive intestinal peptide in gastro-enterology but as voluntary interruption of pregnancy in orthogenic departments). Alternatively, a word may be ambiguous because it has many meanings (polysemy): for example, the word "knee" may represent a joint (dislocation of the knee) or an anatomical angle (right inferior knee of the coronary artery). Polysemy may be eliminated by taking into account the context in which the term is used. Finally, the ambiguity may result from the vagueness of language: "infarction" commonly refers to the heart, but it is more precise to talk about "myocardial infarction" in this case, to differentiate between this type of infarction and mesenteric or cerebral infarction, for instance.

Medical language is also highly expressive and includes many synonyms, i.e. expressions referring to the same object (e.g., myocardial infarction, heart attack, MI).

Natural language is extremely powerful and flexible. It can deal with various degrees of precision and evolution due to changes in knowledge, and it makes it possible to understand the context even when implicit, because humans can interpret and draw inferences from a knowledge of language.

2.3 Normalising the Expression of Medical Concepts in Computing Environments to Ensure Semantic Interoperability

When information is stored in a computer system, the computer "sees" the words simply as a string of characters. The computer system can carry out logical operations on these strings (e.g., it can check whether two words are identical by comparing each character of the textual string).

The storage of information on computers is of value if it provides benefits for the healthcare provider and for the patient. For example, noting that a patient is asthmatic in his computerised medical record should be associated with an automatic reminder to vaccinate the patient against flu, because he has a chronic respiratory disorder.

This requires computerised systems to understand information and, therefore, to make use of the meaning of the information rather than its expression. This is referred to as "semantic interoperability".

Resources are required to limit the ambiguities and imprecision of natural language, and to manage synonymy. These resources must introduce elements of context reproducing the organisation of knowledge necessary for a human to interpret words.

With such resources, it is possible:

- To record clinical data and to store them in electronic patient records with the appropriate level of detail,
- To exchange clinical data between independently developed clinical information systems without human intervention and with no loss of meaning,
- To combine similar data from several independent information systems without human intervention (for example, for health monitoring systems),
- To share decision rules from clinical practice guidelines between hospitals using different, independent information systems without human intervention and to use them with the data stored in these information systems.

Use Cases

2.3.0.1   *Use Case 1*

A patient arrives in the emergency department. The reason for the consultation is entered into the computerised information system and transmitted to the attending physician. The data are imported and integrated into the patient's electronic health record. The diagnosis of acute renal colic is automatically added to the list of the patient's problems, an analgesic prescription is added to the list of treatments and the results of laboratory tests and scanner findings are added to the patient's medical history.

2.3.0.2   *Use Case 2*

A health monitoring system for detecting the exposure of the population to infectious organisms retrieves, each night, the data concerning diagnoses, symptoms and bacteriological results stored in various hospital information systems. It combines these data and analyses them, to detect the emergence of new infectious diseases.

2.4   Terminology Resource Components

An understanding of the structures and utilisation targets of the various terminology resources used in healthcare requires a definition of the lexical assumptions on which they are built. There are two approaches:

The first is based on the onomasiological theory of word formation, which gives names to a meaning, thought or concept. In this case, the different designations or terms mapping to a specific meaning are sought and qualified as synonyms; for example "necrosis of the myocardium after coronary obstruction by a thrombus" can be named "myocardial infarction" or by the acronym "MI".

The second lexicographical approach is based on the collection of different words, from which meanings or concepts are extracted (semasiological approach). In this case, the same term can have several meanings and are said to be homonyms or polysemic, as in "cold" as a level of temperature and the name of a particular illness.

2.4.1   Triangle Concept, Term, Object or Thought, Word, Thing

There are three main framework components based on the Ogden-Richards semiotic triangle (Ogden et al. 1923) and the modified Ogden-Richards semiotic triangle

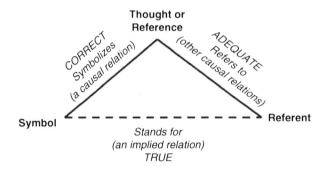

Fig. 2.1 Ogden-Richards triangle

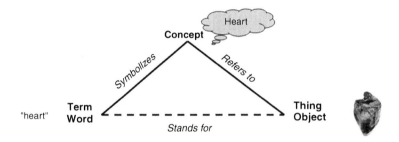

Fig. 2.2 Modified Ogden-Richards triangle

(Campbell et al. 1998): Thing or Object, Thought or Concept, Word (symbol) or Term (Figs. 2.1 and 2.2).

Within a person, the "heart" is a real thing or object. When thinking about this heart there is a thought in the brain referring to heart, which is referred to as a concept. This thought/concept is symbolised by the word/term/symbol "heart" in English.

2.4.2 Definition of Components According to (ISO 1087–1 2000) and (ISO 17115 2007)

2.4.2.1 Concept

A concept is "a unit of knowledge created by a unique combination of characteristics".

It may refer to a material thing (a car) or an immaterial entity (speed). It constitutes the apex of the Ogden-Richards triangle (Figs. 2.1 and 2.2).

It is symbolised by a designation.

2.4.2.2 Designation and Term

Designation is the representation of a concept by a sign, which denotes it. There are three types of designation: symbols, appellations and terms.

A term is the verbal designation of a general concept in a specific subject field, whereas an appellation is the verbal designation of an individual concept.

The term is the lower left point of the Ogden-Richards triangle and corresponds to the object (lower right point) expressed indirectly via the concept (Figs. 2.1 and 2.2).

2.4.2.3 Concept System

Intuitively, concepts can be placed in an organised system: for example, a closed fracture is a type of fracture.

If a terminological phrase is more complex than can be symbolised by a single term, such as "fracture", it is necessary to define several concepts and their relationships.

A concept system is a set of concepts structured as a function of the relationships between them. This set of concepts and relationships is the basis of semantic representation.

There are two main types of relationship: hierarchical and associative.

A hierarchical relationship is a relationship between two concepts that may be either generic or partitive.

An associative relationship is a pragmatic relationship between two concepts having a non hierarchical thematic connection, by virtue of experience, as a causal, site or a temporal relationship. Most concept systems are based on generic relationships (symbolised by IS_A) or partitive relations (symbolised by PART_OF).

A generic relationship is a relationship between two concepts in which the intension (definition) of one concept includes that of the other concept plus at least one additional delimiting characteristic. For instance, the subordinate concept "Talus" has a generic relationship (IS_A) with the superordinate concept "Foot bone". It is a foot bone, but has an additional characteristic (Fig. 2.3).

A partitive relationship is a relationship between two concepts in which one concept is the whole and the other is a part of the whole. For instance, the subordinate concept "Talus" has a partitive relationship (PART_OF) with the superordinate concept "Foot bone structure". It is part of the bone structure of the foot (Fig. 2.3). In a generic relationship, the superordinate concepts named generic concepts have a narrower intension (definition) and lie at the top of the hierarchy and the subordinate concepts are specific concepts that are more precise and located at a lower level of the hierarchy.

In a partitive relationship, the superordinate concept known as the comprehensive concept is connected with co-ordinate concepts, which are at the same level of the hierarchy.

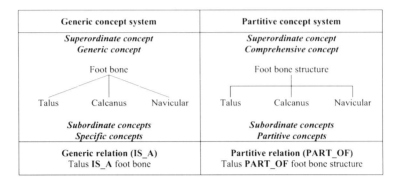

Fig. 2.3 Hierarchical concept systems

2.4.2.4 Definitions

A definition is the representation of a concept by a descriptive statement differentiating it from related concepts. The intensional definition of a concept is a definition describing the intension of a concept by stating the superordinate concept and its delimiting characteristics. (e.g. a femur diaphyseal fracture Is_A a fracture located on the femur diaphysis).

The extensional definition of a concept is the description obtained by grouping together all the subordinate concepts under a single criterion of subdivision. (e.g. noble gas : helium, neon, argon, krypton, xenon, radon).

2.4.3   Compositional Approaches for Concept Representation

Some simple concepts can be combined into a compositional concept representation. Let us take as an example "*Escherichia coli* pyelonephritis". It is possible to identify three categories, classes or axes of concepts: Topography with 'the pelvis or the kidney' (pyelonephr-), Morphology with 'infection' (-itis) and Etiology with '*Escherichia coli*'. The representation of this compound knowledge requires explicit description of the relationships between the components. In our example, "*Escherichia coli* pyelonephritis" can be represented as an infection (morphology) which "has_site" "the kidney" (topography) and which "has_cause" "*Escherichia coli*" (etiology).

The prevention of nonsense representations (such as liver fracture), requires the imposition of constraints between the relationships formalised as semantic links and the authorised components (formalised as categories, classes or axes of characterising concepts). For example, the semantic link "has_site" is authorised only between concepts characterising morphology and concepts characterising topography (Fig. 2.4).

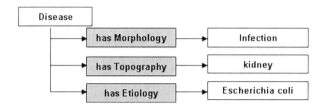

Fig. 2.4 Representation of the concept "*Escherichia coli* pyelonephritis"

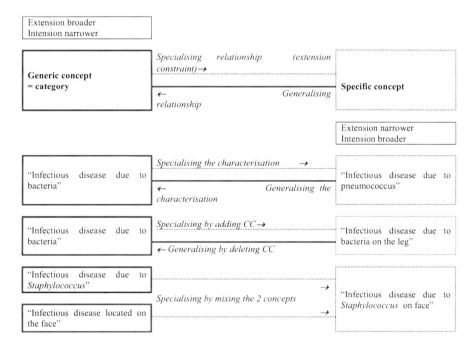

Fig. 2.5 Specialising/Generalising processes in a compositional system (*CC* composite characteristic)

The hierarchy, semantic links and compositional constraints, known as the categorial structure, of a set of elementary concepts defines the conceptual representation field. This field can be used to infer and to subsume automatically the subordinate concepts, as summarised in Fig. 2.5.

2.4.4   Formal Concept Representation

Various knowledge representation tools are available to support compositional approaches to concept representation. Knowledge representation tools are artificial intelligence tools for the representation of knowledge as symbols, facilitating

Fig. 2.6 Semantic network
of C. Peirce

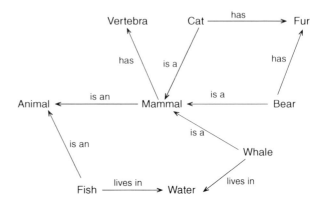

inference from knowledge elements to create new knowledge elements. The first
formal representation proposed was called the Semantic or Frame network and was
described in 1909 by C. Peirce (1909) as shown in Fig. 2.6.

It was developed into a graphical interface of first-order logic known as the
Conceptual Graph by J Sowa (1984) and, more recently, into the Web Ontology
Language, which combines the RDF/XML syntax format and description logic-
based formal representation (Baader et al. 2005; Lacy 2005). An example is
provided by the work of Schulz et al. (2011), using Bio Top upper level ontology:

> PathologicalEntity equivalent to PathologicalStructure or Pathological Disposition or
> Pathological Process

with

> All instances of PathologicalStructure are related to the anatomical objects where they
> occur via the relation PhysicalPartOf or by the more general relation physicallyLocatedIn.
> All instances of PathologicalDisposition are related to their bearers by the relation
> inheresIn.
> All instances of PathologicalProcess are related to the place where they occur by the
> relation hasLocus and to their participating entities by has Participant.

2.5 Terminological System Typology

A terminological system is a system organising the relationships between terms and
concepts in a domain with, when appropriate, any associated rules, relationships,
definitions and codes (EN ISO 1828 2012). The different types are named: terminology, nomenclature, thesaurus, vocabulary, classification, coding system, taxonomy and ontology.

A terminology is a set of designations belonging to a special language (ISO
1087–1 2000) related to the concepts of a specific domain (e.g. Terminologica
Anatomica). Clinical and reference terminologies can be distinguished on the basis
of their use.

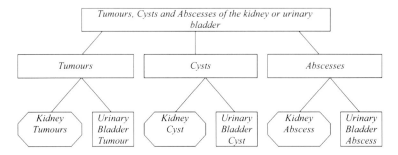

Fig. 2.7 Example of classification

A thesaurus is a dictionary of words in alphabetical order (with keywords and synonyms) organised to facilitate the retrieval and classification of documents, in an index, for example. In a vocabulary, terms are associated with definitions.

A nomenclature is an inventory of terms used to designate objects in a particular field, mostly when the system is based on user-specific rules rather than concepts.

A classification is an organisation of the exhaustive set of concepts of a domain, by necessary and sufficient conditions, such that each concept belongs to only one class. The classes are mutually exclusive and hierarchical (generic or partitive) and exhaustive, due to the creation of residual classes named "Not Elsewhere Classified" and "Not Otherwise Specified" (Fig. 2.7).

A taxonomy is a classification based exclusively on generic hierarchical relations.

A coding system is a combination of a set of concepts, a set of code values, and at least one coding scheme mapping code values to coded concepts (ISO 17115 2007). Codes are used by computers.

Codes may be meaningful, if a human can infer some knowledge from the code: for instance, the ICD 10 Congestive heart failure code I50.0 means that this disease can be found in chapter I, which relates to cardiovascular diseases. However, codes may be meaningless and entirely unlinked to any meaning (e.g. purely alphanumeric strings).

IDC 10 Typology

There are several types of healthcare terminological system. The International Classification of Disease (ICD 10) is a 21-chapter classification based on anatomy and aetiology concepts. Its codes are meaningful.

Volume 2 is an alphabetical index with synonyms, which may be considered to be a thesaurus. Its extension to oncology ICD-O makes it possible to create compositional representations with morphology. It is also a nomenclature. Finally, the coding rules in Volume 3 propose a semi-formal definition of the classes (De Keizer et al. 2000).

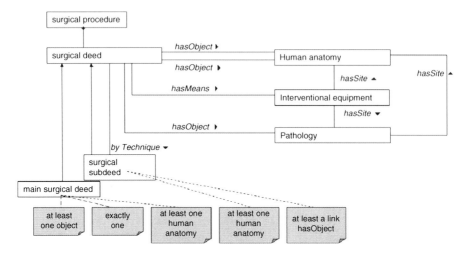

Fig. 2.8 Categorial structure for terminological systems of surgical procedures (Adapted from EN ISO 1828 2012)

Ontology is the study of what exists (the branch of metaphysics dealing with reality). Formal ontologies are theories that aim to provide precise mathematical formulations of the properties and relationships of certain entities.

Three levels of ontologies must be considered:

– Upper-level ontologies, representing the world: BFO, Bio Top, DOLCE, SUMO.
– Reference ontologies in a domain, such as FMA in anatomy, Galen for surgical procedures.
– Applied ontologies, such as SNOMED CT.

Compositional system representations and categorial structure (EN ISO 1828 2012) are semi-formal approaches to ontology. Categorial structures are minimal logic constraints for representing the concepts of a specific domain. For instance, for surgical procedures (Fig. 2.8), according to the categorial structure:

– The principal semantic categories are Anatomy, Deed, Device and Pathology, with two qualifier categories (cardinality and laterality).
– The semantic links are *by_technique*, *has_means*, *has_object*, *has_site* (with inverse):

  – *Has_object* is authorised between deed and anatomy or device or pathology,
  – *Has_site* is authorised between device or pathology and anatomy,
  – *Has_means* is authorised between deed and anatomy, device or pathology,
  – *By_technique* is authorised between deed and deed.

– The minimal constraints required are:

  – A deed and *has_object* shall be present,
  – Anatomy must always be present with either a *has_object* or with a *has_site*,

- The use of pathology is restricted to macroscopic lesions and to cases in which it can distinguish the procedure concerned from other procedures using the same deed and the same anatomy,
- When *by_technique* is used, the deed on the right side of the semantic link must conform to the previous rules.

The categorial structure makes it possible to ensure that new terms describing surgical procedures are associated with a formal definition consistent with a common template.

2.6 Desiderata for Terminological Systems

JJ. Cimino from the Columbian Presbyterian Medical Center in New York has defined 12 characteristics, known as desiderata, for terminological systems used in medical records (Cimino 1998).

1. The content must satisfy the user. To most users "What can be said" is more important than "how it can be said". Omissions are readily observed and timely, formal and explicit methods for plugging gaps are required.
2. The vocabulary must be concept-oriented. The unit of symbolic processing is the concept and each concept in the vocabulary should have a single, coherent meaning.
3. A concept's meaning cannot change and it cannot be deleted from the vocabulary, it is the concept permanence principle.
4. Concept identifiers must be meaningless. Concepts typically have unique identifiers (codes) and these should be non-hierarchical (see code-dependence), to allow for later relocation and multiple classification.
5. The system must be polyhierarchical, to allow multiple classification.
6. Concepts must have a semantic definition. For example, Streptococcal tonsillitis = Infection of the tonsil caused by *Streptococcus*.
7. The system must not have residual categories. Traditional classifications have rubrics that include NOS, NEC, Unspecified, Other, the meaning of which may change over time as new concepts are added to the vocabulary. These are not appropriate for recording data in an electronic health record.
8. The system must have multiple granularities. Different users require different levels of expressivity. A general practitioner might use myocardial infarction, whereas a surgeon may record acute anteroseptal myocardial infarction.
9. Although there may be multiple views of the hierarchy required to support different functional requirements and levels of detail, they must be consistent.
10. There is a crucial relationship between concepts within the vocabulary and the context in which they are used. Cimino defined three types of knowledge:

    - Definitional – how concepts define each another
    - Assertional – how concepts combine
    - Contextual – how concepts are used

11. Vocabularies must be designed to allow for evolution and change, to incorporate new advances in healthcare and to correct errors.
12. Where the same information can be expressed in different ways, a mechanism for recognising equivalence is required. This is redundancy recognition.

2.7   Terminologies in Action

Many terminologies have been designed, each for a specific purpose. They are used:

– To code patient data, in the context of health care, in epidemiological studies or public health;
– To index documents, including biomedical research articles;
– To represent entities in expert systems and decision support systems;
– To serve as an interface for data entry.

Several representations of a given condition may, therefore, co-exist. We will illustrate this phenomenon with the example of haemochromatosis. Haemochromatosis is a disorder that causes the body to absorb and to store too much iron. In the body, iron is incorporated into haemoglobin, which transports oxygen in the blood. Healthy people usually absorb about 10 % of the iron present in the food they eat. People with genetic haemochromatosis absorb about 20 % of the iron they ingest. The body has no natural way to rid itself of excess iron, so extra iron is stored in body tissues, especially the liver, heart and pancreas (source: www. niddk.nih.gov).

The accumulation of iron in body tissues may lead to:

– Osteo-articular symptoms, including joint pain and arthritis;
– Liver disease, including cirrhosis, cancer and liver failure;
– Heart disease, potentially leading to heart failure;
– Abnormal pigmentation of the skin;
– Damage to the pancreas, possibly causing diabetes;
– Impotence.

Symptoms tend to occur in men between the ages of 30 and 50, and in women over the age of 50. However, many people have no symptoms when they are diagnosed.

Genetic haemochromatosis is mostly associated with a defect in the HFE gene. HFE regulates the amount of iron absorbed from food. Two mutations, C282Y and H63D, are known to cause haemochromatosis. The genetic defect is present at birth, but symptoms rarely appear before adulthood.

Haemochromatosis may also be acquired, through blood transfusions, for example.

2.7.1 Terminologies and Their Use to Code Diseases

Many classification systems were designed in the seventeenth and eighteenth centuries. Nosologists tried to do for diseases what botanists had done for plants: to find the natural divisions between diseases, to discover the real essence of the diseases, and to embody this essence in a suitable definition. Thomas Sydenham (1624–1699) was one of the most famous nosologists. He said "It is necessary that all diseases be reduced to definite and certain species, and that, with the same care which we see exhibited by botanists in their phytology."

The classification systems created during this period include Genera morborum (Linnaeus) and Nosologia Methodica (François Bossier de Lacroix also known as Sauvages). Nosology is the key to improving diagnosis and treatment. Sauvages saw nosology as a practical discipline providing practitioners with a compass to chart their voyages through the complex sea of symptoms.

From the nineteenth century onwards, increasing numbers of terminologies were created for practical purposes, such as the reporting of causes of death, particularly in England, for analyses of child mortality and the reporting of cases of plague in London. The need for accurate reporting of the causes of death (William Farr, Jacques Bertillon) led to the development of the International Classification of Diseases.

The International Classification of Diseases is a statistical classification dating back to the eighteenth century that is maintained by the World Health Classification. Its early revisions related exclusively to causes of death. Its scope was extended, in 1948, to include non-fatal diseases. Various versions of the International Classification of Diseases are now used in more than 50 countries, to code diagnoses in the DRG system or equivalent. The tenth revision was released in 1993. It contains 9876 items, allowing the coding of any case, thanks to the "not classified elsewhere" codes (e.g., "other disorders of mineral metabolism") and the "not otherwise specified" codes (e.g. "disorders of mineral metabolism, unspecified"). Granularity varies between items, depending on statistical aspects. Moreover, classification criteria are included: for example E83.1 "Disorders of iron metabolism" is a subclass of E83 "Disorders of mineral metabolism"; it includes haemochromatosis but excludes iron deficiency anaemia. Haemochromatosis does not have its own class and cases of haemochromatosis are therefore simply coded as E83.1 "Disorders of iron metabolism" (Fig. 2.9).

Some diseases may be coded on the basis of their aetiology (B05.0: Measles complicated by encephalitis) or signs (G05.1: Encephalitis, myelitis and encephalomyelitis in viral diseases classified elsewhere).

The 11th version of the ICD will soon be released. Various models have been defined for the reporting of causes of death, the reporting of morbidity, disease coding in DRG systems, and primary care. Traditional medicine, including the various forms of Asian medicine, will also be represented in ICD 11. Moreover, ICD11 will be aligned with SNOMED CT.

Fig. 2.9 Haemochromatosis
in ICD 10

E83 Disorders of mineral metabolism
    Excl.: dietary mineral deficiency (E58-E61)
        parathyroid disorders (E20-E21)
        vitamin D deficiency (E55.-)
  E83.0 Disorders of copper metabolism
  E83.1 Disorders of iron metabolism
    Haemochromatosis
    Excl.: anaemia with iron deficiency (D50.-)
        sideroblastic anaemia (D64.0-D64.3)
  ...
  E83.8 Other disorders of mineral metabolism
  E83.9 Disorder of mineral metabolism, unspecified
E87 Other disorders of fluid, electrolyte and acid-base balance

SNOMED CT (http://www.ihtsdo.org/snomed-ct/) is a clinical reference terminology. It is a comprehensive terminological system for coding clinical information (283,000 concepts, 732,000 terms and 923,000 relationships).

2.7.2 Terminologies for Indexing

There is a need for controlled vocabularies suitable for use in the indexing and cataloguing of biomedical publications. These terminologies must be thesauri containing links showing the relationships between related terms and providing a hierarchical structure facilitating searching at various levels of specificity, from "narrower" to "broader" terms. They correspond to a more or less limited list of terms encompassing synonyms, to facilitate information retrieval.

Medical Subject Headings (MeSH) is the controlled-vocabulary thesaurus used for indexing articles for PubMed. MeSH was designed in 1960 by the US National Library of Medicine. The 2012 edition of MeSH contains 26,581 descriptors (subject headings), which are used to index documents in an unambiguous manner. The descriptors are organised into several modules covering all the domains of biomedicine: Anatomy, Organisms, Diseases, Chemicals and Drugs, Analytical, Diagnostic and Therapeutic Techniques and Equipment, Psychiatry and Psychology, Phenomena and Processes, Disciplines and Occupations, Technology, Industry, Agriculture, Anthropology, Education, Sociology and Social Phenomena, Humanities, Information Science, Named Groups, Health Care, Publication Characteristics, Geographical. MeSH is a directed acyclic graph, in which a term may have more than one parent, for example

- Urinary lithiasis *is a* Lithiasis,
- Urinary lithiasis *is a* Urologic disease.

In MeSH, "Haemochromatosis" is categorised as "Metal Metabolism, inborn errors". In fact, we know that the relationship between "Haemochromatosis" and "Metal Metabolism, inborn errors" should actually be "is generally a" rather than "is a", because haemochromatosis is acquired in some cases, through multiple

> - Nutritional and Metabolic Diseases [C18]
>   - Metabolic Diseases [C18.452]
>   - Metabolism, Inborn Errors [C18.452.648]
>   - Metal Metabolism, Inborn Errors [C18.452.648.618]
>   - Hemochromatosis [C18.452.648.618.337]
>   - Iron Metabolism Disorders [C18.452.565]
>   - Iron Overload [C18.452.565.500]
>   - Hemochromatosis [C18.452.565.500.480]

Fig. 2.10 Haemochromatosis in MeSH

blood transfusions. Such relationships, although not taxonomic, are of interest for information retrieval purposes. For this reason, "Haemochromatosis" is also categorised as "Iron Overload" (Fig 2.10).

2.7.3 Terminologies in Decision Support Systems

Reasoning in decision support systems may be based on taxonomies, in addition to rule-based inferences. Taxonomies support specialisation and generalisation (from the more general to the more specific and vice versa).

Quick Medical Reference (QMR) is a decision support system for assisted diagnosis in medicine, not restricted to a specific domain. It uses a terminology to represent diseases, signs and symptoms. The QMR terminology is organized as a taxonomy, in which haemochromatosis is a subclass of cirrhosis. In this representation, the system focuses on the liver lesions caused by iron overload. Cirrhosis is a complication of haemochromatosis. The relationship between haemochromatosis and cirrhosis is therefore not taxonomic and actually means "may be found when".

2.7.4 Terminologies for Data Entry

Interface terminologies are used to facilitate data entry into electronic medical records. They link user's descriptions to structured data elements in a reference terminology (Rosenbloom et al. 2006).

During the development of a domain-specific interface terminology based on a reference terminology, the domain concepts are identified and mapped to the reference terminology concepts. This creates a subset of the reference terminology (Bakhshi-Raiez et al. 2010). This subset covers the needs of the user and is not directly displayed to users, instead being presented in terms or descriptions familiar to the user.

SNOMED CT may be used as an interface terminology for data entry as it includes many terms for each concept and subsets for data entry are made available.

The VCM iconic language (Visualization of Concepts in Medicine) (http://vcm. univ-paris13.fr/svcm) proposes icons for the graphic representation of the main physiopathological states: patient characteristics (e.g. age, sex), symptoms, diseases, antecedents, risks, various classes of treatment, medical follow-up procedures, health professionals and medical speciality, and medical knowledge. These icons are created by combining several elements from a lexicon of shapes, colours and pictograms, according to simple grammar rules. A health professional can usually learn the VCM lexicon and grammar in a few hours. The VCM language makes the *is-a* relations present in other terminologies visually explicit. For example, a coronary disease is a cardiac disease, but this is not explicit in a visual search because the "coronary disease" term does not include the word "heart" or the prefix "card-". By contrast, the VCM icon for coronary disease makes the relationship with cardiac disease explicit because it includes the heart pictogram. This iconic language is accompanied by a graphical silhouette (Mister VCM) making it possible to organise, in a restricted space, a set of VCM icons corresponding to coded data by anatomy and aetiology.

2.7.5 Conclusions

Each terminology is useful in a given context, but not universally valid. The organisation of the concepts reflects the intended objective. It is useful to find out the aim of a terminology before using it, particularly when hierarchical structures are used in algorithmic process. The underlying question is: Does this resource describe the domain and the concepts needed to achieve the desired goal adequately.

2.8 Aligning Terminologies

2.8.1 Alignment Methods

As detailed above, in health there are almost as many different terminologies, controlled vocabularies, thesauri and classification systems as there are fields of application (Shvaiko and Euzenat 2013).

Given the enormous number of terminologies, existing tools, such as search engines, coding systems and decision support systems, have a limited capacity for dealing with "syntactic" and "semantic" divergences, despite their large storage capacities and ability to process data rapidly. Faced with this reality and the increasing need to allow co-operation with/between the various health actors and their related health information systems, there seems to be a need to link and connect these terminologies, to make them "interoperable". Alignment techniques are of particular importance because the manual creation of correspondences between concepts or between terms is extremely time-consuming. There are two major dimensions for similarity: the syntactic dimension and the semantic dimension. The syntactic dimension is based on lexical methods and the semantic dimension is based on structural and semantic properties of terminologies.

2.8.1.1 Lexical Methods

Lexical methods are based on the lexical properties of terms. These methods are straightforward and constitute a trivial approach to identifying correspondences between terms. The use of such methods to achieve mapping in the medical domain was driven by the similarly of the terms included in many terminologies.

String-Based Methods

In these methods, terms (or labels) are considered as sequences of characters (strings). A string distance is calculated to determine the degree of similarity between two strings. In some of these methods, the order of the characters is not important. Examples of such distances, also used in the context of information retrieval, are: the Hamming distance, the Jaccard distance, the Dice distance. Another family of appropriate measures, known as the "Edit distance" takes into account the order of characters. Intuitively, an edit distance between two strings is defined as the minimum number of character insertions, deletions and changes required to convert one string into another. The Levenshtein distance is one example of such a distance. It is the edit distance with all costs (character insertions, deletions and changes) equal to 1. This measure is also frequently used for spelling errors. For example: asthmma vs. asthma (insertion of one character), astma vs. asthma (deletion of one character) and ashtma vs. asthma (inversion of two characters). However, these methods can only quantify the similarity between terms or labels. They therefore provide low estimates of similarity between synonymous terms with different structures. For example, the words "pain" and "ache" are synonyms. They are thus semantically related and mean the same thing, but none of the distances presented above can identify any links between these two terms. Conversely, these methods find significant similarity between terms that are actually different (false positive), such as "Vitamin A" and "Vitamin B".

Language-Based Methods

In these methods, terms are considered as words in a particular language. NLP tools are used to facilitate the extraction of meaningful terms from a text. These tools exploit the morphological properties of words. Methods based on normalisation processes can be distinguished from those making use of external knowledge resources, such as dictionaries.

Normalisation Methods

Each word is normalised to a standardised form that is easy to recognise. Several linguistic software tools have been developed for the rapid retrieval of a normal form of strings: (i) tokenisation involves segmenting strings into sequences of tokens, by eliminating punctuation, cases and blank characters; (ii) the stemming process involves analysing the tokens derived from the tokenisation process, to reduce them to a canonical form; (iii) stop words elimination involves removing all the frequent short words that do not affect the sentences or the labels of terms, phrases such as "a", "Nos", "of"...etc.

External Knowledge-Based Methods

These methods use external resources, such as dictionaries and lexicons. Several linguistic resources have been developed to identify possible mappings between terminologies. These methods form the basis of the lexical tools used by the UMLSKS API (https://uts.nlm.nih.gov/home.html). They were combined with synonyms from other external resources to optimise mapping to the UMLS. Another external resource largely used in the biomedical field is the lexical data-base WordNet (http://wordnet.princeton.edu/).

Examples

Exact Match The Orphanet disease "Glycogen storage disease type 2" has an exact match with the SNOMED notion "Glycogen storage disease, type II": "Glycogen storage disease type 2" is a synonym of the MeSH descriptor "Glycogen storage disease, type II", which is itself an exact match for the SNOMED notion "Glycogen storage disease, type II".

Alignment by Combination The Orphanet term "Diabetic embryopathy" is aligned with two MeSH descriptors "Diabetes mellitus" and "Fetal diseases": with NLP tools "Diabetic" is matched with the MeSH descriptor "Diabetes mellitus"; and "Embryopathies" is a MeSH synonym of "Fetal diseases" (which is, here, an exact match).

2.8.1.2 Structural Methods

These methods use the structural properties of each terminology to identify all the possible correspondences between terms. They consider terminologies as graphs in which the nodes represent terms and the edges represent relationships between these terms established in the terminology. Most medical terminologies can be represented as graphs. Furthermore, these techniques can also be combined with lexical techniques. Together with the structural properties of each terminology,

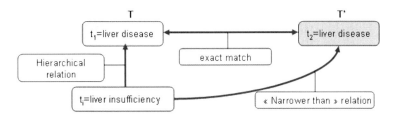

Fig. 2.11 NT alignment

semantic methods also use semantic similarities to find the closest term. The principal technique involves calculating the number of edges between terms, to determine a distance between them. The best known distance estimating similarity is the Wu-Palmer distance, which is defined according to the distance between two terms in the hierarchy and their positions with respect to the root. Unlike these traditional edge-counting approaches, other methods, such as Lin similarity methods, estimate similarity from the maximum amount of information shared by two terms in a hierarchical structure. These similarities can be used to find possible connections between terms or concepts from different hierarchical terminologies, such as MeSH or SNOMED INT, for example. This approach is based on hierarchical relationships and is used to align the remaining terms not mapped by the lexical approach. This mapping provides two types of correspondences:

- Narrow-Mapping: when the remaining term has at least one child (hierarchical relationship narrower than) mapped to at least one term.
- Broad-Mapping: when the remaining term has at least one parent (hierarchical relationship broader than) mapped to at least one term.

"Narrower Than" (NT) Relationships

If a term t_1 from the terminology T has an exact lexical match with a term t_2 from the terminology T', then each term t_i of T narrower than t_1 is narrower than t_2 in T'. A NT relation links t_i in T to t_2 in T' (Fig. 2.11).

"Broader Than" (BT) Relationship

If a term t_1 from the terminology T has an exact lexical match with a term t_2 from the terminology T', then each term t_i of T broader than t_1 is broader than t_2 in T'. A BT relationship links t_i in T to t_2 in T'.

However, an evaluation, generally a manual evaluation, is required to ensure that the alignment is of high quality.

2.8.2   The UMLS

In 1986, US National Library of Medicine (NLM) launched the Unified Medical Language System (UMLS) project. According to Donald Lindberg, Director of the NLM, the objective was to build a vocabulary, a language linking the biomedical literature with observations on the patient, and educational applications in the school, a language connecting these areas.

The UMLS project addresses semantic heterogeneity issues in the biomedical domain. It provides terminological resources that have been made available to the community (Knowledge Sources). Since 1990, the UMLS project has produced annual editions of tangible products that are now regularly used by their intended audience. Assessments of the value of the UMLS products by more disinterested observers are required, but an increasing array of operational systems are making use of one or more of the UMLS Knowledge Sources or lexical programs (Lindberg et al. 1993). The UMLS resources include:

- The Metathesaurus (1990), including a large set of terminologies used biomedical domain and describing relationships between the terms;
- The Semantic Network (1990) is a set of semantic types representing the broad categories of the domain. They are used to categorise the Metathesaurus concepts;
- The Specialist Lexicon (1994) provides the lexical information required for natural language processing. It includes commonly occurring English words and biomedical vocabulary. The Lexicon entry for each word or term records the syntactic, morphological and orthographic information used with associated NLP tools.

2.8.2.1   The UMLS Metathesaurus

The Unified Medical Language System® (UMLS®) Metathesaurus® is a large, multi-purpose, multilingual thesaurus that contains millions of biomedical and health-related concepts, their synonymous names and their relationships. The Metathesaurus includes over 150 electronic versions of classifications, code sets, thesauri, and lists of controlled terms in the biomedical domain. These are the source vocabularies of the Metathesaurus. (http://www.nlm.nih.gov/pubs/factsheets/umlsmeta.html)

The Metathesaurus is organized by concept, or meaning. It links alternative names and views of the same concept from different source vocabularies and identifies useful relationships between different concepts. As such, the UMLS Metathesaurus transcends the specific thesauri, codes and classifications it encompasses.

The UMLS Metathesaurus includes most of the terminologies used in medicine, such as the International Classification of Diseases and SNOMED. It includes different versions of the terminologies, such as ICD9, ICD10, ICD9-CM, and different languages (ICD in French, Spanish, German, Russian, etc.). Each concept has a unique identifier, known as the Concept Unique Identifier (CUI). All the relationships existing between two terms in the source terminologies are represented. All information, terms, concepts and relationships are presented in a unified format.

We can illustrate this with craniostenosis.

- Many source terminologies contain this concept: ICD-10, ICPC, MedDRA, MeSH, OMIM, Read Codes, SNOMED CT
- A definition can be found in MeSH: Premature closure of one or more sutures of the skull;
- Synonymous terms are clustered into a single CUI C0010278 with a preferred label "Craniosynostosis" (Preferred Term). The synonymous terms include Craniostenosis (ICD, ICPC, OMIM, SNOMED CT), Craniosynostosis syndrome (SNOMED CT), Synostosis (cranial) (CRISP), Premature closure of cranial sutures (MedDRA, SNCT), Congenital ossification of cranial sutures, Congenital ossification of sutures of skull, Premature cranial suture closure (SNOMED). We also have abbreviations, such as CRS, CSO and CRS1 (OMIM). In addition, the UMLS provides translations into several languages (e.g. Spanish, German, French).
- The Metathesaurus provides a list of related terms, either more specific, such as Craniosynostosis, type 1 (OMIM), or syndromes such as Hurst syndrome (C0014077) Christian syndrome 1 (C0795794) or SCARF (Skeletal abnormalities, cutis laxa, craniostenosis, ambiguous genitalia, psychomotor retardation, facial abnormalities) syndrome (C0796146);
- The relationships between the concept "Craniostenosis" and other concepts in source terminologies are retained.

The UMLS Metathesaurus contains more than 9,000,000 terms, 2,000,000 concepts and 22,000,000 relationships between concepts. The data correspond to 152 terminologies and 19 different languages.

2.8.2.2   The UMLS Semantic Network

The UMLS Semantic Network consists of (i) about 130 broad categories, or semantic types, providing a consistent categorisation of all concepts represented in the UMLS Metathesaurus®, and (ii) a set of semantic relationships between semantic types (network). The semantic types are of two kinds:

- "Entity" encompasses physical objects (e.g. plants, animals, anatomical structures, chemicals) and conceptual entities (e.g. spatial entities, temporal entities, signs and symptoms);
- "Event" includes activities (e.g. therapeutic procedures, behaviour), phenomena and processes (e.g., biological functions, diseases).

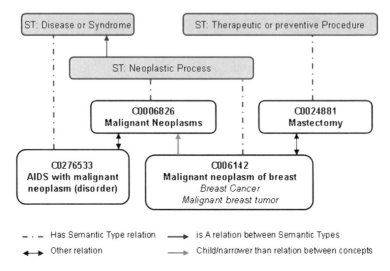

Fig. 2.12 Concept categorisation in the UMLS

Each semantic type is given a definition.

> **Example of Semantic Types**
>
> **UI**: T190
>   **STY**: Anatomical abnormality
>   **ABR**: anab
>   **STN**: A1.2.2
>   **DEF**: An abnormal structure, or one that is abnormal in size or location.
>   **UN**: Use this type if the abnormality in question can be either acquired or congenital abnormality. Neoplasms are not included here. These are given the type 'Neoplastic Process'. If an anatomical abnormality has a pathological manifestation, then it will additionally be given the type 'Disease or Syndrome', e.g., "Diabetic Cataract" will be double-typed for this reason.
>   **HL**: {isa} Anatomical Structure; {inverse_isa} Congenital Abnormality.

The Semantic Network represents 54 relationships. The links between the semantic types provide the structure for the network. The primary link between the semantic types is the "is_a" link, which establishes the hierarchy of types within the Semantic Network. There are also non-hierarchical relationships (e.g., treats, diagnoses) that represent useful and important relationships in the biomedical domain (e.g., drugs treat diseases).

The Semantic Network provides a broad categorisation of Metathesaurus concepts. At least one semantic type is assigned to each concept in the Metathesaurus. For example, mastectomy is categorised as a therapeutic or preventive procedure (Fig. 2.12).

2.8.2.3   SPECIALIST Lexicon

The SPECIALIST Lexicon was developed for the SPECIALIST Natural Language Processing (NLP) System. It consists of a set of lexical entries, with one entry for each spelling or set of spelling variants in a particular part of speech. Lexical items may be multi-word terms if the term is determined to be a lexical item by its presence as a term in general English or medical dictionaries, or in medical thesauri, such as MeSH®. The Lexicon also includes acronyms and abbreviations.

A set of 20,000 words form the core words entered. This core is derived from the UMLS Test Collection of MEDLINE® abstracts, together with words that appear in both the UMLS Metathesaurus and Dorland's Illustrated Medical Dictionary. It also includes words from the general English vocabulary and the 10,000 most frequently used words listed in The American Heritage Word Frequency Book and the list of 2,000 words used in definitions in Longman's Dictionary of Contemporary English. (http://www.nlm.nih.gov/pubs/factsheets/umlslex.html).

The lexicon entry for each word or term records the syntactic, morphological, and orthographic information needed for NLP. For example, the verb "to treat" is associated with "treats", the third person singular form in the present tense, "treated" the past and past participle form, and "treating" the present participle form (McCray et al. 1994).

2.8.3   The Health Multiple-Terminologies Portal

Several terminological resources are aligned in the Metathesaurus® database of the UMLS. These relationships, via CUIs, may be exploited for semantic interoperability between medical applications. However, some terminologies are not included in the UMLS.

The Health Multi-Terminology Portal (HMTP) was created by the CISMeF team at Rouen University Hospital. It allows users to navigate through several terminologies (Fig. 2.13). The MeSH is available online, without login. After free registration on the website, users have access to the hierarchies of the following freely available terminologies and ontologies: ATC, CCAM, CIF, ICD-10, CISP-2, Cladimed, DRC, FMA, IUPAC, LOINC, LPP, MedlinePlus, NCCMERP, PSIP Taxonomy, SNOMED Int., VCM, WHO-ART and WHO-ICPS. The HMTP includes 32 terminologies (11 from the UMLS and 21 non-UMLS), for some of which, such as MedDRA and Orphanet, a licence fee must be paid.

When the user enters a term via the interface (Fig. 2.13), he or she can visualise, for this term:

– Its description ("Description" link);
– The list of the terminologies in which the term exists;
– The different hierarchies in which the term exists ("Hierarchies" link), with a visualisation of the complete trees or reduced trees with only the direct subsumers and subsumees;

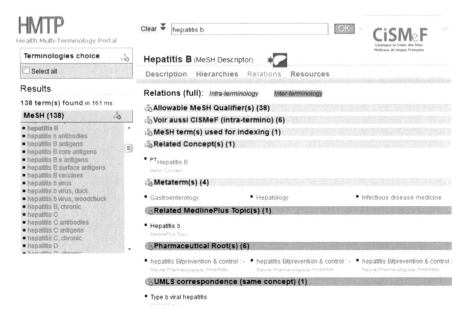

Fig. 2.13 Interterminology and intraterminology relationships for "*hepatitis B*"

- Its intraterminology relationships with other terms in the chosen terminology ("Relations" link);
- Its interterminology relationships deduced with an UMLS Metathesaurus® alignment ("Relations" link);
- Its interterminology relationships deduced with an exact lexical match of the term ("Relations" link).

The HMTP has a multilingual version, the European Health Terminology/ Ontology Portal (http://www.ehtop.eu/). EHTOP includes the 32 terminologies of the HMTP and the same search and navigation functionalities. Only the ICD10 is freely accessible in 11 languages.

2.9   Conclusion

The organisation of diseases into classifications initially facilitated analyses of the causes of death and the first calculations of mortality statistics. With the informatisation of health data, the use of standardised terminologies for the recording of medical information is essential, because it enables the reuse of this information for various goals (e.g. decision support, information retrieval).

Terminological systems have evolved. The first systems organised the terms into lists with or without hierarchical relationships. They were followed by systems allowing the composition of concepts, with a gain in expressivity. The most recent systems use formal definitions of concepts to enable the computer to reason on the basis of these concept definitions and to improve semantic representation. The SNOMED CT ontology is a terminological system increasingly widely adopted by developers of health information systems and is one of the cited terminological references for the implementation of interoperability frameworks (HL7).

2.10 For More Information

This chapter contains many links relating to this topic. We recommend that readers systematically consult the websites mentioned in this chapter.

Many terminological systems exist and only a few have been used as examples in this chapter. Most are accessible via the Internet. We provide a non-exhaustive list of Internet links to access them.

The classification of the world health organisation (WHO http://who.int/classifications/en/)

where you will find:

- International classification of disease, 10th revision (ICD10)
- International classification of disease, 11th revision (ICD11)
- International classification of disease for oncology (ICD-O)
- International classification of primary care, 2nd edition (ICPC)
- International classification of functioning, disability and health (ICF)
- International classification of health interventions (ICHI)
- Anatomical therapeutic classification (ATC) for drugs at http://www.whocc.no/atc_ddd_index/
- WHO Adverse Reaction Terminology (WHO-ART) at http://www.umc-products.com/
- International Classification for Patient Safety (ICPS) at http://www.who.int/patientsafety/en

The reference terminologies
For medicine: SNOMED CT: http://www.ihtsdo.org/
For radiology:
RADLEX: http://www.rsna.org/radlex/
SNOMED Dicom Microglossary: http://www.all-acronyms.com/cat/7/SDM/SNOMED_DICOM_Microglossary/964404

For anatomical pathology
ADICAP: http://www.adicap.asso.fr/
For psychiatry
Diagnostic and Statistical Manual of Mental Disorders IV (DSM): http://www.psych.org/mainmenu/research/dsmiv.aspx
For biological laboratory tests
Logical Observation Identifiers Names and Codes (LOINC): http://loinc.org/

For procedures

Classification Commune des Actes Medicaux (CCAM): http://www.ameli.fr/accueil-de-la-ccam/index.php

Office of Population, Censuses and Surveys Classification of Surgical Operations and Procedures (OPCS4): http://www.connectingforhealth.nhs.uk/systemsandservices/data/clinicalcoding/codingstandards/opcs4

Procedure Coding System (PCS): www.cms.hhs.gov/ICD9ProviderDiagnosticCodes/08_ICD10.asp

For adverse drug reactions

Medical Dictionary for Regulatory Activities (MedDRA): http://www.ich.org/products/meddra.html

For anatomy

Foundational Model of Anatomy (FMA): http://sig.biostr.washington.edu/projects/fm/AboutFM.html

For genetics

Gene Ontology (GO): http://www.geneontology.org/

Human Genome Organisation (HUGO): http://bioportal.bioontology.org/ontologies/45082

The ontology portals

Open Biological and Biomedical Foundation (OBO Foundry): www.obofoundry.org

Bioportal: bioportal.bioontology.org

Exercise

Q1 Search for Haemochromatosis in SNOMED CT, using the following browser: http://vtsl.vetmed.vt.edu/. By looking at the definition of this concept and the set of its antecedents, can you say what haematochromatosis is?

Q2 Do you agree with the SNOMED CT representation of haematochromatosis?

R1 Enter "Haemochromatosis" in the description tab and select the concept "Haemochromatosis (disorder)". Haemochromatosis "has for causal agent" iron.
It is an iron overload, a disorder of iron metabolism, a disorder related to the excess of a trace element. It is a disorder of mineral metabolism, a disorder related to excess intake of micronutrients, a disorder of hyperalimentation, a nutritional disorder.

R2 We can consider whether "disorder of hyperalimentation" is a definition that matches haemochromatosis correctly. This definition was certainly automatically generated (automatic treatment based on concept definitions), hence the need to develop methods for auditing large terminological systems to ensure medical consistency.

References

Baader F, Horrocks I, Sattler U (2005) Description logics as ontology languages for the semantic web. In: Stephan W, Hutter D (eds) Mechanizing mathematical reasoning. Springer, New York

Bakhshi-Raiez F, Ahmadian L, Cornet R et al (2010) Construction of an interface terminology on SNOMED CT. Generic approach and its application in intensive care. Methods Inf Med 49 (4):349–359

Campbell KE, Oliver DE et al (1998) Representing thoughts, words, and things in the UMLS. J Am Med Inform Assoc 5:421–443

Cimino JJ (1998) Desiderata for controlled medical vocabularies in the twenty-first century. Methods Inf Med 37(4–5):394–403

de Keizer NF, Abu-Hanna A, Zwetsloot-Schonk JH (2000) Understanding terminological systems I; terminology and typology. Methods Inf Med 39(1):16–21

EN ISO 1828 (2012) Health informatics – categorical structure for classifications and coding systems of surgical procedures

ISO 1087–1 (2000) Terminology work – vocabulary – part 1: theory and application

ISO 17115 (2007) Health informatics – vocabulary for terminological systems

Lacy LW (2005) Chapter 10. OWL: representing information using the web ontology language. Trafford Publishing, Victoria

Lindberg DA, Humphreys BL, McCray AT (1993) The unified medical language system. Methods Inf Med 32(4):281–291

McCray AT, Srinivasan S, Browne AC (1994) Lexical methods for managing variation. In: Biomedical terminologies, proceedings of the 18th annual symposium on computer applications in medical care, pp 235–239

Ogden CK, Richards IA (1923) The meaning of meaning: a study of the influence of language upon thought and of the science of symbolism. K. Paul, Trench, Trubner/Harcourt, Brace, London/New York

Peirce CS (1909) Existential graphs, MS 514. Eprint of existential graphs MS 514 with commentary by John F. Sowa

Rosenbloom ST, Miller RA et al (2006) Interface terminologies: facilitating direct entry of clinical data into electronic health record systems. J Am Med Inform Assoc 13(3):277–288

Schulz S, Spackman K et al (2011) Scalable representations of diseases in biomedical ontologies. J Biomed Semantics 2(suppl 2):S6

Shvaiko P, Euzenat J (2013) Ontology matching: state of the art and future challenges. EEE Trans Knowl Data Eng 25(1):158–176

Sowa JF (1984) Conceptual structures: information processing in mind and machine. Addison-Wesley, Reading

Chapter 3
Management and Dissemination of Health Knowledge

A. Venot, J. Charlet, S. Darmoni, C. Duclos, J.C. Dufour, and L. Soualmia

Abstract This chapter describes the various types of document available via the Internet for the dissemination of health knowledge. It describes the process for the publication of scientific articles reporting the results of research. The concepts of journal impact factor, citation base and h-index are explained. The Cochrane collaboration for literature reviews and meta-analysis is described. The goals of national and international clinical guidelines are explained and examples are provided of websites providing access to such guidelines. The various types of training materials available from e-learning platforms for medical students are described. This chapter also takes into account drug monographs, which are particularly useful for prescription and disease knowledge bases. Examples of resources useful to members of the public are also provided and we describe the methods for accessing health knowledge, including details of the resource indexing process and the MeSH thesaurus frequently used for indexing. The quality of health information on the Internet is also considered and the principal quality criteria are listed. The HON foundation and certification are described.

A. Venot (✉) • C. Duclos
LIM&BIO EA 3969, UFR SMBH Université Paris 13, 74 rue Marcel Cachin, Bobigny Cedex 93017, France
e-mail: alain.venot@univ-paris13.fr; catherine.duclos@avc.aphp.fr

J. Charlet
CRC INSERM U872 eq 20, Pierre et Marie Curie University, 15 rue de l'Ecole de Médecine, Paris 75006, France

S. Darmoni • L. Soualmia
CISMeF team, LITIS EA4108, Rouen University Hospital, 1 rue de Germont
76031 Rouen cedex, France

J.C. Dufour
Sciences Economiques & Sociales de la Santé & Traitement de l'Information Médicale, UMR 912 Inserm/IRD/Aix-Marseille Université, Faculté de Médecine, 27 Bd Jean Moulin, 13385 Marseille Cedex 5, France

A. Venot et al. (eds.), *Medical Informatics, e-Health*, Health Informatics,
DOI 10.1007/978-2-8178-0478-1_3, © Springer-Verlag France 2014

Keywords Medical journals • Impact factor • Cochrane Collaboration • Clinical guidelines • Drug database • MeSH • HON

After reading this chapter you should:

- Be able to list and describe the various types of medical knowledge resource available.
- Understand the differences between an original scientific article reporting clinical trial results and a meta-analysis of clinical trials on a given topic.
- Understand the differences between systematic reviews of the literature conducted by the Cochrane Collaboration and clinical guidelines for the management of a specific disease.
- Be aware of the various types of educational materials for medicine available from the Internet.
- Know about the contents of drug databases and the way in which knowledge about drugs is represented.
- Be able to list knowledge bases for diseases and explain their various features.
- Be aware of the various types of health resources available to the general public via the Internet.
- Understand the nature of MeSH and how are indexed health documents using this resource.
- Be aware of the principal search engines for accessing different types of medical knowledge.
- Be able to cite the main quality criteria for health information on the Web.
- Know what the HON and HON certification are and be able to identify the HON logo.

3.1 The Various Types of Documents for the Dissemination of Health Knowledge

Web development has greatly changed the management and dissemination of knowledge in the domain of health. New economic models have been devised and implemented, often resulting in the immediate availability of this knowledge, free of charge. Various types of documents have been developed and are available online:

- Scientific articles published in journals and presenting the results of recent research, such as epidemiological studies, clinical trials, studies of diagnostic test performances and imaging tests.

– Literature reviews: compilations of papers presenting the results of studies addressing similar questions (e.g. the efficacy of a new type of hypoglycemic drug for the treatment of type 2 diabetes)
– Clinical guidelines developed by scientific societies or national or international agencies, to support the diagnosis and treatment of specific diseases (e.g.: clinical guidelines for the management of essential hypertension in adults).
– Knowledge presented in a standardised manner, in databases for commercial products, such as drugs, combining scientific, legal and commercial aspects.
– Knowledge on a range of diseases with common characteristics (e.g. rare diseases) stored in databases.
– Medical knowledge in pedagogical documents for the initial and ongoing training of health professionals.
– Medical documents for the general public for which quality criteria and certification procedures have been defined.

These various types of resources and the ways to find and access them are described in this chapter.

3.2 Classification and Description of the Various Types of Resources

3.2.1 Articles and Scientific Journals, Impact Factors and Citation Databases

Scientists conducting research in the field of health generally try to publish their results in journals that are indexed, so that others interested in the same area can find them easily. English is the most widely used language for the dissemination of research results in health.

Several types of journal can be distinguished:

– General journals, which publish articles relating to many areas of health (e.g. Nature Medicine, New England Journal of Medicine, The Lancet, British Medical Journal);
– More specialised journals, focusing specifically on a certain organ (e.g. the heart or kidney) or therapeutic technique (e.g. orthopaedic surgery, organ transplantation), or a particular disciplinary approach (e.g. public health, medical informatics).

All scientific journals now have a website providing access to both the latest issues and an archive of issues from previous years. Access to each item may be free of charge or may require payment (with some journals demanding payment for all articles and others only for recent articles). Some journals are available only in electronic format, whereas others are available in both "paper" and electronic forms. Some journals providing free access to articles require a financial contribution from the authors at the time of publication (e.g. 1,200 Euros). In such cases, readers have free access to the article, which is paid for by the author.

Each journal has a full name (e.g. New England Journal of Medicine) and a standardised abbreviation (e.g. New Engl J Med). It also has an impact factor, which is calculated each year (Callaham et al. 2002). This impact factor is used to determine a hierarchy of importance for these journals and to quantify the scientific output of researchers.

Citation databases have been constructed and are used for the calculation of impact factors. These databases contain the complete references of articles citing a particular article. These data are very interesting for researchers who wish to know the real impact of their work. The actual number of citations of an article may differ significantly from the expected number of citations calculated from the impact factor of the journal concerned, and this may be good or bad news for the authors!

Article citation databases include (Kulkarni et al. 2009):

- Journal Citation Reports ® from ThomsonReuters, http://thomsonreuters.com/products_services/science/science_products/az/journal_citation_reports/
- Scopus, http://www.scopus.com

The Impact Factor of a Journal

The impact factor of a journal is directly related to the number of citations (in all journals) of all the articles published in the journal during the previous 2 years.

If T is the total number of scientific articles published in a journal in 2010 and 2011, and C is the total number of citations of these articles in 2012, the impact factor (IF) for 2012 is $IF = C/T$. The impact factor thus represents the mean number of citations of articles published in the journal 1 or 2 years earlier.

An impact factor for 2012 of 5 indicates that, on average, each of the articles published in the previous 2 years (2010–2011) was cited five times in 2012. The impact factors of major general medical journals may reach values of about 40.

The consultation of these citation databases is generally not free of charge. However, Google Scholar http://scholar.google.fr/ provides free access to citations calculated in an automatic manner, whereas the other databases are generated by manual indexing. These citation databases are also used to calculate indices quantifying the volume and quality of the scientific output of researchers (Nieminen et al. 2006).

Thus, the h factor proposed by Hirsh is defined as follows: an author has an index of h if he or she has h publications cited at least h times, with his or her other publications being cited less than h times. Therefore, a researcher with an h factor of 30 has 30 publications that have been cited at least 30 times, with his or her other publications cited less frequently.

The h factor of an author can be calculated by ranking his or her articles in decreasing order of the number of citations. A graph is them plotted, with the number of citations on the y axis and the rank of each item on the x axis. This generates a curve, the intersection of which with the first bisector gives the author's h factor.

The process of publishing research findings in an international medical journal

For an author, the choice of journal in which to try to publish research findings depends on the match between the supposed quality of a journal (target audience, areas covered, types and quality of the results usually published in the journal) and the quality and interest of the researcher's own results. Major journals with high impact factors are read by a large number of people and their articles often cited, but their requirements in terms of the interest of the results and method quality are higher. It is therefore more difficult to publish in these journals.

Once the target journal has been chosen, the authors must consult its website for "instructions to authors" providing guidelines concerning the form of the article (e.g. concerning the text, figures, tables, references) and how to submit an electronic version of an article.

The authors then write their article in accordance with these guidelines and submit it for publication, pledging not to submit it simultaneously to another journal. The publisher of the journal, or its editorial board, often looks over the article rapidly to ensure that it matches the range of subjects covered by the journal. The article may be dismissed out of hand at this early stage.

If not rejected outright, the article is sent by the editor to several experts of his or her choice in the corresponding field (named reviewers), who are asked to carry out a critical appraisal of the article concerned. The editor provides a summary of their analyses and sends the final decision to the authors: the paper may be accepted directly for publication with no further revision (a very rare event), modifications may be required before resubmission, with the authors having to explain how they have taken the criticisms of the reviewers into account, or the article may be refused. If the article has to be modified, the process may be iterative, with the text passing back and forth between the authors and the reviewers.

It may take several weeks or months for accepted articles to be published. Increasingly, however, the article is posted in electronic form before its publication in the paper version of the journal.

Once published, the article is indexed and appears in bibliographic search engines, such as PubMed. The time between the publication of an article and its indexing in a bibliographic database is variable (from almost immediate to several months).

3.2.2 The Cochrane Collaboration, Literature Reviews and Meta-Analysis

Physicians and healthcare professionals do not have time to read all the literature on a given topic. Moreover, published results may be contradictory or biased. There is therefore a need for accessible systematic literature reviews.

The Cochrane Collaboration was established in 1993 and is an international non-profit organisation present in more than 100 countries (http://www.cochrane.org/). Within this organisation, a network of people (28,000 in 2010) carry out literature reviews (http://www.cochrane.org/cochrane-reviews) with the aim of providing the answers to major questions in medicine, for physicians, patients and policy-makers.

Thousands of literature reviews have already been completed and are available online from the "Cochrane Library" (http://www.thecochranelibrary.com). These systematic reviews concern principally the prevention or treatment of diseases and evaluations of the performance of diagnostic tests.

Each literature review aims to provide answers to precisely formulated questions (e.g., for antibiotics: do these drugs relieve the symptoms of sore throat? How useful are they for treating middle ear infections in children?).

Once the question has been formulated, the work of the Cochrane Collaboration begins with the identification of all articles published on the issue concerned exceeding a predefined quality threshold. Summaries are written, indicating whether there is sufficient evidence to conclude, for example, that a particular type of treatment is or is not effective.

These literature reviews are updated periodically to take recently published findings into account. A specific methodology, that of "meta-analysis", has been developed to determine the level of statistical significance associated with the results of overall analyses of the various studies selected on a given topic.

3.2.3 Clinical Guidelines

Literature reviews can provide only partial conclusions concerning the best ways to manage disease in patients. A higher degree of synthesis, possibly including expert opinion, may be required. The goal is to standardise practices and to inform physicians of the best ways to diagnose and treat a given disease.

Such recommendations are included in clinical guidelines, which are produced by scientific societies, health agencies and private publishers.

In the United States, the Agency for Research and Quality in Health Care has established the National Guideline Clearinghouse (NGC), which holds sets of guidelines based on evidence-based medicine (http://www.guideline.gov/). These documents are available from the Internet. Equivalent sites have been established in the UK (NHS Evidence; http://www.evidence.nhs.uk/) and Canada (CMA

InfoBase; http://www.cma.ca). In France, guidelines are available from the websites of agencies such as the National Health Authority (http://www.has-sante.fr/) and that of French Agency for the Safety of Health Products (http://www.ansm.sante.fr).

3.2.4 Training Materials (Courses, Clinical Cases, Multiple Choice Questions, Videos, Podcasts etc.)

The number of e-learning platforms has greatly increased in recent years, because open source e-learning software under the GNU or LNG, for installation on servers, can now be downloaded and readily customised by universities.

Examples include:

- Moodle (Modular Object-Oriented Dynamic Learning Environment) http://moodle.org/
- Claroline http://www.claroline.net/

These programs allow teachers access to online learning materials in many formats (text, html, pdf, powerpoint, video, podcasts) and to multiple-choice exercises.

Many documents are accessible from the websites of universities, either freely or after university enrolment. These documents are created either by teachers acting individually or by groups of teachers writing in a collegiate manner. In this second case, the content and form of the documents may be particularly useful because they may result from a consensus of several renowned teachers.

3.2.5 Drug Databases

Physicians prescribing drugs to patients require appropriate information to ensure that the drugs are prescribed correctly. The prescribed product must be named and its name spelt without error, and the dose prescribed should be appropriate for the disease or symptom to be treated and for the specific characteristics of the patient, such as age, weight, height and medical history. The physician must also be able to ensure that the prescription is not contraindicated for the patient, taking into account possible allergies, other diseases and treatments.

Thousands of drugs are currently available and it is no longer possible for the physician to be fully aware of properties of all the products they may need to prescribe. This problem led to the publication of various paper-based dictionaries or formularies, to assist physicians with their prescriptions.

The computerisation of health processes then led to the development of computerised knowledge bases for drugs, containing both text information about each commercial product and structured and coded data useful for prescription assistance systems (See Chap. 8 on therapeutic decision support).

Drugs are commercial products with a standardised process of development. The dissemination of knowledge about drugs therefore obeys rules and has specific uses, depending on the country in which it occurs. The information collected during clinical trials before the authorisation of market release, is used to generate monographs, detailing items essential for prescribers, such as:

– The active ingredient: chemical product, molecule principally responsible for the effect of the drug. Information about "therapeutic class" (i.e. the "family" of drugs the active ingredient belongs to) is particularly useful to the physician.
– Indications: clinical situations, mostly diseases (e.g. essential hypertension in adults) and symptoms for which the drug may be useful.
– Contraindications: diseases for which the drug should not be prescribed (e.g. beta-blockers in asthma).
– Doses: the amount of drug to be administered to the patient, together with the periodicity of intake (how many times the drug should be taken per unit time; hourly, daily weekly etc.) and, possibly, the precise moment at which the drug should be taken (with meals or at some distance from mealtimes, in the morning and not at night, etc.). The total duration of treatment may also be given and may differ between indications.
– Adverse reactions: clinical (e.g. nausea) and biological (e.g. increase in transaminase levels) manifestations that may be caused by the drug, taken at the recommended dose. This section often reports the frequency of occurrence of these effects.
– Overdosage: manifestations of a similar nature as those for adverse reactions, but occurring only if a dose exceeding that recommended is given.
– Drug interactions: there may be a list of drugs that are dangerous when administered together with the prescribed medication.

Drug databases have been developed and are maintained in all developed countries. Some are freely available, such as DailyMed (http://dailymed.nlm.nih.gov), which is maintained by the National Library of Medicine in the United States.
Others are commercial products, such as:

• In the United Kingdom, The British National Formulary (http://www.bnf.org),
• In France, The Claude Bernard database (http://www.resip.fr), Theriaque (http://www.theriaque.org)/, Thesorimed (http://www.giesips.org) and Vidal (http://www.vidal.fr)
• In the United States: Firstdatabank http://www.fdbhealth.com, Micromedex (http://www.micromedex.com/) and Multum (http://www.multum.com)
• In Germany, die Rote List (https://www.rote-liste.de)

Health authorities can establish a process for the accreditation of these knowledge bases. The aim is to ensure that the drug databases approved are complete (all medicinal products must be present), accurate and up-to-date, with complete information provided for each drug.

3.2.6 Diseases Databases: The Example of Orphanet in Europe

Orphanet (http://www.orpha.net), created by the INSERM (French National Institute of Health and Medical Research) and the French Directorate General of Health (French Ministry of Health and Sport), is a database on rare diseases, the associated genes and orphan drugs (Rath et al. 2012). This database is available in six languages (English, French, German, Italian, Portuguese and Spanish) and is cross-referenced with the International Classification of Diseases (ICD 10) and OMIM (www.ncbi.nlm.nih.gov/omim). Orphanet also produces an online encyclopaedia, including standardised summaries and review articles written by experts, and a directory of expert services in Europe. In Europe, a rare disease is defined as a disease with a prevalence of less than 1 in 2000 people in the European population. The threshold used to define rarity is not necessarily the same in other countries outside Europe. Besides, the prevalence of a disease may differ between countries or, indeed, between continents. Thus, a disease that is considered rare in France may not be rare in Japan, for example, and vice versa. The diseases referenced in Orphanet are thus diseases that are rare in Europe.

A disease can be sought in the database, by using the alphabetical list, or indicating the exact name, or just part of the name, of the disease in the search box. Orphanet is the only European database providing such complete information for rare diseases: most of the diseases are linked to a review article or a summary text, and to a list of specialised clinics, diagnostic laboratories, research projects, patient registries, clinical trials, networks of professionals, and patient organisations identified for the disease.

The review articles of the encyclopaedia are written by experts, in a standard format, to make them easier to read. They comprise the following sections: summary, key words, name of the disease and synonyms, differential diagnosis, prevalence, clinical description, genetic aspects, prenatal diagnosis, management, reference list.

The texts are written in English and are subjected to a rigorous review process, managed by a European editorial committee. All the review articles are translated into French, and most are also available in the other languages of the database. Most diseases are also linked to a selection of relevant websites.

Information about expert services is currently collected in 35 European countries, under the responsibility of national scientific committees, according to common selection criteria. All information on the site is validated by the scientific committees in each country, with a complete update performed annually. The database is also subject to continuous quality control procedures. Orphanet is freely accessible and free of charge for all audiences, and the same information is available for both patients and healthcare professionals alike. Orphanet is funded by a number of sources including, notably, the INSERM, the French Directorate General of Health, and the European Commission.

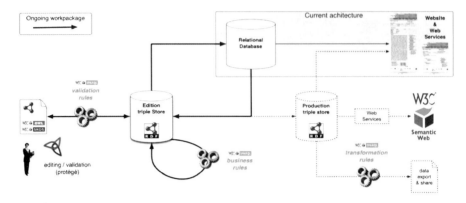

Fig. 3.1 The Orphanet editorial workflow

In 2010, Orphanet launched a process for overhauling the referencing system for rare diseases, to improve the workflow for the updating of the knowledge base, particularly classifications. Thus, the thesaurus of rare diseases is currently being transformed into an ontology and classifications useful for the referencing of medical knowledge (e.g. a classification of genetic diseases) will be generated from this ontology, which will become the reference resource of the system. Knowledge about diseases will thus be updated in a single ontology, with all changes automatically reflected in the classifications. As concerns the display of classifications, the availability of the ontology and classifications will improve the standing and interoperability of the other knowledge bases linked to rare diseases, such as biobanks or HPO.[1] Finally, the use of Semantic Web[2] technologies will, in the long run, make it possible to extract data from knowledge bases in interoperable formats, such as Sparql Endpoint (see Fig. 3.1).

3.2.7 Resources Available to the General Public

Many sites provide members of the general public with information about diseases and drugs in everyday language understandable to the layman (Santana et al. 2011).

For instance, Medline Plus (http://www.nlm.nih.gov/medlineplus) is a database designed specifically for the general public but also accessible to healthcare professionals. It was created by the NLM (National Library of Medicine) in the US and provides selected, validated and updated information for more than 700 diseases (see "Health topics") and drugs (see "Drug Information"). This

---

[1] *Human Phenotype Ontology* http://www.human-phenotype-ontology.org/.

[2] http://www.w3.org/2001/sw/

database is free and contains no advertising links. It is available in English and Spanish, and offers two search modes: by keyword and alphabetical index (Olney et al. 2007).

In France, Doctissimo http://www.doctissimo.fr/ provides another example. This database provides information on many health topics, including diseases and drugs.

HONcertified sites (a quality label explained later in this chapter) are of particular interest.

3.3 Methods for Accessing Health Knowledge

The most appropriate way to obtain health knowledge is to make use of specialised websites and servers. However, this approach is not the most prevalent, with general search engines, such as Google or Yahoo being the most used. Web 2.0 has become yet another way to transmit and exchange health information, through moderated forums developed for health professionals and for patients and their families.

Specialised websites and servers provide access to health information and knowledge through directories indexed alphabetically or by theme. Access can also be obtained via search tools, in which the user can enter a query in natural language, as in Google, for example. The search tool associated with the website or server then makes use of predefined algorithms to retrieve a set of documents corresponding to the user's query. However, an initial document indexing step is required to allow the search tool to match the query with the document index.

Depending on the search tool, the query may be simple (i.e. simple words in free text) or more sophisticated (i.e. with a query language and its associated syntax and a set of keywords from a controlled vocabulary). In this second case, indexing is performed manually, thus reducing the set of documents (as in Yahoo for general categories, or in MEDLINE, NGC for health). Conversely, with automatic indexing (as performed by GoogleBot, the Google crawler), the number of indexed pages may reach several billions, but with a lower level of precision, as the process is entirely automatic, with no human control.

3.3.1 Resource Indexing

An index is a surrogate for the content of a document. It may have several different types of component. It may be a single word (e.g. *asthma, hepatitis, diagnosis*), a term consisting of several simple words (e.g. *yellow fever, pH dependence*), an entry term in a thesaurus (i.e. a list of terms and descriptors from a controlled vocabulary; e.g. *status asthmaticus; type II diabetes; hepatitis B, chronic*).

An index is difficult to extract, either manually or automatically (Huang et al. 2011). It may also describe the container of the document, with content and container represented by metadata. Metadata provide information about documents and objects or structured information about documents. For the Internet, metadata provide descriptive information about Web resources and represent the content, structure and logistical information for any information object, including electronic resources. The Dublin Core Metadata Initiative (DCMI) is a metadata element set designed to facilitate the discovery of electronic resources (http://dublincore.org). Originally designed for the author-generated description of Web resources, the DCMI is now used by museums, libraries, government agencies and commercial organisations alike. The Dublin Core has 15 elements, including *descriptors, title, authors, publication date and language*.

3.3.2   The MeSH

The MeSH® thesaurus (Medical Subject Headings http://www.nlm.nih.gov/mesh/) was developed by the US National Library of Medicine (NLM) in 1960. It is maintained, with an annual update, by experts. In its 2012 version, it contains 26,581 descriptors, which represent medical concepts, organised into 16 categories (including [A] Anatomy, [N] Health Care, [V] Publication Types and [Z] Geographic locations). About 200–300 descriptors are added each year. During the update to generate the 2012 version, 454 descriptors were added (e.g. *exome, endophytes*), 41 descriptors were modified (e.g. *Technology, Medical* replaced by *Medical Laboratory Science*; *Mental Retardation* replaced by *Intellectual Disability*) and 15 descriptors were deleted (e.g. *Neuroleukin*; *Laboratory Techniques and Procedures*). It is available in more than 20 different languages.

3.3.2.1   Descriptors (or MeSH Headings)

The MeSH® descriptors are organised into hierarchies, with the most general descriptor at the top of the hierarchy and the most specific descriptor at the bottom. A hierarchical tree first describes general categories (e.g. *Diseases*) and then more specific notions (e.g. *Lyell syndrome*), at different levels of the tree. Each hierarchy contains no more than 11 levels and each descriptor is identified by a unique number and one or more "MeSH Tree Numbers", depending on its position in the tree.

For example, *chronic hepatitis* has the MeSH Tree Number C06.552.380.350, which is more specific than *hepatitis* C06.552.380. A descriptor may belong to several trees, as the MeSH is a multiple-hierarchy thesaurus. A descriptor may have several siblings and a sibling may have several ancestors.

3.3.2.2 Qualifiers (or MeSH Subheadings)

The qualifiers make it possible to specify which particular aspect of a descriptor is addressed in a citation, and then to focus on a subfield of the descriptor. The qualifiers are also organised into hierarchies. Each qualifier must be used contextually. In its 2012 version, the MeSH contains 83 qualifiers. For the indexing of a scientific article and citation, an NLM indexer may associate a qualifier with a MeSH descriptor. For example, the association of the descriptor *hepatitis* with the qualifier *diagnosis* (noted *hepatitis/diagnosis*) restricts the hepatitis to its diagnostic aspects.

Extract of Abstract Indexing of a Citation in MEDLINE

Horrocks JC, McCann AP, Staniland JR, Leaper DJ, De Dombal FT. - Computer-aided diagnosis: description of an adaptable system, and operational experience with 2,034 cases. Br Med J. 1972 Apr 1;2(5804):5–9.

　　PMID – 4552593 *(PubMed identifier: each indexed citation a single and unique identifier)*

　　DP – 1972 Apr 1 *(Date of publication of the article)*

　　TI – Computer-aided diagnosis: description of an adaptable system, and operational experience with 2,034 cases *(title)*

　　PG – 5–9 *(page numbers)*

　　AB – This paper describes ... Experience in this setting suggests that computer diagnosis may be a valuable aid to the clinician. *(abstract)*

　　FAU – Horrocks, J C *(full author name)*

　　AU – Horrocks JC *(author name in the citation)*

　　. . .

　　LA – eng *(language of the citation)*

　　PT – Journal article *(publication type)*

　　PL – ENGLAND *(place of publication)*

　　TA – Br Med J *(journal title abbreviation)*

　　JT – British Medical Journal *(journal title)*

　　MH –Abdomen *(MeSH term)*

　　MH – Abdomen, Acute/etiology

　　MH –*Computers *(major MeSH term)*

　　. . .

　　EDAT – 1972/04/01 *(edition date of the citation in the PubMed database)*

　　MHDA – 1972/04/01 00:01 *(The date MeSH terms were added to the citation)*

　　SO – Br Med J. 1972 Apr 1;2(5804):5–9. *(source: bibliographic information)*

3.3.3   Search Engines in the Domain of Health

3.3.3.1   PubMed/MEDLINE

MEDLINE (Medical Literature for Analysis and Retrieval System Online) is a bibliographic database of references in the biomedical field. It covers all disciplines, including medicine, pharmacy, biology and public health. Produced by the NLM, it provides various types of access, the most widely used of which is PubMed (http://www.ncbi.nlm.nih.gov/PubMed), which has been available free of charge since 1997.

More than 5,500 biomedical journals are listed in the database, with over 18 million article references indexed manually, dating back to 1960 (or 1945 for OldMedline, which contained about 480,000 articles in May 2011). There are also more than 800,000 recent articles referenced in PubMed but not yet indexed manually. Finally, PubMed contains about 675,000 referenced articles that are not, strictly speaking, in the MEDLINE database. In total, PubMed contains over 20.7 million references. Most of the journals concerned are in English.

From the outset, the MeSH thesaurus was used to index the various MEDLINE resources. PubMed queries can be carried out by inserting the name of an author or an illness in the search box on the homepage. However, more complex and precise queries are also possible. The MEDLINE indexers (of which there are about 50) manually index scientific articles, each taking about 15 min, with an average of 15 MeSH descriptors, with two levels of indexing: Major (marked with an *; see below) and Minor. The Major MeSH terms are considered more important than the Minor terms.

PubMed Query with the Appropriate Syntax

A query language is associated with the PubMed search tool. The generic syntax is as follows: Term [field] operator
　　The Boolean operators are indicated in upper case in the query: AND, OR.
　　Fields can, for example, take the following values:
　　[AD] Affiliation
　　[ALL] All fields (research in all fields of manual indexing)
　　[AU] Author Name
　　[TA] Journal name
　　[MH] MeSH term
　　[MAJR] Major MeSH term
　　[DP] Publication date
　　[PT] Publication type
　　[SH] Qualifier, or MeSH subheading
　　[TW] Text (words in the document text)
　　[TIAB] Search in titles and abstracts

PubMed Query with the Appropriate Syntax (continued)

The PubMed tool is not case-sensitive (upper-case and lower-case letters are treated in the same way). The following queries are equivalent: hepatitis [mh] = Hepatitis [mh] = HEPATITIS [MH].

By default, this greatly increases the number of documents retrieved: a query of "hepatitis" will return all documents indexed by the descriptor "hepatitis", but also all documents indexed by derivatives of the term "hepatitis". This "explosion" of findings can be disabled in the query with the fields [mh: noexp] or [MAJR: noexp].

Qualifiers can be affiliated to the descriptors in queries, with the following syntax: MeSH Term/Qualifier [Fields]. The qualifier name or its abbreviation can be used: hepatitis/diagnosis [MH] = hepatitis/di [MH]

Several qualifiers for the same descriptor can be combined, using the OR operator:

Hepatitis/di [MH] OR Hepatitis/th [MH]

3.3.3.2 National Guideline Clearinghouse

Created in 1998, this website is an initiative of the US health agency AHRQ (Agency for Healthcare Research and Quality). Its purpose is to provide standardised access to clinical guidelines in English. The current recommendations are identified, summarised, described and classified if they meet stringent criteria for inclusion. This site currently includes more than 2,500 recommendations.

3.3.3.3 CISMeF

The Catalogue and Index of French-speaking Medical Sites (CISMeF http://www.chu-rouen.fr/cismef/) identifies major sites and medical documents from public sources (health agencies, scientific societies, universities, foundations etc.) or associations. Created in February 1995, at the initiative of the University Hospital of Rouen, this database focuses on health and medical sciences.

Information is selected and indexed with the French version of the MeSH thesaurus (Darmoni et al. 2012). Since 2007, CISMeF has made use of a multiple terminology for manual (already 12 % of the catalogue) or automatic indexing and for information retrieval.

Each resource is associated with a descriptive note, in a standardised format making us of 11 of the 15 Dublin Core elements (http://www.dublincore.org), including title, author, description, website publisher, date, URL, language, keywords, resource type.

Many improvements have been made to the MeSH thesaurus since 1995: 20,000 synonyms have been added to MeSH terms with ambiguous acronyms managed by the search engine, 6,000 definitions have been translated (MeSH Scope Note), approximately 12,000 MeSH Supplementary Concepts have been translated, with over 6,000 synonyms for these MeSH SC. Two concepts have been added: 121 metaterms (or super-concepts) making it possible to search for information within a medical discipline (e.g. Dermatology), and 296 types of resource (e.g.: Forum List diffusion), extending the publication types available and able to contain descriptions of health resources available via the Internet.

Example of a Metaterm: Dermatology

Metaterms correspond to a biological or medical speciality associated with one or more keywords (keywords or trees) or qualifiers. In MeSH, the descriptor "Dermatology" (H02.403.225) is described simply as a health profession (H02) in the hierarchy of disciplines and professions [H].

In CISMeF, the metaterm "Dermatology" was created to improve information retrieval by combining a number of descriptors from different trees, and the corresponding subtrees:

Dermabrasion (E02.218.210; E04.680.250)

Dermatology (H02.403.225)

LSD (C02.256.743.494; C22.196.497)

Dermoscopy (E01.370.350.515.184)

Hygiene of the skin (E02.547.800)

CTCL (C04.557.386.480.750.800; C15.604.515.569.480.750.800)

Skin diseases (C17.800)

Three priorities have been identified: resources for education, recommendations for good clinical practice and documentation specifically for patients and the public. Hospitals, health centres and clinics, institutions, medical libraries, electronic journals, databases and associations (professional and patient) are also identified. Almost 91,000 resources are accessible in various ways: alphabetical, subject index, access by resource type (e.g. hospitals, universities, databases), together with the search engine, introduced in 2000, Doc CISMeF, which is primarily destined for use by health professionals. The catalogue can be searched in English or French.

3.4 The Quality of Health Information Available from the Internet

3.4.1 Why Do We Need to Pay Attention to the Quality of the Information About Health Available from the Internet?

The health information available from the Internet is of heterogeneous quality. Some of this information is of high-quality, but misinformation, outdated, incomplete, intentionally or unintentionally misleading information may also be posted and maintained on the Internet. This heterogeneity of information quality results principally from the variability of the process leading to the dissemination of health information via the Internet. This process may be rigorously controlled, as, for example, when a medical journal publishes scientific articles online that have previously passed through validation and review stages before acceptance for publication.

By contrast, this process may be devoid of knowledge validation stages if, for example, an ordinary user provides medical information on an unmoderated forum, a personal site or blog, or via a social network. Between these two extremes, there are various modes of publishing online, resulting in the dissemination of health information of various degrees of reliability (e.g. a wiki allowing the collegiate and instantaneous correction (rightly or wrongly) of information by self-appointed qualified Internet users).

The quality of the health information found on the Internet is, therefore, highly heterogeneous. This is a potentially dangerous situation, because most users have no background medical knowledge enabling them to identify and consider only the relevant, high-quality information.

This is a growing problem because the number of Internet users searching for medical information is increasing. In April 2010, a study conducted by IPSOS (http://www.ipsos.fr) on behalf of the College of Physicians showed that more than 70 % of French people have used this medium to search for health information. In the US, data collected by the Pew Internet & American Life Project have shown that searches for health information rank third in the hierarchy of online activities (after the use of e-mail and querying search engines). The vast majority of Internet users make use of generalist directories or common search engines, such as Google, Yahoo or Bing, rather than specialised search tools based on reliable resources.

One of the problems with the use of health information available from the Internet is that the quality of the information must be systematically assessed on the basis of a number of objective criteria.

3.4.2 The Main Quality Criteria for Health Information

Many initiatives have been developed, since the mid-1990s, for the definition of quality criteria for health information disseminated via the Internet. Most of this work has led to the publication of multicriterion scoring grids and/or the appraisal of resources (sites or documents) available via the Internet. Some of these grids have led to the creation of labels or mechanisms of accreditation or certification by a trusted third party, to facilitate the identification of resources satisfying some or all of the quality criteria of these grids. Finally, others have helped to establish "codes of conduct" or "practical recommendations" for the dissemination of health information.

One conclusion that can be drawn from these studies is that the quality of health information can be assessed with several criteria of two main types:

Table 3.1 The main quality criteria for health information on the Internet

Source	Is it possible to identify the institution producing and/or publishing the information?
	Is it possible to identify the authors and their areas of expertise?
	Is there an editorial board and/or a scientific committee?
	Have the original sources of information (citations) been given?
	Is it possible to identify the promoter and the funding source behind the information? Is there a conflict of interest?
Content	Is the creation date and/or updating of information mentioned? Is it acceptable?
	Is the public targeted by the information indicated?
	Is the editorial content well arranged (coherent plan, spelling, grammar, translation)?
Interface	Is it possible to obtain feedback (report any errors to the author, write to the person responsible for managing the online publication, etc.)?
	Are the accessibility of information, the organisation and navigability of the site satisfactory?

- Criteria relating to the "form" of information, focusing on the "container", the environment in which the information is placed (the website itself).
- Criteria relating to the "substance" of information itself, the "content" and intrinsic qualities of the information, regardless of the format in which that information is disseminated.

The table above presents a number of quality criteria for health information available from the Internet, which are easy to assess without having to read the entire document or site. These quality criteria can also be applied to websites containing information relating to domains other than health (Table 3.1):

Whatever the type of information, the consultation of Internet resources requires a critical mind and an ability to select, sort, prioritise and cross-check various sources of information. This can only be achieved by taking the quality criteria into account and giving higher priority to items of information than to others, depending on the context and nature of the information (e.g. you are likely to have less stringent requirements concerning the date on which a human anatomy diagram, which is unlikely to change, was last updated than for aspects of the treatment of a particular disease, which is likely to benefit from continual advances in science).

3.4.3 The HON Foundation and HON Certification

The Health On the Net Foundation (HON) http://www.hon.ch is one of the initiatives directly contributing to the development of criteria for the quality of health information on the Internet (Boyer et al. 1998). The HON has set itself the task of guiding Internet users to online sources of health information that are reliable, understandable and relevant. HON was created in 1995. Since 2002, it has been recognised as a non-governmental organisation by the United Nations. It has its headquarters in Geneva and is funded by the Canton of Geneva, the European Commission, the French National Health Authority and the Provisu Foundation.

Fig. 3.2 'Clickable'
HONcode-certified logo,
giving the most recent date of
certification for the website

In France, under a partnership agreement with the national agency for health, the "Haute Autorité de Santé", the HON has been the official organisation responsible for certifying health websites since 2007. Health websites seeking HON certification must conform to the eight HONcode principles established when the HON foundation was created. The HONcode has spread worldwide and, over the last couple of decades, has become the reference in the field of health website certification.

The HONcode is based on a grid of eight quality criteria that sites must agree to meet for certification:

1. Authoritative: The qualifications of the authors must be indicated.
2. Complementarity: Information should support, not replace, the doctor-patient relationship.
3. Privacy: Respect for the privacy of patients and the confidentiality of personal data submitted to the site by the visitor.
4. Attribution: Citation of the sources of the information published and its date, for medical and health pages.
5. Justifiability: The site must back up claims relating to benefits and performance.
6. Transparency: Accessible presentation, accurate e-mail addresses for contact.
7. Financial disclosure: Identification of funding sources.
8. Advertising policy: Advertising clearly distinguished from editorial content.

Any website containing health information for the general public and/or professionals, the content of which may or may not be strictly medical, can request certification from HON. This request, which is free and voluntary on the part of website publishers, is followed by an assessment by reviewers from the foundation. The reviewers then check that each of the eight criteria are met by the website and that there are elements explicitly allowing these criteria to be checked (provision of a privacy policy, the editors of the website and sources of funding listed, etc.). If the website meets all the criteria, without exception, it is certified and allowed to display a dynamic logo (Fig. 3.2), on which users can click to be redirected to a dedicated page on the HON website on which the certificate of HONcode compliance is confirmed and dated. This provision of access to the certificate on the HON Foundation website allows users to check that the certificate is not being misused and that the site does not display the HONcode logo without having successfully passed the certification.

A certified website is periodically inspected by HON reviewers. Certification of the website may be suspended if it no longer complies with the HONcode or renewed if the criteria are still satisfied. Certification visits occur at intervals of less than 12 months. Reviewing may occur sooner, on the basis of internet users' reports or the findings of the automated reporting tools of the foundation.

However, it would be presumptuous to assert that HONcode certification guarantees the accuracy of the information published by certified websites and that the HONcode liberates visitors entirely from the need to analyse health information critically. Indeed, erroneous information, whether intentionally so or otherwise, may not necessarily be identified by the foundation's reviewers. However, the HONcode provides a guarantee of the quality of the editorial policy and the commitment of the publishers to maintaining this quality.

3.5   For More Information

This chapter contains many links relating to this topic. We recommend that readers systematically consult the websites mentioned in this chapter.

The reader should spend time learning to query PubMed with the "advanced" and "limited" options. The reader can, for example, construct queries to find an answer to Question 2 of Exercise 1 initially without using these options and, then, using them.

The reader may consult the website of the Cochrane Collaboration to obtain more information about its organisation, its objectives and the resources that can be found on the website.

The reader should browse the MeSH tree on the website of the NLM, to become familiar with the various keywords that can be used in some medical search engines.

Exercises

Q1 Here are four questions a doctor may have to answer:

Q11: Is acupuncture effective for treating migraine?

Q12: What publications have there been on the treatment of type 2 diabetes in the last 3 years?

Q13: What are the treatment options for a 74-year-old patient with type 2 diabetes?

Q14: What are the contraindications for the prescription of antibiotics, such as amoxicillin and potassium clavulanate?
In which types of resource can the answers to these questions be found? Which websites may the doctor consult to find the answers? (Give separate answers for each question). Find answers to the four questions. Note the time

Exercises (continued)
you take to find the answer to each question. Draw conclusions on access to medical knowledge, particularly in clinical situation (i.e. during the consultation).

Q2 A patient wants more information about bradyopsia and the genes involved in this disease. What options does he/she have?

Q3 A healthcare professional wants to consult as many documents related to insulin therapy as possible. Propose different ways to obtain access to these documents.

Q4 Search recommendations and consensus conferences in MEDLINE.

Q5 Search the term "gated SPECT" (single-photon emission computed tomography), eliminating references published in J Nucl Med.

R1

R11 The answer can be found in a literature review, see the Cochrane Library.

R12 The answer can be obtained by querying PubMed.

R13 Find access to a clinical practice guideline focusing on the management of type 2 diabetes and read the chapter specifically relating to old people.

R14 Find a drug database, such as DailyMed. Select the drug amoxicillin and potassium clavulanate and read the contraindications in the monograph.

R2 The patient may have access to this information via Orphanet website. If he/she does not know the exact spelling of the disease, he/she may browse the alphabetical index rather than the search engine within this website. A link to the genes (two known in 2011) and a link to PubMed (query "*bradyopsia* OR *prolonged suppression response electroretinography*") completes the available information.
Access to English documents via MedlinePlus might have been possible, but owing to the rarity of the disease, no resource destined for use by patients is available on MedlinePlus.

R3 One possible route of access would be to use the National Guideline Clearinghouse, selecting "Guidelines" and using the key words "insulin therapy" for the query. Alternatively, this information is accessible via PubMed and the MEDLINE references.

R4 ("guidelines"[MeSH Terms] OR "Consensus Development Conferences"[MeSH Terms] OR "consensus development conference"[Publication Type] OR "consensus development conference, nih"[Publication Type] OR "guideline"[Publication Type] OR "practice guideline"[Publication Type])

(continued)

Exercises (continued)
The search can be completed by consulting the following sites: http://www.
guideline.gov/ et http://rms.nelh.nhs.uk/guidelinesfinder/ (UK).

R5 (gated spect [All Fields]) NOT (j nucl med [Journal Name])
The expression gated SPECT is not recognised by PubMed, because it is not
available in MeSH. It is therefore necessary to search with the exact expression.
This can be achieved by carrying out the research on [All Fields] i.e. all the
fields including MEDLINE titles and abstracts.

References

Boyer C, Selby M, Scherrer JR et al (1998) The health on the net code of conduct for medical and
    health websites. Comput Biol Med 28(5):603–610
Callaham M, Wears RL, Weber E (2002) Journal prestige, publication bias, and other
    characteristics associated with citation of published studies in peer-reviewed journals. JAMA
    287(21):2847–2850
Darmoni SJ, Soualmia LF, Letord C et al (2012) Improving information retrieval using Medical
    Subject Headings Concepts: a test case on rare and chronic diseases. J Med Libr Assoc 100(3):
    176–183
Huang M, Névéol A, Lu Z (2011) Recommending MeSH terms for annotating biomedical articles.
    J Am Med Inform Assoc 18(5):660–667
Kulkarni AV, Aziz B, Shams I et al (2009) Comparisons of citations in Web of Science, Scopus,
    and Google Scholar for articles published in general medical journals. JAMA 302(10):
    1092–1096
Nieminen P, Carpenter J, Rucker G et al (2006) The relationship between quality of research and
    citation frequency. BMC Med Res Methodol 6:42
Olney CA, Warner DG, Reyna G et al (2007) MedlinePlus and the challenge of low health literacy:
    findings from the Colonias project. J Med Libr Assoc 95(1):31–39
Rath A, Olry A, Dhombres F et al (2012) Representation of rare diseases in health information
    systems: the Orphanet approach to serve a wide range of end users. Hum Mutat 33(5):803–808
Santana S, Lausen B, Bujnowska-Fedak M et al (2011) Informed citizen and empowered citizen in
    health: results from an European survey. BMC Fam Pract 12:20

Chapter 4
Representation of Patient Data in Health Information Systems and Electronic Health Records

M. Cuggia, P. Avillach, and C. Daniel

Abstract With the development of IT, more and more hospitals and health facilities are currently using electronic health records (EHR) in replacement of the paper-based patient record. The main goal of an EHR is to improve the health care process. Moreover, EHRs make easier the reuse of patient data for other purpose like research studies or management. In this chapter, we first discuss the added value of EHRs. Then we present their main categories and the different ways to represent and coding data in such systems. The place of interoperability standards is critical to integrate EHRs in Health information system. Therefore we present and discuss the two main semantic standards (HL7 and OpenEHR) used to structure and code clinical data in EHRs as well as the initiatives encouraging vendors to implement them.

Keywords Patient record • Electronic health record • Data representation • Semantic interoperability

M. Cuggia (✉)
Laboratoire d'Informatique Médicale, Faculté de Médecine, INSERM U936,
2, av. Léon Bernard, Rennes 35043, France
e-mail: marc.cuggia@univ-rennes1.fr

P. Avillach
Biomedical informatics and public health department, University Hospital HEGP, AP-HP,
Paris, France

INSERM UMR_S 872 team 22 : Information Sciences to support Personalized Medicine,
Université Paris Descartes, Sorbonne Paris Cité, Faculté de médecine, Paris, France

C. Daniel
CCS Domaine Patient AP-HP, 05 rue Santerre, Paris 75012, France

A. Venot et al. (eds.), *Medical Informatics, e-Health*, Health Informatics,
DOI 10.1007/978-2-8178-0478-1_4, © Springer-Verlag France 2014

After reading this chapter, you should:

- Be able to define what a patient record is and its main objectives
- Be able to identify the added value of an electronic health record (EHR) in regard to its paper version
- Know the rationale and principles of EHR certification
- Know the different types of EHR
- Know the main modalities of data organization in an EHR
- Be able to explain why the expression of the context is important to represent the medical data
- Be able to explain why semantic interoperability of EHRs is essential in the medical domain, and to know its main principles

4.1 Aims of the EHRs

The patient record is a collection of documents that provides an account of each episode in which a patient visited or sought treatment and received care or a referral for care from a health care facility.

According to F. Roger France (Roger et al. 1978), the patient record (PR) is "the written memory of all information about a patient, continuously updated, and its utilization is both individual and collective."

The PR is an indispensable tool for the medical practice. It is intrinsically linked to the health care process. PR allows the written collecting on a physical medium (paper) of the data generated during this process all along the patient's life. It reflects the medical state of each patient, and the diagnostic and the therapeutic actions taken.

The PR contains different kinds of data coming from different origins (administrative data, medical data, paramedics data), produced, inferred and collected by all the actors involved in the health care process. These data are the relevant facts corresponding to the different decisions and actions that have been taken to treat the patient.

The administrative part of a PR provides the information for his/her identification and the socio-demographic data are continuously updated during the patient's life (identity, health insurance status, employers, etc.).

The quantity and the complexity of the data contained in the PR are constantly increasing with the development of the medical specialties and their technicalities.

Designing a single PR meeting all the needs of the whole healthcare actors is still an issue. Even for a same profession, the needs could be very different from a medical specialty to another: e.g. a specialist in allergy collects the allergy history of a patient with an extreme precision, while an orthopedist does not usually need this level of details.

The patient record also includes the manner in which the patient was informed about the diagnostic and therapeutic strategy. It contains his/her consent for the cares and for the secondary use of clinical data in other settings such as clinical research or public health.

Indeed, beyond the individual use of its data, the PR is also used collectively in order to characterize a population of interest for biomedical or public health research.

4.2 From the Paper Based Patient Record to the EHR

According to the International Organization for Standardization (ISO) definition, an EHR is a "repository of patient data in digital form, stored and exchanged securely, and accessible by multiple authorized users" (ISO/TR 20514:2005; Häyrinen et al. 2008). It contains retrospective, concurrent and prospective information and its primary purpose is to support continuing, efficient and quality integrated health. Electronic patient records are used both in hospitals and in general practice. Most of the time, the computerization of the patient record is a complex and a progressive process. Some of the EHR contain only the main relevant documents such as discharge letters, post operative reports, histologic observations, etc. As the EHR is a container, progressively more information can be integrated, such as therapeutic prescriptions, lab data, daily clinical charts or radiologic data. Cohabitation with the paper based PR is often unavoidable, at least during the first years of the EHR deployment.

4.2.1 Added Value of the PR Computerization

The objectives of an EHR are to:

- Decrease the access and the delivery time
- Share the data between actors involved in the care process
- Meet the needs of security, audit trail, and avoid the medical errors
- Provide data for biomedical or public health research, for teaching purpose or for management

Table 4.1 compares advantages and drawbacks between paper-based PR and EHR.

4.2.1.1 Health Information Exchange

With the EHR, patient data are immediately available, in real time, at the bedside and also remotely (for instance at the doctor's home). The data are also available for

Table 4.1 Comparing the functionalities between a paper based PR and an EHR (Degoulet and Fieschi 1996)

Features	Type of patient record	
	Paper based	EHR
Delivery and sharing patient data between professionals		
Data integration (including multimedia data)	+	+++
Readiness	+	++
Completeness	+	+++
Access	Sequential	Simultaneous
Data availability	Local	Global
Remote access	0	+++
Linking the episodes of care	+	+++
Security, data protection		
Data security	+	+++
Confidentiality	++	+
Decision support		
Reminder and alarm	0	+++
Diagnosis and therapeutic decision support	0	++
Multimedia	0	++
Secondary data reuse		
Professional practice assessment	+	+++
Biomedical and epidemiologic research	+	+++
Management	0	++
Learning and training activities		
Process of care	+	+++
Acceptance of care protocols	+	++
Link to knowledge database	0	+++

the other components of the information system. For example, the integration of the information flow between the EHR and the laboratory management system improves the speed of the data processing (transmission of the orders from the bedside to the laboratory and reception of the results in real time).

The electronic data can be shared at large scale, beyond a single hospital, for instance at regional, national or international scales.

By improving the data sharing, the EHR supports the continuity of care and provides a better security for the patients (e.g. sharing patient history or allergies data).

4.2.1.2   Data Protection, Security and Traceability

The EHR provides better data traceability and the activities around the patient. All the actions of the users are stored (e.g. the access to the record). Moreover, each data entry is time stamped and signed.

4.2.1.3 Decision Support

A national report, published in 2000, by the American Institute of Medicine (Kohn et al. 2000) had estimated that around 100,000 American citizens died each year of medical errors. These errors are due from a great part of them, either to a lack of information, or wrong data in the patient record. Decision support systems connected to the EHR can help the physician by generating reminders and alerts (e.g. to detect drug adverse effects) or by suggesting, from the patient data, a diagnosis or a therapeutic strategy. Using an EHR avoids typing or capture data errors. Data are captured once for all, from the source, if necessary with some entry controls. The data are then re-used and shared all along the process of care. For instance, when a physician types the drug prescriptions, this information is automatically re-used for the nurse care plan, for the pharmacy department, for the billing system, etc. Moreover the readability of computerized data is much better than the handwritten data.

4.2.1.4 Secondary Reuse of the Patient Data

The EHR is source of information for management, evaluation and research purposes. Compared to the paper-based version, computerized data makes easier the secondary re-use of the data. Data extraction and statistical analysis can be carried out for studies or to provide indicators and dashboards for different domains, like for instance clinical research (e.g.: to find eligible patient for clinical trial), epidemiologic survey (e.g. for infectious diseases) or evaluation of professional practices: (e.g. adequacy to clinical guidelines) (Jensen et al. 2013; Meystre 2007).

4.2.1.5 Teaching Activities

The EHR is also a tool for teaching and training students. Patient data can be extracted to create pedagogical resources. The functionalities of the EHR can be used to teach how to collect systematically clinical facts for optimizing the diagnostic and therapeutic strategies.

4.2.2 Accreditation, Certification

The EHR is a key element for health care, supporting some critical processes that might jeopardize the patient life (e.g. drug prescription). Therefore, health care organizations must ensure the quality and reliability of their EHR. As in the aeronautic sector, some initiatives exist to deliver a software certification or accreditation.

Certification refers to the confirmation of certain characteristics of an object, person, or organization. This confirmation is often, but not always, provided by some form of external review, education, assessment or audit. The accreditation is a specific organization's process of certification. From the vendors' point of view, obtaining such label is usually considered as one of the best way to promote their products. Two relevant initiatives exist in this domain:

The European Institute for Health Records or EuroRec Institute (Eurorec 2013) is a non-profit organization founded in 2002 as part of the ProRec initiative. The institute is involved in the promotion of high quality EHR systems in the European Union. One of the main missions of the institute is to support, as the European authorized certification body, EHRs certification development, testing and assessment by defining functional and other criteria.

The HL7 EHR System Functional Model provides a reference list of functions that may be present in an EHR. The function list is described from a user perspective with the intent to enable consistent expression of system functionality. This EHRS Functional Model, through the creation of Functional Profiles for care settings and realms, enables a standardized description and common understanding of functions sought or available in a given setting (e.g., intensive care, cardiology, office practice in one country and primary care in another country).

4.2.3   Organizational and Ethical Factors

The way that an electronic record is used is definitely different from a paper-based record. An EHR might improve the efficiency of the working tasks but most of the time, some significant changes occur in terms of processes.

4.2.3.1   Taking into Account the User Needs

Human factors can lead to resistance to change and jeopardize the EHR adoption. The main factors are computerized entry mode, lack of computer culture, lack of incentives from administrative or medical authorities, lack of training activities or information about the added value of the EHR and fears on data protection and system security.

The adoption of an EHR requires that users and decision makers perceive its added value. Patients expect an improvement of the care process. Physicians need that the EHR makes the data access easier, and supports them to adopt the best diagnostic and therapeutic strategy according to updated evidences and the specific characteristics of the patients. Decision makers expect to reuse the data for management and planning purposes, and to get, as soon as possible, a financial return of investment.

The EHR has to provide different views of the same data and for the different users, according to their needs.

The EHR functionalities as well as the access rights have to be specified according to the role and the profile of each category of users. This is only possible if a comprehensive analysis of the business processes, the working environment and the needs of sharing or exchanging data has been initially carried out.

4.2.3.2 Regulatory Aspects of EHRs

Personal health data are critical. Their special status imposes enhanced protection with the aim of ensuring their confidentiality (see Chap. 11). Many European countries are currently trying to resolve this issue related to security and privacy of health data in EHR. The European Union enacted in 1995 (95/46/EC) and 2002 (2002/58/EC) directives on the protection of personal data collected via the services and public communication networks whose objective is to make easier movements of personal data, emerged as necessary for the establishment and functioning of the common market, while ensuring respect for fundamental rights and freedoms of individuals.

In most countries a regulatory framework for sharing health data based on European regulation is designed to implement the terms of respect for human rights. This is to ensure that the information provided to the patient is clear, complete and prior to enable it so that the patient can exercise its rights and particularly the right of access. The regulatory framework also aims at ensuring that explicit and informed consent to data sharing is collected.

In most European countries, EHR shall be declared to a regulatory body. This declaration is intended to explain the purpose and scope of the use of clinical data, the functionalities of the system, the specifications and the rules of physical and logical security applied to computer processing implemented.

Hosting health data requires the implementation of a policy of security and confidentiality to provide patients and health professionals the necessary level of access to health data while ensuring a maximum protection against all kinds of abuse. The aim is to ensure the user authentication for accessing reading or writing to the clinical information system and to guaranty that only the relevant data is accessible to authorized people or organizations. Secure access to data is a challenge in complex organizations such as an academic hospital.

In many countries, the implementation of security policies of health information systems is still carrying questions about the foundations of legal frameworks, as well as the technical regulation in place to ensure the effective protection of rights.

4.2.3.3 Current Stake of EHR Is About the Secondary Reuse of Data for Research Purposes

These new utilizations will only develop if the confidence of health professionals and patients is reinforced by the implementation of new infrastructures (such as Clinical Data Warehouse and information technology platforms dedicated to translational research) that take into account legal, privacy and security aspects (see Chap. 18). De-identification of datasets and protection of genetic information is still challenging (Kushida et al. 2012).

4.3 Typology of EHRs

We can distinguish different categories of EHRs (ISO 2004):

4.3.1 EHRs in Primary Care Facilities

In primary care, General Practioners (GP) use EHR for managing their activities, for exchanging or sharing data with healthcare networks or for transmitting information to the billing system of health care insurance. The barriers to adoption of EHR systems by primary care physicians can be attributed to the complex workflows that exist in primary care physician offices, leading to nonstandardized workflow structures and practices (Ramaiah et al. 2012).

4.3.2 EHRs in the Hospital Information System

EHR have been recently and widely adopted by private and academic hospitals. Indeed, the functional coverage of the HIS concerned initially the administration, the support departments, and laboratories or imaging departments. Deploying an EHR in clinical wards has a strong impact on the organization and on the care process. Most of the countries, for instance in Europe, have developed an active incentive politic based on large funds and regulation to encourage hospitals to use EHRs.

4.3.3   Shared EHRs and Personal Health Records (PHRs)

Sharing and exchanging information between health professionals, beyond a hospital or a GP surgery becomes a critical stake. Most of countries in the world are developing initiatives to support healthcare networks with IT infrastructures. For instance, in United State, Regional Health Information Organizations (RHIO), also called a Health Information Exchange Organization, are a multi stakeholder organization created to facilitate a health information exchange (HIE) – the transfer of healthcare information electronically across organizations – among stakeholders of that region's healthcare system. The ultimate objective is to improve the safety, quality, and efficiency of healthcare as well as access to healthcare through the efficient application of health information technology.

Beside EHRs for providers, personal health records (PHR) for patients have been developed. PHR is a longitudinal record, containing patient medical history and critical data to support it. While the EHR is created and maintained by healthcare professionals, the patient is responsible of the access and the content management of his/her own PHR (see Chap. 13) (Evans and Kalra 2005).

4.3.4   Pharmaceutical Record

In some countries, pharmaceutical records are developed in order to secure drug prescription and dispensation. The pharmaceutical record gives to the pharmacists an access to the history of the drugs delivered to the same person in the whole of the dispensaries during a given period of time, in order to avoid the drug interactions and/or redundant prescriptions. It is created upon request and consent of the individuals involved and information is disclosed only to those authorized by the owners.

4.4   Organization and Content of the EHR

4.4.1   Healthcare Processes and EHR

A comprehensive hierarchical breakdown of the capabilities of EHR systems has been developed by HL7 as the EHR System Functional Model (EHR-S FM) (Benson 2012; HL7 2013).

At the top level, there are three groups of functions:

- Direct care functions used to provide direct health care, or self care, to one or more persons

- Supportive functions that use EHR data to support the management of healthcare services and organizations (Administration support); inputs to systems that perform medical research, promote public health and improve the quality of care at a multi-patient level (Population Health Support)
- Information infrastructure which are backbone elements of security, privacy, registry, interoperability and terminology.

The EHR-S FM is not a list of specifications for messaging, implementation, or conformance but is a valuable resource for industry, healthcare providers, governments, and other organizations to use as a common language for discussing the functionality of electronic health records.

4.4.2   Global Organization of EHRs

Each of these processes produces data throughout patient care management. In its broadest sense, the patient record is the repository of data from these business processes. A variety of formats are used to organized clinical data. Two main models are the Source Oriented Medical Record in which the information is organized according to the "source" of the information and the Problem Oriented Medical Record in which the information is organized according to "problems" (Coiera et al. 2003).

4.4.3   Source Oriented Medical Record

This model is based on the organization of most paper records. The data are organized according to the "source" of the information (i.e. data issued from the physician, the nurse, or data collected from an imagery department) Data are recorded usually in chronological order into specific components:

- The administrative component contains information about patient identity, address, telephone, entry and discharge dates, etc.
- The clinical documentation component contains daily progress notes from healthcare professionals (physicians 'notes, nursing assessments, problem lists, medication lists, discharge summaries, etc.)
- The order entry component contains orders for both diagnostic and procedures (radiology tests, laboratory tests, consultant requests, nursing orders, medication orders, etc.)
- The tests and imaging component contains the images, the results of lab tests and their interpretation
- The care plan component contains information related to future monitoring and treatment administration as well as patient education

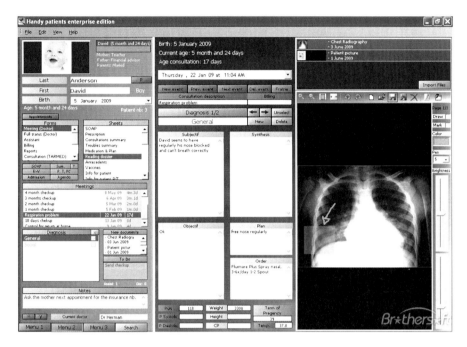

Fig. 4.1 Copy screen of an EHR http://en.wikipedia.org/wiki/Electronic_health_record

This mode of organization silos promotes an analytical representation of the data, at the expense of a synthetic visualization. However, information can be reorganized into specific views in order to meet the needs of each medical specialty or even each clinician (Fig 4.1).

4.4.4   Problem Oriented Medical Record

The Problem Oriented Medical Record was first described by Larry Weed in 1968 as an alternative to the chronologically organized medical record (Weed 1968). It was one of the first and most influential attempts to structure the patient record.

Following this model, the record is divided into two parts. The first part covers the patient's social, family and past medical history. The second main section is the progress notes. Progress notes are organized into a series of sections, each of which is given a heading which broadly can be called a 'problem' either by the patient of the clinician. A problem is anything that causes concern (not only a diagnosis) such as stomach pain or hypertension for example. The problem list is a list of all the patient's problems indicating those that are 'Current', 'Dormant' or 'Resolved'. The label associated with the heading may change as the problem or its understanding develops.

Progress notes can be clinical notes, test results, medication records or even images. They don't change over time and are grouped under one or more of the problem headings.

Each progress note has a problem heading and four sub-headings, using the acronym SOAP:

- S – Subjective, meaning the information provided about history and symptoms by the patient or relative.
- O – Objective, meaning information obtained by direct examination of the patient or from clinical investigations (laboratory, radiology etc.)
- A – Assessment, meaning the clinicians assessment about what is the matter with the patient (diagnosis), prognosis etc.
- P – Plan, meaning the plan of future action, including investigations and treatment (drug prescriptions, physiotherapy, surgery and so on).

Drugs prescribed are also listed in a separate medication list.

In US, PROMIS, a problem-oriented electronic patient record system was implemented for many years on medical and gynecological wards at the University of Vermont.

Most GP in the UK systems are using problem-oriented EHRs linking medication and tests with problems.

The main advantages of POMR are:

- Improved readability of the record especially for problems with few entries.
- A systematic approach for patient care management.

The main limitations of POMR are:

- The simple approach of POMR makes clinical information quick to pick up but POMR can be complex to maintain.
- Not all headings are 'problems' in the strict sense of the word. For example, the heading of 'Vaccinations' is used commonly to denote where all the entries related to a vaccination history may be found.
- A 'single problem consultation' is rare. Many different issues may be discussed within a single consultation for example respiratory infection in a lung cancer patient.
- A clinical or paraclinical statement may legitimately belong under more than one problem heading. In this case, the data must be recorded several times. For example, the finding of high blood pressure could appear under for example 'Adult screening ' or 'Essential Hypertension' or even 'Anxiety State'.
- Different clinicians, and indeed anyone with the right to view the clinical record, need different amounts of information from the record as well as different views.
- Some problems are complex and hence difficult to read.
- There is no easy way to express causal relationships between problems (pneumonia following a fall causing fractured neck of femur).

In the real world some problems tend to go in cycles, getting better and then worse over a long period (for example, Asthma) and therefore cannot be easily classified in 'Current', 'Dormant' or 'Resolved'. This model is suitable for only some conditions like surgery (a "current" problem like appendicitis is "solved" definitively by an appendectomy).

4.5 Healthcare Information models

The data entered in an EHR can be complex and includes sometimes implicit meaning as in the following example:

Mr. Alain DuPont

Came on Friday 19 h for an emergency!
Another angina!
Red throat
BP: 12/8
Clamoxyl 1gx2 6 days
No certif this time

In this example, the reader quickly deduced that:

- The physician was not very pleased to see this patient for this emergency and in addition at the end of the day;
- This is a priori a new episode of angina;
- Clinical examination is really minimal;
- The antibiotic treatment is recorded as amoxicillin 1 g twice a day for 6 days but the dosage form and route of administration are not recorded;
- Support for this episode is not the same as during previous consultations since no medical certificate is provided.

However, the EHR is not designed to be the exact mirror of current paper-based records. EHRs provide the opportunity of organizing, structuring, standardizing and encoding healthcare information so that this information can be processed and provide added-value services to the healthcare professionals and most importantly to the patient through decision support and broad information communication for better patient care coordination.

In order to automatically process the data that are collected in EHRs, it is necessary that their meanings are explicitly represented in a format that can be interpreted by computers as well as by humans.

4.5.1 Healthcare Data Structures

We distinguish three levels according to the structure of clinical information:

- Level 1 (unstructured data): the format of the clinical documentation is human-readable free-text
- Level 2 (semi-structured data): the format of the clinical documentation consists in one or more structured sections. Each section contains a single narrative block.
- Level 3 (structured data): the format of the clinical documentation allows each section to include machine-processed clinical statements at almost any level of granularity.

Thus, it offers the benefits of both human-readable and machine processed documents.

The lower levels of clinical documentation provide rather low technical barriers to adoption, while providing a migration route towards structured coded records.

Structuring clinical information is a complex task but although there are many different and competing information models of electronic health records, high-level principles can be stated. Reference Models defined by Standard Development Organizations (mainly CEN TC251, HL7, CDISC) for the electronic health record were designed in the context of exchanging clinical information between parties. These models represent the clinical content as a hierarchy of useful clinical information to be exchanged/shared in specific use cases.

As an example, the ISO 13606 Reference Model defines (CEN/ISO 2013):

- A top-level container of all or part of the electronic health record (EHR) of a single subject of care (patient) ("Extract").
- Sets of information committed to one EHR by a clinician relating to a specific clinical Encounter ("Compositions"). Progress notes, laboratory test reports, discharge summaries, clinical assessments and referral letters are all examples of Compositions.
- Clinical Statements ("Entries") recorded in the EHR as a result of a single clinical action, observation, interpretation or intention. It may be thought of as a line in the record. Examples include the entries about a symptom, a laboratory result, a diagnosis or a prescribed drug.
- "Elements" are the leaf nodes of the EHR hierarchy which are single data values, such as systolic blood pressure, a drug name or body weight.

Related Elements may sometimes be grouped into Clusters. For example, systolic and diastolic blood pressures are separate Elements, but may be grouped into a Cluster (e.g. 140/90), which represents one Item in an Entry.

Entries may be grouped together in Sections. A Section is a grouping of related data within a Composition usually under a heading such as Presenting History, Allergies, Examination, Diagnosis, Medication, Plans. Sections may have sub-Sections.

4.5.2 Healthcare Data Elements

Clinical statements correspond to the recording of clinical actions. We can distinguish different clinical actions such as Observation, Procedure, Substance Administration, Supply and Patient Encounter (Velde and Degoulet 2010).

Observations often involve measurement or other elaborate methods of investigation, but may also be simply assertive statements, such as a diagnosis. Many observations are structured as attribute-value pairs.

Procedures cover diagnostic procedures (e.g. imaging, laboratory investigations), therapeutic procedures (surgical procedures, prescribing a medicine, physical treatment such as physiotherapy) and also administrative procedures, counseling.

Substance administration refers to a medication that has been administered to a patient.

Supply involves provision of a material by one entity to another (for example, dispensing a medicine).

Patient Encounter is defined as an interaction between a patient and a care provider for the purpose of providing healthcare-related services. Examples of Patient Encounter include inpatient and outpatient visits, home visits and even telephone calls.

Clinical statements refer to healthcare entities such as Living Subject, Material, Place and Organization.

Observation value contains the information determined by the observation action. The data type is usually constrained to a specific data type, such as physical quantity or a code.

Data types are the basic building blocks used to construct or constrain the contents of each element. Most EHRs use only a small number of common data types.

Simple data types contain just a single value, while complex data types may contain more than one sub-element, each of which has its own data type. The data type of a component can also be a complex data type. Complex data types reflect associations of data that belong together, such as the parts of a person's name, address or telephone number, or linking identifiers with their issuing authority.

Simple data types include:

- Date: represents a date in format: YYYY[MM[DD]]. For example, 2 August 2008 is represented as 20080802.
- Formatted text: allows embedded formatting commands, bracketed by the escape character.
- String: used for short strings up to 200 characters.
- Text: used for longer texts up to 64 K characters.

The most commonly used complex data types are for example coded concepts, names and addresses, composite quantity which has sub-components (quantity and units), specimen source (covering information about specimen type, body site,

Table 4.2 Examples of data elements and data types

Element	Attribute	Value	Data type
Medical history	Medical history	Paragraph	Text
Presence/absence of hypertension	Hypertension	Yes/no	Boolean
Type of myocardial infarction	Type of myocardial infarction	List of terms or coded expressions (ICD-10) (I210) anterior myocardial infarction (I211) inferior myocardial infarction	Coded data
Systolic blood pressure	Systolic blood pressure	Decimal unit: mm Hg	Physical quantity
Birth date	Birth date	Day/month/year (+/− hour/ second)	Date/hour

collection method, additives etc.), timing/quantity (allowing the specification of the number, frequency, priority etc. of a service, treatment or test), etc.

Table 4.2 illustrates data elements and simple data types.

Examples below illustrate more complex data structures corresponding to vital signs (Table 4.3) and susceptibility tests (Table 4.4).

In complex data structures such as vital signs or susceptibility tests, a rich set of relationships between entries needs to be defined to reflect the structure of clinical information and links between different healthcare data elements.

Examples of relationships are:

- Causal relationships: Used to express that an item caused another item, such as substance administration (e.g. penicillin) caused an observation (e.g. a rash), or observation (e.g. diabetes mellitus is the cause of kidney disease).
- Compositional relationships: Used to show that an item is a component of another more complex item (e.g. hemoglobin measurement is a component of a full blood count).
- Evidence relationships: Used to show that an item provides supporting evidence of another item (for instance, "possible lung tumour" is supported by "mass seen on chest -xray").

4.5.3 Encoding Clinical Information

Natural language is the most common way to express clinical information. Although information systems cannot easily automatically exploit free text, recent progresses in Natural Language Processing resulted in solutions able to encode unstructured textual clinical information.

Free text in EHRs can be annotated with controlled vocabulary either manually _ for example, the clinician will encode diagnoses stated in discharge summaries

Table 4.3 Vital signs

Attribute	Value	Date/hour	Author
Temperature	37.5 °C	15/09/2011–11:15	Isabelle Nurse
Pulse	80/min	15/09/2011–11:30	Isabelle Nurse
BP			
Systolic BP	132 mmHg	15/09/2011–11:45	Isabelle Nurse
Diastolic BP	86 mmHg	15/09/2011–11:45	Isabelle Nurse

Table 4.4 Antibiogram

Antibiogram				
Sample #1 (E.coli)			Sample #1 (S. aureus)	
B Lactamin				
Ampicilin	>16	Resistant		
Cefalotin	<= 1	Intermediate	<= 2	Sensible
Quinolon				
Ciprofloxacin	<= 0.5	Sensible		
Ofloxacin	<= 2	Sensible	> 4	Resistant
Aminosid				
Amikacin	<= 0.5	Sensible		
Gentamicin	> 8 R	Resistant		

selecting relevant ICD-10 codes _ either automatically using Natural Language Processing (NLP) techniques. Due to current limitations of such techniques with regards to ambiguous or complex or polysemous (use of a word with multiple meanings) data, a manual validation of the automatic encoding process is still usually required.

When clinical information is collected through a questionnaire as a structured set of elements, the value of some structured observation may be represented by coded concepts. Coded values need to be uniquely identified, but there is always the problem that two different coding schemes use the same code value. The solution is to explicitly identify both the coding scheme and the code value.

Therefore, for example the complex data type "Coded Descriptor (CD)" is used to represent not only the wording of the observation value (e.g. display Name = "Acute myocardial infarction") but also its code (I21) in a coding system whose name and version are also explicit (code System 2.16.840.1.113883.6.3; System Name ICD-10) and possibly the original expression that served as the basis for encoding, which preserves the meaning of the data before encoding.

Terminology binding is the process of establishing links between elements of a terminology and an information model. We often find situations where there are several possible ways to express the same meaning, due in part to the overlap between information models and terminologies. The terminology should be used for specifying specific concepts and value sets and simple semantic relationships, such as laterality, post-coordinated expressions at various levels of nesting.

The structural information model should be used for specifying information and meta-data for any clinical statement such as dates and times, people and places, numbers and quantities; grouping and organization of the record framework including the record structure, the way that items should be grouped together and anchors for terminology components, such as codes.

4.6   Context of the Healthcare Information

The EHR is not a collection of facts, but a set of observations about a particular patient, which have been made by healthcare professionals, each at a specific time and place for some purpose. Because each clinical statement is an observation, it is quite possible for two statements about the same event to disagree with each other, but this can be resolved if the context or provenance of each statement (who stated it, when and where) is recorded.

In the patient record, the representation of the context in which data is collected is crucial for its interpretation. Whatever the formal character of the structure used, patient data are usually hierarchically organized and "nested" to each other. It is this structure that allows expressing context.

Some medical data can be very intricate and interdependent. For example, a positive bacteriological examination of urine refers to organisms found on culture media. Each seed is tested by a susceptibility test referring to a list of antibiotics and resistance values.

The context information needed to interpret health data are of different types:

- The "compositional" context refers to the relative positioning of a given element relative to others in the folder. For example, the information "recurrent tonsillitis" has no intrinsic meaning. Indeed, is it the patient's diagnosis or a history? This is because it is positioned in a field or topic explicitly, that this data is contextualized and becomes explicit.
- The "provider" context conveys important information about the source of the data (author of the data, its role and responsibility, the institution and the department concerned, the date the data is entered, or control or production and release).
- The context of the medical reasoning. For example, it allows expressing the presence or absence of a diagnosis, its degree of certainty, data from the literature to substantiate scientific reasoning.
- The context of the care process: in general, the patient data is captured throughout his care. The new data is often related with previously recorded data. For example, it is common that the establishment of a new diagnosis is actually related to a pre-existing problem. The context of the care process thus establishes such a link and allows, among others, to express a cause and effect with an episode of care with existing or a step in a therapeutic protocol.

4.7   Sharing, Exchanging and Processing Clinical Data

The degree of structuring and coding of patient data directly affects the possibilities of data exchange/sharing as well as automatic data processing and therefore of providing advanced features for patient care coordination but also research or public health.

However, since methods of collection and representation of the information through encoded data structures is not natural for health professionals, we must find the right balance among unstructured data enabling professionals to speak with all the necessary medical discourse, and structured and coded data to add value to the EHR (diagnostic aid, statistical processing, production indicator, etc.) (Jensen et al. 2012).

"Interoperability" is defined as the ability of information systems to enable health stakeholders (healthcare professionals, patients, citizens, health authorities, research organizations or training, etc.) to exchange or share electronic data.

The ISO standard "Open Systems Interconnexion" (OSI) explains how two systems can communicate (Table 4.5). The OSI model consists of seven layers (physical, data link, network, transport, session, presentation, application). Interoperability is achieved when the transmitter and receiver systems for each layer share a common language.

The lower layers describe, for example, the physical medium that will convey information (this may be an electrical cable, optical fiber, or electromagnetic waves). The upper layers describe how the data will be structured and organized. The last layer – application level (level 7) is specific to each domain (financial, health, etc.) and concerns syntactic and semantic interoperability. It is at this layer that standards shall be defined (information models and coding procedures) to ensure that the meaning of the data is explicitly transmitted from one system to another.

According to this model, in its report "Semantic Interoperability for Better Health and Safer Healthcare; Research and Deployment Roadmap for Europe ", in January 2009, the European Commission distinguishes technical interoperability, syntactic interoperability that both enables the sharing of dematerialized free text health data interpretable by humans (for example, an hospitalization discharge report) and finally semantic interoperability that enables the sharing of dematerialized structured and coded health data using a system reference code (e.g., diagnoses coded in ICD-10, lab test results encoded in LOINC, etc.).

The challenge of semantic interoperability is to define the rules for sharing/exchanging health data that are interpretable not only by humans but also by computers (Kalra and Blobel 2007).

How to share information like "25 August 2011, the value of the patient's systolic blood pressure is 160 mmHg"? In a paper-based record this information may be registered in the physician's note as: "Today: 160."

In the area of patient care coordination, standard-based structured and coded clinical data facilitates access to data and knowledge that are relevant in the context

Table 4.5 Simplified view of the OSI model representing different interoperability layers to implement for the sharing and exchange of medical data.

Today (25 August 2011) SBP: 160 (On August 25, 2011, the value of the patient's systolic blood pressure is 160 mmHg)

Semantic	The semantics of the information is interpreted by the sending and receiving systems because it is coded using a coding system shared (e.g. SNOMED CT)	<observation>... <code = '271649006' displayName = 'Systolic blood pressure' codeSystem = '2.16.840.1.113883.6.96' codeSystemName = 'SNOMEDCT'/>
Syntax	The syntax of the information is interpreted by the sending and receiving systems because it is consistent with a shared model (template) corresponding to the observation "systolic blood pressure". This model includes the parameter name (e.g. " systolic blood pressure "), its value, its unit (in case of physical quantity) and the date of observation	<observation> <templateId root = '1.3.6.1.4.1.19376.1.5.3.1.4.13'/> <Original text: systolic blood pressure <effectiveTime value = 20110815/> <value xsi:type = PQ value = 160 unit = mmHg/> </Comment>
Message	Message structure	Start-Body-End
Security	Encryption	O # * 22 * # * @ @
Transport Network	Supporting physical and logical transport (Ethernet cable, optical fiber, Wifi)	

of the patient (access to care protocols, rules, alerts: reminders for vaccination date, drug-indications, etc.). In the field of clinical research, epidemiology or public health, standard-based structured and coded clinical data can be used to support patient recruitment in clinical trials, adverse event detection as well as clinical trial execution (see Chap. 17).

4.8 Standardization Organizations and Initiatives in Health Informatics

Standardization bodies in the field of health informatics ensure, within the framework of their mission, the production of health informatics standards for the exchange, sharing and use of health data.

Some standards organizations such as HL7, CEN TC 251, TC 215 ISO, DICOM or CDISC (in the field of clinical research informatics) produce information models whose objective is to describe how data elements are organized within messages and documents that are exchanged or shared by HIS. These models have a crucial role in the representation of health information, including not only health data but the context in which they are produced and should be interpreted.

Other standardization bodies, such as WHO and IHTSDO, produce terminologies or ontologies used to "encode" the data in health information systems such as ICD-10 or SNOMED CT (See Chap. 2).

4.8.1 Health Level Seven HL7

Health Level 7 (HL7) is an organization that defines an eponymous set of technical specifications for electronic interchange of clinical, financial and administrative data between information systems. These specifications are variously integrated into formal American (ANSI) and international (ISO) Standard Development Organizations.

Initially American, HL7 specifications tend to become international standards. HL7 defines the structure of particular messages or documents shared or exchanged in XML format.

All models of messages or documents are defined from a reference model ("Reference Information Model" (RIM)) which is an ISO standard (ISO/HL7 21731:2006). HL7 has also developed a specification of the health care data types (ISO/HL7 21090:2009 standard).

HL7 has also defined a common architecture for clinical documents called "Clinical Document Architecture" (CDA) (standard ISO/HL7). It is possible to build from the CDA model (Fig 4.2) document models adapted to most medical specialties and in most contexts of use such as, for example, models for discharge summaries, for multidisciplinary meeting reports, lab results, etc.

A CDA document consists of a header and a body. The header is always structured and specifies the author of the document, the producing organization of the document etc. Three levels of interoperability are described for the body:

- Level 1: the body content of the CDA document is free text and is therefore not "readable" by computers.
- Level 2: the body of the CDA document is divided into sections like history, history of disease, diagnosis, etc. These type of sections are coded according to a controlled vocabulary (e.g. LOINC), but the content remains free text.
- Level 3: the body of the CDA document is machine-readable.

Note that regardless of the level of CDA (1, 2 and 3), a narrative block is always present, to allow a document to be still readable by humans. HL7 and ASTM worked together to define a minimum set of key clinical data to share or exchange between health professionals in a context of continuity of care ("Continuity of Care Document" (CCD)).

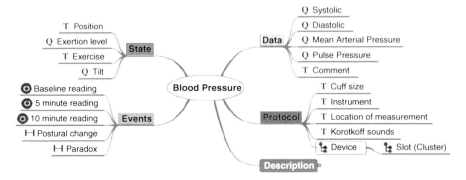

Fig. 4.2 Graphical representation of the archetypal blood pressure. The set of data elements necessary and sufficient to capture this information is represented (OpenEHR 2013)

4.8.2 *The European Committee for Standardization, CEN TC 251*

EN13606 is a standard information model developed by CEN largely reflecting the work of OpenEHR (OpenEHR 2013 an international research initiative).

The main feature of EN13606 and OpenEHR is that they define an information model comprising two levels: the reference model and archetypes.

The reference model represents the general characteristics of the data elements of a medical record. This model defines a set of classes that constitute the main blocks and sections found in an electronic file (e.g. classes such as Observation, Assessment, Instruction and Action).

The reference model is very generic and insufficient to define in more detail the data elements required in the medical field (e.g., systolic blood pressure).

Specific data elements are specified through archetypes developed by groups of medical experts. Archetypes derive from classes of the reference model by adding constraints on these classes expressed using the ADL language (Archetype Definition Language).

All data elements necessary to capture information from a field or topic must first be modeled. Thus, for example, the blood pressure archetype (Fig. 4.2), expresses the notions of systolic blood pressure, diastolic blood pressure, but also the average position of the patient, the type and size of the cuff, etc.

4.9   Conclusion

Medicine, healthcare, and information technology are elements of a world in which more and more office records and health information are being made available in electronic format. This trend opens the door to endless uses for information captured in EHRs–from biosurveillance to decision support.

Nevertheless, these systems are still not fully interoperable despite a huge effort to develop standards of interoperability. Indeed, medical data are definitely different from other data (like financial data). They are complex, sometime ambiguous, and they change very quickly as new knowledge is discovered. Representing formally medical data and their context is still an open issue. It is although one mandatory condition to reuse them automatically. The future of these systems will take benefit of semantic web technologies, using ontologies for data indexing and reasoning.

4.10   For More information

The MeSH expression "Electronic Health Records" was added in 2010. It is the only instance of the generic term "Medical Records Systems, Computerized" that inherits the concept "Medical Records". An example of Medline query is: ("Electronic Health Records"[Majr] AND (Review[ptyp] AND "2008/02/12"[PDat]:"2013/02/09"[PDat] AND "humans"[MeSH Terms] AND English [lang]))

Consult the Web sites:

HL7: http://www.hl7.org/

Interop health: http://www.interopsante.org

IHE (Integrated Health Enterprise) http://www.ihe.net/

OpenEHR: http://www.openehr.org/

Congress website HIT (Health Information Technology): http://www.health-it.fr/

Congress website HIMSS (Healthcare Information and Management Systems Society): http://www.himss.org

Canadian site highway/infoway: https://www.infoway-inforoute.ca/lang-fr/

Exercises

Q1 Regarding the contribution of computerization of the patient record: What are the correct answers?

A Decrease access and delivery times of medical information

B Share data between healthcare professionals

C Do not reduce the size of the patient records in paper form

(continued)

Exercises (continued)
D Reply to safety and traceability requirements
E Conduct studies or produce indicators

Q2 Concerning medical decision support from a computerized patient record:
What are the correct answers?
A Helps prevent medical errors
B Refers only the diagnostic approach
C Must be implemented
D Allow the patient to choose a physician
E Generates alerts to prevent drug adverse effects

Q3 The Personal Medical record (DMP) contains medical data indicated by:
What are the correct answers?
A Hospital doctors
B General practitioner
C Physiotherapist
D Pharmacists
E Occupational physicians

Q4 Structured data from a computerized patient records can:
What are the correct answers?
A Be encoded using terminology
B Be a paragraph of 20 lines of text
C Come from a drop-down menu
D Be a checkbox
E Be an image scanner

Q5 The main limitations of structuring a problem-based patient record are:
What are the correct answers?
A Number of problems can be identified entangled
B Chronological data representation
C A data organization clearer to medical reasoning
D An incentive for the user to adopt a systematic approach to problem
E A symptom, clinical or paraclinical can be attached to various problems

R1: ABDE

R2: AE

R3: ABC

R4: ACD

R5: ABE

References

Benson T (2012) Principles of health interoperability HL7 and SNOMED, 2nd edn. Springer, London

Coiera E, Magrabi F, Sintchenko V (2003) Guide to health informatics, 2nd edn. CRC Press, Boca Raton

Degoulet P, Fieschi M (1996) Introduction to clinical informatics. Springer, New York

EuroRec: European Institute for Health Records [Internet]. cited 2013 Jan]. Available from: http://www.eurorec.org/

Evans MG, Kalra D (2005) Healthcare computer systems–global approches. Lancet 365:10–11

Häyrinen K, Saranto K, Nykänen P (2008) Definition, structure, content, use and impacts of electronic health records: a review of the research literature. Int J Med Inform 77:291–304

HL7 Standards Product Brief – HL7 Electronic Health Record-System (EHR-S) Functional Model (FM), Release 1 [Internet]. [cited 2013 Jan]. Available from: http://www.hl7.org/implement/standards/product_brief.cfm?product_id=18

ISO/TR 20514:2005 ISO/TR 20514:2005 Health informatics – Electronic health record – Definition, scope and context

Jensen PB, Jensen LJ, Brunak S (2012) Mining electronic health records: towards better research applications and clinical care. Nat Rev Genet 13:395–405

Kalra D, Blobel BGME (2007) Semantic interoperability of EHR systems. Stud Health Technol Inform 127:231–245

Kohn LT, Corrigan JM et al (eds), Committee on Quality of Health Care in America I of M (2000) To err is human: building a safer health system. The National Academies Press

Kushida CA, Nichols DA, Jadrnicek R et al (2012) Strategies for de-identification and anonymization of electronic health record data for use in multicenter research studies. Med Care 50(Suppl):S82–S101

Meystre S (2007) Electronic patient records: some answers to the data representation and reuse challenges. Findings from the section on Patient Records. Yearb Med Inform 2007:47–49

openEHR – Homepage [Internet] [cited 2013 Jan 20]. Available from: http://www.openehr.org/

Ramaiah M, Subrahmanian E, Sriram RD, Lide BB (2012) Workflow and electronic health records in small medical practices. Perspect Health Inf Manag 9:1d

Roger FH, De Plaen J, Chatelain A et al (1978) Problem-oriented medical records according to the Weed model. Med Inform (Lond) 3:113–129

Velde RV, Degoulet P (2010) Clinical information systems: a component-based approach, Softcover reprint of hardcover 1st ed. 2003. Springer

Weed LL (1968) Medical records that guide and teach. N Engl J Med 278:593–600

Chapter 5
The Processing of Medical Images: Principles, Main Applications and Perspectives

L. Legrand

Abstract This chapter provides an introduction to the basic concepts of medical digital image processing. We kept the mathematical complexity of the chapter at a low level so that life science students can understand the underlying principles behind the methods of analysis and computer-assisted diagnosis.

After a brief description of the way in which the main medical image modalities are produced, the differences between image processing and image analysis are specified. Then the sampling and quantification processes, and their effects on image quality are explained. Grey level images and colour images are then introduced, as well as histogram-based image segmentation. The advantages and drawbacks of different lossy and lossless image compression techniques are given, and popular image formats are depicted, including the DICOM standard.

The principal image processing techniques are described: from pixel to pixel operations to spatial convolution techniques. Filtering in the frequency domain and mathematical morphology are rapidly described. So are shape recognition and classification, image registration and merging. Finally, Picture Archiving and Communication Systems (PACS) are briefly presented.

Throughout the chapter, examples are provided in order to ease the readers' understanding.

Keywords Sampling • Quantification • Pixel • Histogram • Lossy compression • Lossless compression • DICOM • Convolution • Segmentation • Frequency domain • Binary mathematical morphology • Shape recognition • Registration • Merging • PACS

L. Legrand (✉)
Laboratoire LE2i – UMR CNRS 6306, Université de Bourgogne UFR Médecine, 7 Bd Jeanne d'Arc, Dijon Cedex 21079, France
e-mail: Louis.Legrand@u-bourgogne.fr

A. Venot et al. (eds.), *Medical Informatics, e-Health*, Health Informatics,
DOI 10.1007/978-2-8178-0478-1_5, © Springer-Verlag France 2014

5.1   Introduction

After reading this chapter, you should:

- Have understood the meaning of important terms frequently used in the processing of medical images: segmentation, compression, DICOM, filtering, morphological processing, registration, PACS
- Have understood the basics and objectives of the principal techniques of image processing and analysis
- Be able to cite the principal image formats and their advantages
- Know how to explain the principles of image compression and its limits in medicine
- Be able to distinguish between lossy data compression and lossless compression.
- Be familiar with the DICOM standard of medical images
- Be able to define the notion of the segmentation of medical images
- Know the main principles of mathematical morphology
- Be able to cite applications of the different techniques in the field of medical biology

Medical imaging is evolving steadily thanks to contributions from the world of physics, chemistry, mathematics, engineering sciences and medicine. Over the last three decades, there have been considerable advances in medical imaging. New imaging techniques in two dimensions (2D), three dimensions (3D) and more have become extremely important in radiology. Today, establishing certain diagnoses or assessing the severity of certain diseases would be impossible without the help of medical imaging.

Though medical imaging has become one of the most important domains of scientific imaging, it owes this position to the rapid and continuous progress in computed tomography, and to the concomitant developments in methods of analysis and computer-assisted diagnosis the principles of which are presented in this chapter.

5.1.1   Imaging in the World of Medicine

Medical imaging dates back to 1895, when Roentgen discovered X-rays, which are at the origin of radiography, the only imaging method available at the start of the last century. The radiograph was obtained by exposing a film to X-rays that had passed through the human body. The result is a two-dimensional analogue image, which is the projection onto a radiographic film of organs in three dimensions. Today, the quality of the images obtained by radiography of the chest or mammography are often sufficient for a reliable diagnosis and for low-cost screening.

Scintigraphy, CT scan, sonography and MRI appeared one after the other during the twentieth century. These imaging techniques are today used to obtain anatomical, physiological, metabolic and functional information about the human body (see examples of images obtained with the different techniques on: http://www.inserm.fr/thematiques/technologies-pour-la-sante/dossiers-d-information/imaging-functional-biomedical).

They use computers to create and display numeric images that are getting better and better. Thanks to computers, multidimensional numeric images of physiological structures can be calculated and processed to see hidden elements that could be useful for the diagnosis. These elements would be difficult, even impossible to distinguish in two-dimensional imaging. The computerized processing and analysis with the different imaging techniques provide considerable aid for medical diagnosis. Artefacts have been reduced and the quality of images improved by taking into account knowledge about the physical processes involved in creating medical images and about human physiology; the images are thus easier for doctors to interpret.

5.1.2 How Medical Images Are Produced

The general aim of medical imaging is to provide information about the physiological processes of the body or organs by using external or internal sources of radiation, or a combination of the two. In practice, there are four principal commonly-used imaging techniques in clinical medicine as well as in research: CT-scan or tomodensitometry, scintigraphy and positron emission tomography (PET) used in nuclear medicine, sonography, and magnetic resonance imaging (MRI).

5.1.2.1 Radiography and CT Scan

Radiography and CT-scans use X-rays as the source of external energy to create very precise anatomical imaging. X-rays are ionizing radiation that is more or less attenuated depending on the density of the tissues they pass through. The images obtained measure the attenuation coefficients of the X-rays. Iodine-based contrast agents may be necessary to see certain organs.

Radiographic film is being replaced more and more by electronic detectors, which are more sensitive and make it possible not only to reduce the doses of radiation received during the examination, but also to digitize the radiography directly. Radiography can be used, for example, to detect fractures or to see lungs damaged by disease.

A CT-scanner has a series of emitters and receptors arranged in a circle around the patient, which provides a succession of two-dimensional images of slices of the body calculated by the computer. Computerized digitalization and processing of the images provide a three-dimensional representation of the organs of the body from these 2D slices. Because of the great precision of CT-scan images they are often used to see structural anomalies or volume modifications (haemorrhages, tumours, embolism. . .).

5.1.2.2 Sonography

Ultrasound is another example of an external energy source. They are non-ionising acoustic waves imperceptible to the human ear. A probe is pressed against the skin as close as possible to the organ being studied. The probe emits ultrasounds that go through the tissues, and are reflected back to the probe in the form of echoes. The echoes are analysed by a computer and transformed in real time into an ultrasound image. Among other uses, sonography is particularly useful in obstetrics, urology and hepatology.

In Doppler sonography, the echoes are transformed into sounds, curves or colours that reflect the speed of blood flow in arteries and veins.

5.1.2.3 Scintigraphy and Positron Emission Tomography (PET)

The imaging techniques in nuclear medicine use an internal source of energy to generate images of the human body. Small quantities of radioelements are injected into the body to interact with the target organs or the tissues being studied. Then a gamma-camera coupled with a computer collects and analyses the gamma rays emitted by an isotope.

This principle is applied to scintigraphy, Single Photon Emission Tomography or SPECT and PET (Positron Emission Tomography). These imaging techniques in nuclear medicine provide metabolic information on the physiological functions of organs. Scintigraphy is notably used in exploration of the thyroid, the skeleton, the heart and frequently in oncology. There are many applications of PET in oncology including lung cancers, ENT (Ear, Nose and Throat) cancer and breast cancer.

5.1.2.4 Magnetic Resonance Imaging (MRI)

Magnetic resonance imaging uses an external magnetic field, produced by a magnetic coil placed around the patient, to excite the target atomic nuclei. The excited hydrogen protons become an internal source of energy that will provide electromagnetic signals through the process of relaxation. These signals are captured and converted into images.

MRI provides high-resolution images of soft tissues in the human body (the brain, spinal cord, muscles, tendons, viscera) with excellent characterisation. It's an expensive examination, but is justified to characterise certain lesions more precisely when there are doubts following other examinations.

Recent progress in MRI has made it possible to generate functional images of the brain notably using perfusion images.

5.1.2.5   Other Biomedical Imaging Techniques

Biomedical images, notably those obtained using optical microscopes and scanning electron or transmission microscopy are also very important in medicine, with innumerable fields of application.

In this category, we can include fluorescence imaging. This technique, which uses an external source of ultraviolet energy to stimulate internal biological molecules, is becoming more and more widespread. These molecules absorb ultraviolet energy and become internal sources of energy, which they emit in visible electromagnetic wavelengths.

5.1.3   The Main Categories: Image Processing and Image analysis

5.1.3.1   Image Processing

Image processing comprises a wide range of methods that act on an image to generate another image. The aim of the changes made by these methods is generally to improve the visibility of the characteristics and details of the images, or to facilitate the analysis of these images, or even to improve image quality for printing purposes.

5.1.3.2   Image Analysis

Image analysis makes it possible to obtain digital data from images, using a combination of image processing techniques and measurements. These quantified data can then be analysed statistically for decision-making purposes and graphic representations.

5.2   Digital Images

5.2.1   Digital Images: Sampling and Quantification

Natural scenes of the real world contain continuous variations of colours and shadows. They send a continuous stream of colour images; in the words of physics, they can be described as analogue signals. To be processed by computers, these continuous images must be digitised, that is to say transformed into discrete signals using acquisition systems (digital cameras, or medical imaging systems, for example). Digital images of the real world are thus obtained.

The transformation of an analogue signal into a digital signal requires two steps. The first step is discretization in space; called *sampling*. When it is performed in two dimensions (2D), sampling determines the size of the image, that is to say its height and width. The second step consists in the discretization of grey levels for monochrome images, or components for colour images; this is called *quantification*.

The result of these two steps is a digital image made up of *pixels* (contraction of *picture elements*). The number of pixels in an image is determined by the sampling step. The number of values that a pixel can have depends on the quantification (Fig. 5.1).

A black and white or monochrome image has grey levels. Most of the time, the value of a pixel is a whole number or a vector. For an image in grey levels, the value of pixels is, in general, coded on 8 bits; it ban thus have $2^8 = 256$ different values, and vary from 0 (black) to 255 (white). In medical imaging, to increase the number of grey levels that a computer can distinguish, the coding can go up to 16 bits and the pixels therefore have $2^{16} = 65,536$ different grey levels.

The *histogram* of an image (Fig. 5.2) shows the distribution of pixels in the image. It is a two-dimensional graphic with on the x axis the values that pixels can have (from 0 to 255 for monochrome images coded on 8 bits), and on the y axis, for every level of grey possible, the number of pixels presents in the image. The histogram immediately shows if the picture is rather dark, rather bright, slightly or highly contrasted. It can be used to determine the initial pre-processing necessary to facilitate subsequent analysis. However, it provides no information on the spatial distribution of pixels because two completely different images can have the same histogram.

5.2.2   Colour Images and False Colours

A colour image can be decomposed into three components. Decomposition into primary colours RGB (Red, Green, Blue) is the most common. This is called an additive system of colours because any colour can be obtained by adding different proportions of these three primary colours (Fig. 5.3).

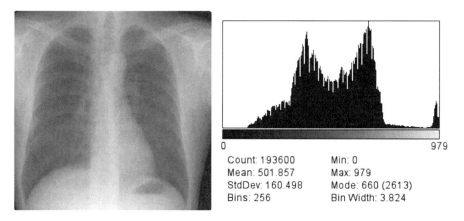

63	93	132	160	155	141	147	159
67	100	137	162	159	151	159	165
70	98	131	157	163	164	173	170
67	85	114	148	159	164	172	167
59	64	86	123	142	152	165	165
50	48	60	88	114	132	146	155
44	51	56	65	87	101	102	121
42	55	58	53	73	86	77	103

Fig. 5.1 MRI image, and pixel values of the enlarged part of the image (Image provided by Dr Alain Lalande, Dijon CHU)

0   979

Count: 193600   Min: 0
Mean: 501.857   Max: 979
StdDev: 160.498   Mode: 660 (2613)
Bins: 256   Bin Width: 3.824

Fig. 5.2 Digitized radiography of the thorax and its histogram (Source http://barre.nom.fr/medical/samples/)

The value of a pixel in a colour image is a vector with three components (one for each primary colour). A total of $2^8 * 2^8 * 2^8 = 16,777,216$ different colours can therefore be represented.

Other decompositions are possible. They can be linked to those above using linear or non-linear equations, for example, HSB decomposition (Hue, Saturation, and Brightness).

In printing, the absorption properties of colours with the CMYK decomposition (Cyan, Magenta, Yellow, and Black) are used. This is called subtractive synthesis.

In diagnostic medical imaging, the images show in grey levels the physical properties of the body through which the rays pass in. CT-scan images represent the

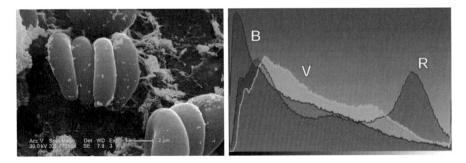

Fig. 5.3 Colour image of red blood cells (Source phil.cdc.gov/phil/home.asp), and the histograms for the *red* (R), *green* (V) and *blue* (B) components

attenuation of X-rays when they pass through the body being examined. It may be useful to add colour so as to discern the characteristics of an image more clearly. To achieve this, false colour are added to Doppler ultrasound images, for example, to colour the blood flow in red or in blue depending on whether the blood is moving towards or away from the transducer. The more intense the colour, the faster is the blood flow.

5.2.3   Image Compression

The resolution of digital images is becoming greater and greater, because the capacity of detectors and of systems to generate these images is better and better and becoming more widespread. High-resolution images are therefore more and more common in widely-used apparatus like mobile phones. Even though the capacities of computer hard discs and the bandwidths for networks are increasing, image compression is becoming necessary to meet growing needs for storage and transmission.

In medical imaging, a huge number of high-resolution images are generated every day in imaging departments. It is estimated that several Tera octets of images are generated every year at a single imaging platform.

5.2.3.1   Properties of Compression Algorithms

The principal property of a compression algorithm for the purpose of storing images is the *compression ratio* it can achieve. Different definitions of compression ratio are currently used. One simple definition gives the compression in the form of a ratio. Suppose that an image of 256×256 requires 65,536 octets in its uncompressed form and only 16,384 octets once compressed, the compression rate in the form of a ratio is therefore 4:1. Some authors prefer to express the compression power as a percentage of the space saved, 75 % in this case.

The *speed* of compression and of decompression is also a major criterion, especially if the aim is to transmit images; in this context, compression is only useful if transmission of the uncompressed image takes more time than the compression, the transmission of the compressed image and its decompression.

There are two families of compression algorithms: lossless compression algorithms, and lossy compression algorithms. By definition, the image that results from lossy compression is different from the uncompressed image.

5.2.3.2   Lossless Compression

Lossless compression techniques produce compressed files from which the original image can be reconstructed exactly. *Huffman* coding is the basis for many compression algorithms. The principle is to code frequent information with fewer bits than used for infrequent information. The length of the code attributed by the algorithm varies according to the frequency of the elements to be coded.

Lempel-Ziv (LZ77 and LZ78) coding is a type of lossless coding that attributes codes of fixed length to groups of pixels of variable length. It is very effective for images that present large homogeneous zones. It can be used free. It is used, for example, by gzip, file compression software (files in.gz).

However, the *Lempel-Ziv-Welch (LZW)* variant is protected by patents held by Unisys. LZW decoding can be used free of charge, but not the coding, which explains why the latter is not always available in software.

In information theory, there is a limit beyond which an image cannot be compressed more without losing information. This limit corresponds to the entropy of the image. The power of image compression for medical images achieved using lossless techniques is in general relatively low (less than 50 %) because they rarely include repetitive patterns.

5.2.3.3   Lossy Compression

To achieve a greater compression ratio, it is necessary to use techniques that result in the loss of part of the information contained in the image. The information lost is less important for the interpretation and analysis of the image. The distortion measured between the compressed and uncompressed image is a major criterion when choosing which method to use.

The *JPEG (Joint Photographic Expert Group)* format is very often used for digital cameras and on the Internet. It allows the 24-bit coding of colour images (capable of representing 16.7 million colours) with variable compression ratios or without losing information.

The principle of JPEG coding is to subdivide the original image into blocks of 8×8 pixels. Each block is then transformed using a *discrete cosine transform (DCT)*. The result is a block of 8×8 quantified coefficients. Data loss and the variation in the compression ratio occur during this quantification. The coefficients obtained after quantification are read from the strongest to the weakest, and then processed using lossless Huffman coding.

In fact, JPEG compression is not particularly effective when lossless is used. The standard has therefore evolved to the standard *JPEG2000* in which the DCT has been replaced by the *wavelet transform*. The result is more effective in the lossless configuration than with JPEG, and it is also more robust. In addition, there are no longer the artefacts that appear in the form of squares in JPEG compressed images at certain compression ratios.

In medical imaging, compression of a single image with different ratios inside and outside a region of interest may be useful. Indeed, as medical images are centred on the organs being studied for the diagnosis, the region of interest must be compressed without error. However, the large areas around the region of interest, which have no diagnostic value, can be compressed with a high proportion of error.

5.2.4   Other Widely-Used Image Formats (GIF, TIFF, PNG...)

The storage of images on discs is achieved using standardised formats that make use of compression algorithms. The principal formats that present advantages with regard to image processing are as follows:

The *GIF* format *(Graphic Interchange Format)* is a format for 8-bit coded colour images (256 colours possible). Developed by CompuServe, this format systematically uses the *LZW* algorithm to compress images. A lot of image processing software allows *GIF* images to be read, but not saved in this format, because of patents on *LZW* coding. The only interest of *GIF* lies in the fact that it is more efficient than *JPEG* for coding drawings with large areas of uniform colour in 256 colours.

The *TIFF* format *(Tagged Image File Format)* is a format that manages 24-bit coded colour images (16.7 million colours). It is suitable for printing images because it can record images in the CMYK format. Almost every type of image processing software recognises the standard *TIFF* format, but very few are able to manage the compressed *TIFF* format, which requires patented LZW coding.

The *PNG* format *(Portable Network Graphics)* is one of the most versatile formats. It can manage images coded on 48 bits, that is to say 2.8E + 14 colours. At the same time, it is able to save drawings in 256 colours of better quality than those using the *GIF* format. Another advantage: compressed images in the *PNG* format use the *LZ77* algorithm, which is free of royalties.

5.2.5   The DICOM Format

The previous imaging formats do not meet all the complex needs of image management in the world of medicine. The *DICOM* (*Digital Imaging and Communications in Medicine*) format, developed on the initiative of medical imaging equipment makers and American radiologists, (http://medical.nema.org) is used. More than a format, *DICOM* is a standard used worldwide in medical imaging systems.

A *DICOM* file contains on the one hand a heading (which stores data concerning the patient, the type of image, the equipment used, the image acquisition conditions...) and on the other hand the image itself, which can be in more than two dimensions (in 3D or in movement). To reduce the size of the image, it is possible to compress it in the lossy or lossless JPEG format.

DICOM was designed to allow interoperability with various imaging systems developed by different companies. As such, it also includes communication standards that comply with the *OSI* (*Open Systems Interconnection*) model of interconnection of open systems, which ensures that two different implementations can communicate effectively. *DICOM* is used in the production, the storage, the transmission, the interrogation, the recovery and the display of medical images. All of the disciplines that produce medical images are concerned, from radiology to cardiology, from oncology to neurology. Choosing imaging systems that comply with the *DICOM* standard guarantees that they will be able to communicate with other equipments, whether current or future, and whoever the manufacturer. On-going developments aim to improve security, performance and flow management.

5.3   Image Processing Methods

In image processing, it is said that we are in the spatial domain when we work directly on the pixels of the image.

After a Fourier transformation (http://en.wikipedia.org/wiki/Fourier_transform) we obtain a new image in the frequency domain. Images can be processed in the spatial domain or in the frequency domain. In the latter, an inverse transformation is performed after the processing in the frequency domain to return to the spatial domain.

In the spatial domain, there are two different cases: in the first, the value of a pixel in the resulting image depends uniquely on that of the same pixel in the original image; this is called pixel to pixel transformation. In the second case, the value of the pixel in the resulting image depends on that of pixels in the immediate vicinity of the pixel in question in the original image; this is called neighbourhood transformation.

5.3.1 *Image Improvement Using Pixel to Pixel Transformations*

Improvement methods use techniques destined to produce images that are easier to interpret or to process. The objective is generally to enhance contrast or the dynamics of the image, or to reduce or eliminate noise. Noise essentially appears during the image acquisition phase. It is often random, but can be periodic. Noise can be characterised by sudden random transitions in grey levels.

5.3.1.1 Pixel-to-Pixel Transformations

Pixel-to-pixel transformations normally use a transcoding table (*Look Up Table* or *LUT*). For an 8-bit image, the transcoding table has 256 values. The LUT can be used to display an image differently on a screen, by changing the distribution of pixel values, without changing the values recorded on the disc. In this table, the index or the address corresponds to the value of the pixel in the original image, and the value contained at this address is the value displayed by the pixel in the image.

The histogram of an image with a dynamic range lower than that available (256 values in this example) can be stretched so that the contrast in the image displayed is better.

5.3.1.2 Histogram Equalisation

A usual method to improve images in the spatial domain is histogram equalisation. In this operation generally performed automatically, we transform the histogram of the original image into a flat, uniform, so-called equalised histogram in which every pixel value is represented the same number of times in the resulting image. In general, the resulting image has an improved appearance (Fig. 5.4).

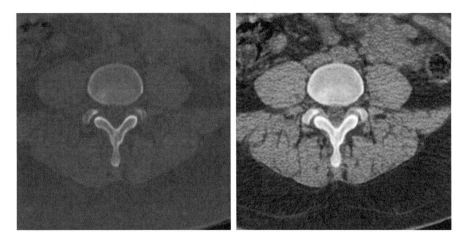

Fig. 5.4 Original image and image after histogram equalisation (Source of original image: http://barre.nom.fr/medical/samples/)

5.3.2   Filtering in the Spatial Domain

In the previous paragraph, the pixels were considered independently, without taking into account the neighbouring pixels. We will tackle here a set of processing strategies that take into account the immediate neighbourhood of the pixels to be treated. This processing involves two types of filters: those which use spatial convolution filtering, and rank filters.

5.3.2.1   Spatial Convolution

Image convolution is an operation that can be performed in the spatial domain with convolution matrices, also called convolution kernels (Fig. 5.5). It can also be performed in the frequency domain as we will see in the next paragraph. A *convolution kernel* is a matrix, often 3×3, which generally contains whole coefficients (Fig. 5.6). The coefficients are combined with the pixels of the original image in the following manner:

- The kernel is centred on a pixel of the original image
- The pixel p_i of the image is multiplied by the coefficients w_i of the corresponding mask,
- The results are added together, and the total is divided by the sum of the coefficients of the mask.
- To preserve the original image, the final result is written in another image at the position corresponding to the place of the original pixel.

This operation is performed for all of the pixels of the original image by reading them from left to right and from top to bottom. The bigger the filter ($5 \times 5, 7 \times 7$ ...), the greater its effect.

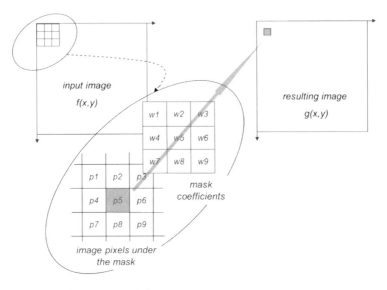

Fig. 5.5 Principle of spatial convolution

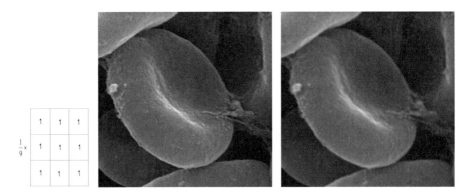

Fig. 5.6 Mean filter: mask, original image (Source phil.cdc.gov/phil/home.asp), and filtered image

The principle of spatial convolution is quite general. Objectives as different as filtering noise or detecting contours can be achieved by choosing appropriate coefficients for the convolution kernel.

5.3.2.2 Spatial Filtering by Convolution

The information contained in an image can be characterised by the concept of spatial frequency. Homogeneous areas of the image where there is little variation in grey levels have low spatial frequencies. In contrast, areas with rapid variations in grey levels, with contours or noise, contain high spatial frequencies.

0,2098	0,458	0,2098
0,458	1	0,458
0,2098	0,458	0,2098

Fig. 5.7 Gaussian filter: mask, original image and filtered image

To accentuate or attenuate a frequency band, low frequencies or high frequencies, it is possible to filter the image using spatial convolution. The only things that change are the coefficients of the convolution mask used.

5.3.2.3 Low-Pass Spatial Filtering

A low-pass spatial filter attenuates or eliminates information on the contours contained in the high frequencies in the image, and preserves the information contained in the low frequencies.

Generally, the mask coefficients used are all equal to 1, and the result is divided by the sum of the coefficients (by 9 for a 3 × 3 mask). This filter is called a '*mean*' filter: the pixels contained in the immediate neighbourhood of the treated pixel adopt the mean value (Fig. 5.7). It attenuates the noise in the homogeneous zones, but smoothes the contours of the image which becomes more blurred.

An alternative to the *mean filter* is the *Gaussian filter* (Fig. 5.7). Its coefficients are calculated from a 2D Gaussian function; the periphery of the kernel is smaller and because of this, the Gaussian filter deteriorates the contours of the image to a lesser extent.

5.3.2.4 Median Filter and Other Rank Filters

In rank filters, the neighbouring pixels are not combined according to a mathematical formula that takes account of the coefficients of a convolution kernel. They are ranked in ascending or descending order, and the value of the resulting pixel is that of the pixel of a chosen rank. For the median filter, the value of the resulting pixel is that of the median pixel.

Median filters can be used to attenuate the noise without blurring the contours. They are very effective for impulsional noise or salt-and-pepper noise (Fig. 5.8). They are non-linear filters.

The maximum filter and the minimum filter are other rank filters that adopt the maximum and minimum values, respectively, of the neighbouring pixels. The minimum filter enhances dark zones, while the maximum filter enhances bright zones.

Fig. 5.8 Median filter: original image, image with salt-and-pepper noise, filtered image

5.3.2.5 Enhancement and Detection of Contours

Enhancing an image is the opposite of smoothing it. Enhancement aims to accentuate the details and contours contained in the images. The detection of contours is an essential step in the recognition of objects or organs in an image.

Contour detection algorithms search regions of the image where grey level intensity varies rapidly, that is to say where the gradient is greater than a certain threshold. The contours correspond to high spatial frequencies.

5.3.2.6 Unsharp Masking

Unsharp masking is a hybrid technique that uses a smoothed image to enhance the contours of images (Fig. 5.9). It consists in two steps: first of all smoothing the image, then subtracting from the original image the result of the smoothing multiplied by a coefficient less than 1. The result below was obtained with a multiplying coefficient of 0.6.

5.3.2.7 High-Pass Spatial Filter

A high-pass spatial filter accentuates the details of the image, contained in the high frequencies, and attenuates or suppresses low frequencies (Fig. 5.10). However, it may also increase noise, which therefore has to be reduced as much as possible before high-pass filtering. The kernel of the high-pass convolution filter is shown in Fig. 5.10.

5.3.2.8 Detection of Contours by the First Derivative

Contours are characterised by rapid changes in the grey levels that surround them. They correspond to a gradient or a first derivative that is greater than a certain threshold. Contours can be detected by derivation or differentiation. In its discrete

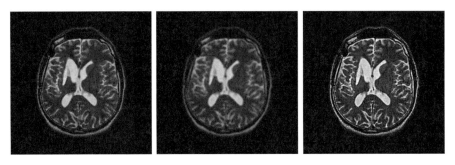

Fig. 5.9 "Unsharp masking": original image, blurred image, difference between two images (Source of original image http://irmtpe.wordpress.com/lirm/)

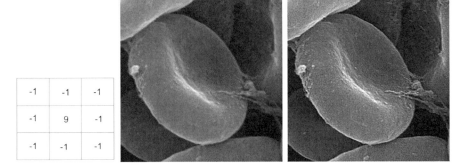

Fig. 5.10 High-pass filter: mask, original image and enhanced image

form, differentiation is performed by subtracting two values, and then by dividing the result by the step that separates the two values. The division is superfluous if the step is equal to one.

From this simple principle, several convolution masks were created to detect contours in an image. To create the derivative in the sense of the x axis and then the y axis, in general, the operators of the first derivative use at least two masks one after the other. These masks are such that the sum of their coefficients is zero, so that they give a value of zero in zones without contours. The most widely used are the masks of Roberts, Prewitt and Sobel (Figs. 5.11 and 5.12).

Roberts masks, 2×2, are sensitive to noise and are quite weak at detecting contours, unless they are very marked.

Prewitt masks, 3×3, present the advantage of having a more clearly defined centre than 2×2 masks, but they remain sensitive to noise.

Sobel masks compensate for this drawback: they include weighted smoothing when the horizontal and vertical derivatives are calculated, and they are thus less sensitive to noise.

Other operators have been developed to improve the results of contours detection even further.

Fig. 5.11 The 2 masks of Roberts, Prewitt, and Sobel

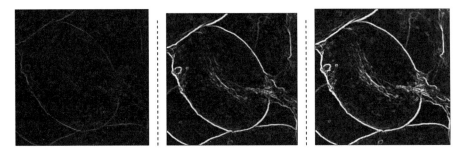

Fig. 5.12 Image processed by Roberts, Prewitt and Sobel filters

The *Kirsch* operator, for example, includes eight convolution masks based on the first derivative, each corresponding to one of 8 directions defined in the image. The Kirsch filter retains the highest of the 8 values.

5.3.2.9 Detection of Contours Using the Second Derivative

A given contour corresponds either to a local maximum of the first derivative of a function, or the passage through zero of the second derivative of the same function. In the discrete form, it has been shown that the 2 s derivatives in the x and y axis can be obtained by using a single mask, the *Laplacian* mask (Fig. 5.13).

In practice the *Laplacian* mask is little used alone because, as it uses the second derivative, it is extremely sensitive to noise. One can cite the "Mexican hat" or "sombrero" operator as example of using the *Laplacian* operator with a Gaussian function.

5.3.2.10 Optimal Filtering

Certain filters have been developed to achieve an optimal result by chaining several steps. This is the case of the *Canny* filter, in which the first step consists in filtering the image with a Gaussian mask. The contours are then detected using a *Prewitt*-type operator. The algorithm then detects the maximum locations of the contours found and suppresses points that are not included: the result is one-pixel-thick

Fig. 5.13 Original image, image filtered by the *Laplacian* and by the *extended Laplacian mask.*

contours. Double thresholding is then used to prevent fragmentation of the contours.

5.3.3   Filtering in the Frequency Domain

In certain types of medical images, like CT-scan and MRI images, the calculations are carried out in the frequency domain after Fourier transformation. Wavelet transformation is a more recent technique that allows a time-frequency representation; it will not be covered here.

5.3.3.1   Notion of Frequency

Images are made up of pixels with varying grey levels, from dark to light grey. *Spatial frequency* measures the rate of variation in these grey levels in an area of the image. High frequencies correspond to rapid variations that are found near contours and in details of the image. Low frequencies correspond to almost homogeneous areas in which the grey levels hardly vary at all.

5.3.3.2   Linear Filters in the Frequency Domain

The Fourier transformation of a spatial image generates *a spectral image* or spectrum, which represents all of the frequencies contained in the original image. The first step of filtering in the frequency domain consists in calculating the spectral image of a spatial image. The frequency components that we wish to remove are thus eliminated while at the same time the frequency components that we wish to keep are preserved. We then return to the original spatial space using an inverse Fourier transformation.

The linear filters described in the paragraph on filtering in the spatial domain can be used in the frequency domain, which is not the case for non-linear filters such as the median filter which cannot be used in the frequency domain.

Low-pass frequency filters are equivalent to spatial smoothing filters. They are obtained by preserving the low frequencies of the spectral image and by suppressing the high frequencies.

The effect of enhancing spatial filters is obtained in the frequency domain by high-pass filtering, that is to say preservation of high frequencies of the spectral image and suppression of low frequencies.

An intermediate effect is obtained by using band-pass filters: they suppress extreme frequencies, high and low, and preserve intermediate frequencies.

5.3.3.3   Interest of Processing in the Frequency Domain

Any image can be processed in the frequency domain. One interest of processing in the frequency domain is that filtering by convolution, which is relatively complex in the spatial domain, is a simple multiplication in the frequency domain. This is the *convolution theorem*.

Filtering in the frequency domain requires calculation of the direct and inverse Fourier transformations of the image, which may appear complex. However, the number of operations of convolution in the spatial domain increases considerably with the size of the convolution kernel. For a given filtering, the overall calculation time is shorter in the frequency domain than in the spatial domain as long as the convolution kernels of the spatial domain are greater than 9×9.

Another advantage of processing in the frequency domain is that it is possible to remove or enhance particular frequencies of the image. In particular, periodic patterns of an image in the spatial domain can easily be enhanced; the periodic noise can be more easily targeted and removed in the frequency domain, which is not possible in the spatial domain.

5.3.4   Mathematical Morphology

Mathematical morphology has produced a large number of morphological operators, which are particularly useful in image analysis. This paragraph is devoted to presenting their properties when they are applied to binary images, that is to say images with two grey levels noted 0 or 1. A pragmatic algorithmic approach is focused on here rather than complex mathematical formulae. The general application of morphological operators to greyscale images is not tackled here.

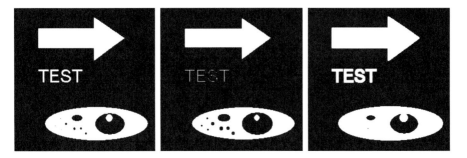

Fig. 5.14 Original image, erosion, dilation

5.3.4.1   Structuring Elements

Morphological operators combine images with small matrices, made up of 0 or 1, called *structuring elements*. Structuring elements are analogues of convolution masks.

The algorithm is as follows: the centre of the structuring element is positioned above each pixel of the input image. The value attributed to the corresponding pixel in the resulting image depends on the comparison of the structuring element with the pixels of the input image.

The two principal mathematical morphology operators are erosion and dilation (Fig. 5.14). They have opposite effects. The operating mode is described here by considering binary images with white objects in the foreground on a black background.

5.3.4.2   Erosion

If at least one pixel of the structuring element coincides with one pixel of the background of the initial image, the corresponding pixel of the resulting image is set at 0. Otherwise it is unchanged. All of the pixels close to the background with a value 1 are therefore set to zero.

Erosion is an operation that diminishes the size of the white objects in the foreground and increases the area of the black background. Erosion depends on the size and the shape of the structuring element. It eliminates isolated pixels or narrow regions.

Erosion is used to eliminate details smaller than the size of the structuring element, and can be used to create one pixel wide contours of objects by subtracting the eroded image from the original image. It can also be used to separate and count objects that touch or cover each other in an image.

5.3.4.3 Dilation

In dilation, if at least one pixel of the structuring element coincides with one pixel in the foreground of the original image, the corresponding pixel of the resulting image is set to 1. In Otherwise it remains unchanged. The pixels in the background that are close to objects in the foreground are set to 1.

The effect of dilation is thus to increase the size of the white objects in the foreground and fill the small black holes inside objects. The enlargement depends on the size and the shape of the structuring element. Objects in the foreground tend to approach each other, or even merge. Dilation can also be used to create one pixel wide contours of objects by subtracting the original image from the dilated image.

5.3.4.4 Opening and Closing

Other morphological operators can be defined using erosion and dilation. Their effects depend on the size and the shape of the structuring element. Among the most widely used, opening and closing can be used to clean binary images before analysis (Fig. 5.15).

Opening is erosion followed by dilation using the same structuring element. It preserves the white objects in the foreground that have the same shape as the structuring element, and tends to eliminate those that have a different shape. At the end of the operation, the objects in the foreground keep the same size, but the contours are smoother.

Closing is dilatation followed by erosion using the same structuring element. The aim is to fill small holes.

5.3.5 Shape Recognition and Classification

One of the aims of image processing is to separate the image into components or into regions that are meaningful for a given analysis. This is called image segmentation. It can be used, for example, for counting and measuring a certain type of cells in a blood sample. It can also involve recognising a tumour in an image of the brain so as to measure the position and dimensions extremely precisely, or to outline the white and grey matter of the brain not affected by the tumour.

Before measuring the characteristics of objects in an image, they first have to be labelled, that is to say recognise all of the pixels that belong to a given object. A particular colour can be attributed to each labelled object (Fig. 5.16). We can also identify objects and determine their dimensional properties.

The classification consists in grouping the objects measured in classes. It is often the last step of image analysis. Attempts are being made to automate the process as

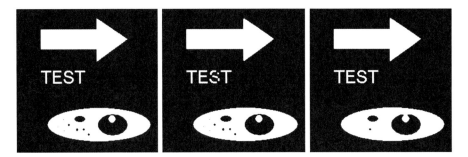

Fig. 5.15 Original image, after opening, after closing

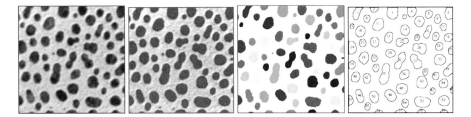

Fig. 5.16 Original image, segmented image, labelled image and analysed image (Source of original image: http://imagej.nih.gov/ij/images/)

much as possible to help in medical diagnosis. Statistical techniques, neuronal networks and genetic algorithms are often used.

5.3.6   Registration and Merging

Different medical imaging techniques provide complementary information, which is very often useful to include in order to improve the diagnosis. It is possible, for example, to merge the functional information of a PET image with the anatomic precision of a CT-scan image. Indeed, as cancer cells have an increased consumption of glucose, it is possible to use PET with ^{18}F-FDG, a sugar that is analogous to glucose labelled with fluorine 18, to identify cancerous zones. The merged image below (Fig. 5.17) combines the functional image obtained by PET of a breast tumour with the CT-scan anatomical image, which is clearer and more precise.

Merged images require that the images to be combined be defined according to the same geometric landmarks. This is never the case for images acquired using different techniques. It is therefore necessary to estimate the relative spatial transformation, which would make it possible to pass from the geometric landmarks of one image to that of the other. This is called image registration.

In medical imaging, registering an image is a frequently needed operation:

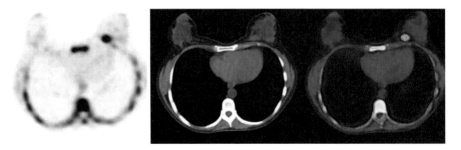

Fig. 5.17 PET image, CT scan image, merged image (Images supplied by Dr Alina Berriolo-Riedinger, CGFL Dijon, FR)

– When images of the same patient acquired using different techniques need to be merged,
– When it is necessary to measure the evolution of a disease using images acquired using the same technique in the same patient but at different times
– When one wants to superimpose images of one patient or several patients in an atlas.

5.4   Imaging Networks, Archiving (PACS)

In a hospital setting, the constantly increasing role of medical imaging in diagnosis has led to a steady rise in the production of digital images. Accessing, sharing and archiving these images must be encouraged. Hospital Information Systems (HIS) are incorporating more and more specialised networks called PACS (*Picture Archiving and Communication System*) for images or RIS (*Radiology Information System*) for other data.

PACS, archiving systems and image transmission, require computers, networks and imaging systems devoted to the storage of, access to and the communication of medical images in the DICOM format. They manage digital images from a variety of origins including US scans, CT-scans, MRI, X-rays . . .

The PACS is incorporated into the RIS, the radiology information system, used by radiology departments to manage the administrative information for patients (identification, requests for examinations, billing. . .)

5.5   Conclusion

The effective digital processing of medical images requires considerable knowledge of the techniques used for image acquisition. The characteristics of the data collected in a specific environment (cardiac MRI, for example) must be clearly understood to select or develop methods for processing, analysis and precise and relevant interpretation. In medicine, a considerable number of methods for processing and image analysis have been developed for different applications of segmentation, analysis and shape recognition. These will make it possible to improve and extend the field of non-invasive methods for diagnosis.

5.6 For More Information

The bibliographic references below will enable you to go deeper into the concepts presented here (Berry 2008; Bovik 2005; Burger and Burge 2008; Castelman 1996; Dhawan 2011; Dougherty 2009; Forsyth and Ponce 2011; Giardina and Dougherty 1988; Gonzalez and Woods 2008; Haralick and Shapiro 1992; Jähne 2005; O'Gorman et al. 2008; Rangayyan 2004; Russ 2011; Russ and Russ 2008; Sonka and Hlavac 2007; Umbaugh 2010; Webb 2003). Some of the documents cited can be used for practical learning as they make use of free biomedical image analysis software such as ImageJ (http://rsbweb.nih.gov/ij)

Some books cited in this chapter contain a CD (Jähne 2005; O'Gorman et al. 2008; Russ and Russ 2008; Umbaugh 2010) or provide the address of a site (Gonzalez and Woods 2008; Jähne 2005; Umbaugh 2010) where it is possible to download programs that illustrate the concepts.

TWorks that deal with these questions are given in the references section:

To go even further, the reader can also consult the many journals in the field such as IEEE transactions on Medical Imaging, IEEE transactions on Image Processing, IEEE transactions on Pattern Analysis and Machine Intelligence, Computer Vision, Computer Vision and Image Understanding, Pattern Recognition, Pattern Recognition Letters, as well as the proceedings of specialised conferences.

Exercises

Q1 How many different grey levels can be represented in an image of (a) 8 bits (b) 12 bits (c) 16 bits?

Q2: What would the histograms of the following images look like?

Q2.1 A set of objects of the same level of grey placed on a background of a different but uniform level of grey

Q2.2 A set of relatively bright objects placed on a relatively dark background, the objects and the background are made up of pixels with similar but varied grey levels.

Q2.3 An overexposed X-ray

Q2.4 An underexposed X-ray.

(continued)

Exercises (continued)

R1 (a) 256 grey levels (b) 4,096 grey levels (c) 65,536 grey levels.

R2: the histograms of the images described above would be made up of:

R2.1 two peaks, one corresponding to the level of grey of the objects, the other to the level of grey of the background

R2.2 two modes: the one closest to zero corresponding to the background of the image, the other to the objects

R2.3 Predominantly made up of bright pixels on the right-hand side of the histogram

R2.4 Predominantly made up of dark pixels on the left-hand side of the histogram

References

Berry E (2008) A practical approach to medical image processing. Taylor & Francis, New York

Bovik A (ed) (2005) Handbook of image and video processing, 2nd edn, Communications, networking an multimedia. Academic Press, San Diego

Burger W, Burge MJ (2008) Digital image processing. Springer, Berlin

Castelman KR (1996) Digital image processing. Prentice Hall, Englewood Cliffs

Dhawan AP (2011) Medical image analysis, 2nd edn. Wiley-IEEE Press, Hoboken

Dougherty G (2009) Digital Image Processing for Medical Applications. Cambridge University Press, Cambridge, UK

Forsyth DA, Ponce J (2011) Computer vision a modern approach, 2nd edn. Prentice Hall, New Jersey

Giardina CR, Dougherty ER (1988) Morphological methods in image and signal processing. Prentice-Hall, Englewood Cliffs

Gonzalez RC, Woods RE (2008) Digital image processing, 3rd edn. Pearson Prentice Hall, Upper Saddle River, NJ

Haralick RM, Shapiro LG (1992) Computer and robot vision, vol 1 and 2. Addison-Wesley, Reading

Jähne B (2005) Digital image processing, 6th edn. Springer, Berlin

O'Gorman L, Sammon J, Seul M (2008) Practical algorithms for image analysis, 2nd edn. Cambridge University Press, New York, NY

Rangayyan RM (2004) Biomedical image analysis. CRC Press, London

Russ JC (2011) The image processing handbook, 6th edn. CRC Press, Boca Raton

Russ JC, Russ JC (2008) Introduction to image processing and analysis. CRC Press, Boca Raton

Sonka M, Hlavac V (2007) Image processing, analysis, and machine vision, 3rd edn. Brooks/Cole Publishing, Pacific Grove, CA

Umbaugh SE (2010) Digital image processing and analysis: human and computer vision applications with CVIPtools, 2nd edn. CRC Press, Boca Raton

Webb A (2003) Introduction to biomedical imaging. Wiley, Hoboken

Chapter 6
Enhanced Medical Intervention: Surgetics and Robotics

A. Moreau-Gaudry and P. Cinquin

Abstract The aim of a medical or surgical intervention is to improve and, where possible, restore the patient's health in compliance with societal rules. To meet this objective, new approaches to increase the Quality of interventions have been identified in recent years. They are based on concepts that are illustrated through two specific field: Surgetics and Robotics. These innovative approaches have led to the general notion of "Enhanced Medical Interventions", that may be seen as new ways of management of patient care in the context of improvements in the quality of interventions. Although these approaches based on innovative technology are very promising, these new types of interventions are meaningful only if associated with considerations related to the assessment of Medical Benefit, such assessment being indeed essential to identify the usefulness of these new approaches towards public health policy.

Keywords Enhanced medical intervention • Surgetics • Robotics • Innovative technology • Health technology • Health technology assessment • Development cycle • Medical benefit • Health information

A. Moreau-Gaudry (✉)
Laboratoire TIMC-IMAG, Domaine de la Merci, La Tronche Cedex 38706, France
e-mail: Alexandre.Moreau-Gaudry@imag.fr

P. Cinquin
Laboratoire TIMC-IMAG, équipe GMCAO, Institut d'Ingénierie de l'Information de Santé, Faculté de Médecine, 38706 La Tronche cedex, France
e-mail: Philippe.Cinquin@imag.fr

A. Venot et al. (eds.), *Medical Informatics, e-Health*, Health Informatics,
DOI 10.1007/978-2-8178-0478-1_6, © Springer-Verlag France 2014

After reading this chapter you should:

- Be able to specify and explain the developments made in recent decades to help increase the quality of surgical operations
- Be able to define and know how to explain the concept of "physician enhancement"
- Be able to define surgetics and explain its contribution to the enhancement in assisting the surgeon or operator
- Be able to describe the various systems of robotic assistance that can be used in surgeon enhancement and explain the ways in which they contribute to this
- Be capable of explaining the key role of information in enhanced medical interventions
- Know the main categories of players involved in the development cycle of a technical innovation in the field of healthcare
- Know the main stages in the development cycle of a technical innovation in the field of healthcare
- Understand the particularities of the field of enhanced interventions in medicine

6.1 Introduction

The aim of a medical or surgical intervention is to improve and, if possible, restore the patient's health, and in a wider context to improve and restore the health of the population undergoing the procedure, in compliance with societal rules. In the first part of this chapter, "Increasing the Quality of interventions", we describe the principal challenges identified in recent years that are being addressed to meet this objective, with examples in several medical and surgical fields. The second and third parts, titled respectively "operator enhancement and surgetics" and "operator enhancement and robotics" present the concepts underlying these two approaches that have been developed in response to these challenges. The fourth part is focused on "Enhanced Medical Interventions" and is meant to introduce new approaches to the management of patient care in the context of improvements in the quality of interventions. Finally, this chapter concludes with a section dedicated to the "Assessment of Medical Benefit" and presents considerations of how these innovative technologies can be evaluated so as to demonstrate their usefulness towards achieving the overall goal of optimal public health introduced above.

6.2 Improving the Quality of Interventions

Improving the quality of medical and surgical interventions is an interdisciplinary activity requiring collaboration between both fundamental and applied researchers in many disciplines. Aspiring to this goal, in a context of technological progress, leads us to constantly push the boundaries of what exists and create new challenges that are ever more ambitious. Among the many challenges identified in recent decades that may increase the quality of surgical operations, two seem most pertinent, namely to reduce the invasiveness of operations by the use of innovative instrumentation and to maximize the benefit/risk ratio using new intraoperative approaches. These are presented below.

6.2.1 Reducing Invasiveness

By "reduction in the invasiveness of an intervention" we mean a lesser degree of disturbance of the tissue structures when we perform the operation. This is a rapidly growing trend due to the expected benefits for the patient and thanks to technological advances, particularly in terms of miniaturization. A typical and widely applicable example in the field of surgery is presented below. It focuses on giving priority to the importance of providing the surgeon with appropriate information so that he/she can operate more efficiently and with greater precision.

In recent years the field of surgery has been shaken up by an unprecedented revolution due to the advent of minimally invasive endoscopic surgery. Unlike "conventional" surgery for which large incisions are required so as to expose the target organs, minimally invasive surgery is characterized by making several small incisions of a few centimeters in length (for example, in the abdomen or around a joint), through which trocars are placed. A trocar is a hollow tubular structure which passes right through the tissue barrier that surrounds the target organ or cavity and through which different instruments (cameras, scalpels, etc.) can be introduced. The instruments introduced into the surgical cavity are intended to be maneuvered so as to view, manipulate and section the anatomical structures. This type of minimally invasive approach has many benefits for the patient. Among these, are reductions in intraoperative bleeding, fewer transfusions, a shorter duration of the operation, and shorter hospital stays.

The boom of minimally invasive surgical techniques, with the beneficial consequences they provide for the patients, has only been possible thanks to advances in instrumentation that combine efficient surgical procedures with minimally invasive approaches. Nevertheless, one of the underlying challenges of these new surgical methods is that of providing the surgeon with information about the surgical site, in a manner that he/she can easily and rapidly interpret and use. More specifically, choosing a minimally invasive surgical approach implies displacement of the surgeon from the site of surgery. Indeed, the surgeon accesses the surgical site, not directly, but via his instruments; he/she no longer sees the organs involved and cannot touch, feel or palpate them. His perceptions, essential for the realization of the procedure, are thus totally dependent on the instruments used at the site.

In this new minimally invasive surgical approach, the first challenge is to be able to give the surgeon a full view of the operative site, i.e. to restore sensory perceptions (primarily sight and touch) as best as possible, so that he/she has a "complete" view of the organ/tissue he/she is operating on. Responses to this first challenge led, among others, to the development of high-tech endoscopic cameras. These can film and display real-time HD views of the operative site, and even restore the perception of depth using a stereoscopic technique. A second challenge, more advanced, is to go beyond the limits of human perception, and to give the surgeon a new "augmented" view of the operative site with information that was not previously accessible intraoperatively. This new way of viewing of the operative site is a rapidly evolving field, especially given the advances in imaging technology that make possible the acquisition of increasingly accurate multimodal information (X-ray images, magnetic resonance images, and others) and hitherto was not possible. The principal challenge is to fuse these different types of information into a single patient reference frame (image referential) and to be able to extract an overall summary of the site that is readily accessible and can be exploited by the surgeon when making decisions during the operation.

To pursue the example presented earlier of endoscopic cameras, one of the current issues is to introduce at the operative site, not just one but several miniaturized cameras. These could potentially be multi-spectral in order to simultaneously acquire different complementary views of the surgical site and even to synthesize all these views into a single "augmented" view so that the surgeon has better and enriched visual control of the site.

Lastly, we should mention the recent emergence of "Natural Orifice Translumenal Endoscopic Surgery" (NOTES) (Auyang et al. 2011; Clark et al. 2012). This type of endoscopic (or keyhole) surgery is another step toward reducing the invasiveness of operations, in which the instruments are brought as close as possible to the target site using the natural body orifices (digestive tract, female reproductive tract or urinary tract).

Having pertinent and precise information concerning the target site (as well as the means for acting on this information) raises new issues for which the answers will be strategic for the future evolution of this new surgical technique.

6.2.2   Improving the Efficiency/Morbidity Ratio

While new medical devices have been developed to reduce the invasiveness of surgical procedures, an effort to improve the efficacy/morbidity ratio is the basis of other innovative approaches aimed at facilitating and optimizing the use of these devices. Two major challenges have been identified, namely the planning of the operation and its implementation.

6.2.2.1   Improving the Planning of the Operation

The first challenge is planning of the procedure, i.e. "the method of choosing the precise goals and proposing the means to achieve them". In some surgical fields, planning has been gradually improved by the introduction of computer programs. These programs allow simulation preoperatively of what will take place intraoperatively and may predict the consequences of a particular operative strategy, thus permitting the surgeon to optimize his strategy. The originality of these approaches is to integrate in a intraoperative setting, pre- and per-operative patient information and knowledge derived from various fields (geometric, kinematic, mechanical, functional, etc.) presented so as to allow a better understanding of the surgical procedure. To illustrate this approach, we take the example of reconstructive surgery of the Anterior Cruciate Ligament (ACL).

The ACL is a knee ligament responsible for the stability of the joint. This ligament may be ruptured during a trauma. ACL reconstruction surgery involves implanting a graft within the joint to try to restore its full functionality. One of the difficulties of this reconstruction is the choice of the exact 3D position of the femoral and tibial insertions of the ends of the graft. In recent decades, hardware and software simulating the intraoperative setting have been developed to provide the surgeon with new tools for both pre- and intraoperative planning and decision making. These innovations are examples of computer assisted medical interventions or CAMI. These are not described in this chapter, but there is a wealth of literature on computer assisted ACL reconstruction (see section "for further information"). Among other things, these methods enable the ends of the femur and the tibia to be geometrically modeled preoperatively in 3D using the real anatomic data of the patient's knee with an accuracy of less than a millimeter. Moreover, it is possible to establish an unambiguous real-time relationship between the anatomic reality of the joint and the geometric modeling. Thus, relative movements of the tibia with respect to the femur can be recorded using the medical device and displayed virtually in real time, with of course, the ability to examine all possible configurations of the joint. The availability, both before and during the procedure, of a virtual environment representing the anatomic reality of the patient enables the surgeon to optimally position in 3D the femoral and tibial ends of the graft so as to minimize the criterion of joint stability during flexion and extension (anisometry). The surgeon can then proceed with his operation as planned.

6.2.2.2 Improving the Accuracy of Surgical Gestures

The second challenge that must be considered concerns the performance of the surgery itself. Indeed, offering the surgeon the opportunity to plan the operation in an extremely precise way is meaningless unless he/she has adequate means to implement this strategy as accurately and optimally as possible.

New strategies have been proposed to address this problem. Improving the performance of the gesture, in terms of accuracy, is commonly obtained by the use of material assistance. As an illustration, in the field of orthopedics, the surgeon will frequently use an ancillary device attached in advance to bony anatomical landmarks, to introduce screws more accurately or to make bone incisions with greater precision. These strategies also include robot assisted procedures that allow human limitations (such as tremor) to be overcome. For example, in the context of laparoscopic surgery, the endoscopic camera is usually held by an operating room assistant who manipulates it and tries to obtain the best view of the surgical site for the surgeon, who meanwhile is manipulating the surgical tools. To assist the surgeon in maneuvering the camera, a voice-controlled robotic camera holder, the "Light Endoscopic Robot" or LER, has been developed by the TIMC-IMAG laboratory at Grenoble (Long et al. 2006, 2013). A modified version of this system, called ViKY® is currently marketed by Endocontrol® (http://www.endocontrol-medical.com).

Improving the performance of the operation also includes increasing the dexterity of the surgeon's gestures. Thus, new devices have been designed to restore the dexterity of hands, restricted due to the severe constraints imposed by laparoscopic operating conditions. As an example, and still in the context of laparoscopic surgery, the surgeon is limited by the instrumentation when trying to make stitches. He/she manipulates the needle using a needle holder that can only be moved about to a very limited extent due to the position of the needle holder with respect to the wall of the digestive tract, which is offset from the site of the stitches. To overcome these difficulties, new medical devices are under development to increase the number of degrees of freedom of the tip of the needle holder and thus to maximize the mobility despite the restricted space for maneuver. An example is the JAIMY® system, currently being developed by Endocontrol® (http://www.endocontrol-medical.com/jaimy.php, http://youtu.be/A-0cTxv3TOY).

Of course, other robotic approaches are being developed to meet the challenges of improving the performance of surgical gestures under laparoscopic conditions. One example is the Da Vinci robot® produced by Intuitive Surgical® since 2003 (http://www.intuitivesurgical.com/products/). This robotic system is based on the concept of manipulator arms (3 or 4 depending on the model). Each arm has seven degrees of freedom and allows for greater precision when performing the surgical gesture. A manipulator arm can hold either an endoscopic camera which provides a three-dimensional view to the surgeon, or a surgical instrument such as an electrocautery.

Finally, another approach to improve the performance of surgical procedures involves the training of surgeons. Innovations that allow the surgeon to gain experience of these increasingly complex surgical procedures, without the slightest risk to patients, appear very attractive. In recent years the use of "simulators" has become more and more popular. Nevertheless, the problem of this approach lies in their ability to simulate, as closely as possible, the conduct of an operation. It is easy to appreciate the importance of detailed knowledge bases for each surgical field, essential to achieve these objectives.

Different Approaches to Improve the Quality of Surgical Interventions

Without being exhaustive, the desire to reduce the invasiveness of these interventions, and taking advantage of technological advances in instrumentation, have led to the development of minimally invasive laparoscopic surgery and more recently to the development of NOTES (Natural Orifice Translumenal Endoscopic Surgery). The search for "optimal" operating procedures has led to the development of new medical devices (robotic or otherwise) for assistance in the planning and implementation of surgery. Nevertheless, whatever the proposed approach, there remains a challenge shared by all. It can be schematically summarized and stated by means of four actions drawn from the field of health informatics: define, acquire, process and make available to physicians the most pertinent information possible, so as to facilitate the performance of his/her actions and thus improve the benefit/risk balance associated with the operation.

6.3   Enhancement of the "Operator" and Surgetics

Our desire to increase the quality of operations has led us to identify key challenges for which innovative solutions have been developed. Some are being used daily in the operating room. Nevertheless, whatever the solutions, a common factor can be identified. This is the "enhancement" of the operator. This enhancement can take many different forms depending on the particular application. This section presents a first approach to operator enhancement, with an emphasis on health information (including both anatomical and functional), its definition, acquisition, processing and use by the surgeon using the system (robotic or not). The most illustrative example is that of "surgetics", which is based on surgical navigation systems.

6.3.1   Surgetics

The term *"Surgetics"* comes from the words "surgery" in English and "informatique" (computer science) in French and arose from an awareness of the importance of cooperation between man and machine for performing medical and in particular surgical procedures. These procedures, increasingly more complex, have benefited from the use of technology (computers, robots), which, with data processing (semi-automatic) of increasingly rich and voluminous amounts of information, have helped change the way in which surgery is done.

Here we should emphasize the contribution of information technologies in processing data. We use the term processing in its broadest sense i.e. the passage from low-level acquisition data captured by sensors to the provision of detailed new high-level knowledge. Indeed, it is principally these technologies that enable us to supplement human analysis, which is potentially flawed or simply impossible. An example is the matching of two 3D images of the same organ but made in two different ways. Matching these two different modalities visually can prove to be extremely difficult (or impossible), especially when the image of the target organ is distorted during one of the two imaging processes. A resort to methods used in the field of signal processing allows the realization of the merger. By using such an approach, the surgeon can, with the use of informative intraoperative information of relatively poor quality (3D ultrasound imaging, for example), realign in an « elastic » manner more highly informative better quality images obtained preoperatively (such as MRI or CT scans). It is from these more informative images that he may be able to perform a better more appropriate intervention.

Such an approach is currently proposed using Urostation®, a device developed by Koelis® (http://www.koelis.com) with which biopsies of the prostate are guided using multimodal imaging (ultrasound and MRI) (Baumann et al. 2012). MRI imaging, performed pre-operatively, enables the surgeon to identify potentially cancerous areas. The MRI images are then realigned with ultrasound images obtained in quasi-real-time and used for sampling, with the possibility of targeting zones where malignancy is highly suspected (thanks to the alignment of the pathological MRI zones with the ultrasound image).

The principles of "surgetics" are based on a now well established trilogy of "Perception", "Reasoning", and "Action" (Cinquin et al. 1995). At the "perception" stage, all technological solutions (coupled with potential biotechnological approaches) are used to acquire relevant information for the planning the operation. In the "Reasoning" stage the provision of explicit and usable information allows the surgeon to establish an optimal treatment strategy. Finally, technical support for actually carrying out the planned operation is proposed. This assistance may make use of a robot or not. In fact, these principles, described in the literature, are implicitly or explicitly reiterated throughout this chapter.

Before addressing the next section on the principles of "GPS for the surgeon", it is necessary to introduce the notion of coupling technological and biotechnological solutions, in the context of providing information that is pertinent to improving the patient's health. Indeed, this notion is essential and perfectly illustrates the necessity for cooperation between different fields so as to acquire "useful" information. The scientific basis schematically shown below illustrates our point. It is now possible to develop biocompatible molecular probes that target, with high specificity, the organic components characteristic of a tissue and/or its quality (malignant or benign). In addition, these probes can be labeled with a fluorophore, which, when excited at the appropriate wavelength, emits photons at a known wavelength. Using a suitable sensor, this feature allows tissues displaying the fluorophore or "rich" in the fluorophore, be visualized and consequently the tissues targeted by molecular probes can be "seen". Thus, approaches based on this principle are being developed, in particular, to assess, intraoperatively, the quality of the surgical borders during surgical prostatectomy.

Schematically, when total resection of the prostate is indicated, ideally the procedure performed should result in the complete removal of all prostatic tissue at the site of the operation. Intraoperatively, after removal of the prostate, the edges of the site are brushed with molecules that bind specifically to prostatic tissue and that are labeled with a fluorophore. After rinsing, excitation at the appropriate wavelength of the fluorophore should not produce a fluorescence signal. Indeed, such a signal would indicate the persistence of prostatic tissue and therefore an incomplete resection at the site, which would then require a complementary intraoperative procedure.

This technique combines the fields of pathophysiology (identification of a specific marker of prostatic tissue), sensor development (sensors sufficiently sensitive and specific to detect the fluorescence signal and to be used intraoperatively) and chemistry (attaching a fluorophore to the biological molecule that will target the residual prostate tissue). It is a perfect example of a transversal approach that provides "useful" information to assist the surgeon during the operation.

6.3.2 Surgical Navigation: GPS for the Surgeon

Surgical navigation works on a similar principle to that of GPS navigation for cars. A locator (magnetic, optical and/or mechanical) enables the surgeon to monitor in real time the position of an object (a surgical tool, a sensor or even an anatomical structure) with usually sub-millimetre precision. Following a plan made before the operation, based, for example on MRI data, and after transfer of this plan to the intraoperative referential by data fusion, the system allows the surgeon to visualize on a screen the real position of the surgical instrument and compare it to its planned ideal position and/or with respect to pre-recorded anatomical reference points (landmarks). The surgeon uses this positional information to control his surgical gestures and follow the planned strategy.

With these systems, the operator is "augmented" thanks to his/her ability during the operation to follow the strategy virtually (this virtual world is linked unequivocally to the very real anatomy of the patient), the information about the patient is often multimodal, and moreover the surgeon can follow his gestures in real time, from information displayed on the screen.

Medical devices based on these concepts are currently marketed and in daily use. One example includes the recent navigation system for interventional radiology during CT scans developed by Imactis® (http://www.imactis.com).

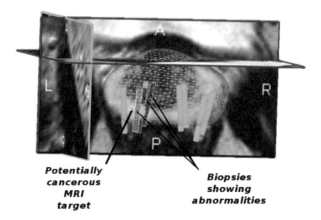

Potentially
cancerous
MRI
target

Biopsies
showing
abnormalities

Fig. 6.1 Visualization on the MRI image of the 3D localization of biopsies guided and performed intraoperatively by ultrasound imaging augmented by MRI. The potentially cancerous "MRI" region to be sampled is shown in 3D (*sphere*) in the MRI image. The *cylinders* represent the ultrasound-MRI-guided biopsy punctures actually performed. Biopsies showing abnormalities on histological examination are located near the area identified by MRI as being potentially cancerous

Use of New Information Analysis Technologies to Improve the Quality of Medico-Surgical Procedures

Technological advances in the last few years now allow us to use fused multimodal patient data intraoperatively. The surgeon can thus better target his/her gestures. To illustrate this point, in order to diagnose prostate cancer it is necessary to perform biopsies of the organ, which when analyzed histologically can confirm (or not) the diagnosis. One difficulty is in performing these biopsies, which are guided by 3D ultrasound imaging but which cannot discriminate between abnormal tissue (potentially cancerous) and normal tissue. The company Koelis® offers radiologists and urologists a guidance and mapping system for prostate puncture-biopsy based on the fusion of 3D ultrasound and MRI images. The latter provides the first with high resolution information on the potentially cancerous regions (see Fig. 6.1). The operator can thus guide the sampling needles towards those regions, identified by MRI, then scaled and fused on the real-time ultrasound images, and see the result of his gesture.

6.4 Enhancement of the "Operator" and Robotics

Here we present a complementary but different strategy to "enhance" the performance of the operator. Whereas above we had stressed the importance of information for planning the operation and its execution, here we present the main types of robotic systems used to achieve greater precision in the operator's gestures. This presentation describes systematically the various robotic assistance systems that have been recently developed or are currently under development.

6.4.1 What Is a Robotic Assistance System?

Before going further, it is necessary to clarify what we mean by a "robotic assistance system." From the French definition of the term "robot" proposed in Wikipedia (http://fr.wikipedia.org/wiki/Robot), a robotic assistance system is "a mechatronic device" that combines three components, mechanical, electronic and computers, to perform automatically tasks that are dangerous, arduous, repetitive or impossible for humans; or simpler tasks by doing them better than they would be done by a human. All the systems listed below meet the above definition.

6.4.2 Semi-Active Systems

As the name suggests, interaction is needed between the robotic system and the operator, this teamwork is sequential. To illustrate this concept, we present an example in the field of orthopedic surgery. When a knee prosthesis is implanted, the femoral bone must be resected along various cutting planes so as to embed the prosthesis, as best as possible, into the bone. One frequently encountered difficulty is to optimally perform these resections, i.e. respecting the relative angular positions of the cutting planes, imposed by the geometric characteristics of the prosthesis. Nevertheless, meeting these criteria is not so easy (possible wobble between each cut responsible for prejudicial angular disparities). To overcome these difficulties, the robot "Praxiteles" was developed by the laboratory TIMC-IMAG (Plaskos et al. 2005; Koulalis et al. 2011). The robot iBlock® is the commercial version of this (PRAXIM ® – http://www.praxim.fr). In a first step, the robot, laterally attached to the femoral bone, approaches the cutting guide (without the cutting tool) to its target position. Once positioned optimally, a mechanical lock prevents any undesired movement of the cutting guide relative to the bone. The cut can then be made with confidence by the surgeon following the cutting guide. When the cut is completed, the process is repeated (positioning of the cutting guide in the new cutting plane, fixing it in position and then and then cutting), until the necessary cuts have been made in all planes. The advantage of this approach is that it considerably simplifies the security problems inherent in the use of robots in the operating room. Indeed, the most dangerous phase in the use of a robot is when it is moving, and special care has to be taken to monitor and control the performance of "invasive" maneuvers, by way of the addition of an immobilization step in the guided system. Here, the performance of the surgeon is "enhanced" by the use of a cutting guide that is optimally positioned for him according to the pre-established plan of the operation.

6.4.3   Synergistic Systems

The surgeon and robotic system interact simultaneously; the robot holds the surgical instrument, but cannot maneuver it directly. Movements are initiated and controlled by the surgeon. Specifically, the surgeon maneuvers mechanically at the external end of the robot. He exerts forces to this end that result in the instrument being moved such as to engage with the patient. The robot simply "filters" the movements of the surgeon. It "allows" forces that will move the instruments in a manner consistent with the predefined strategy of the operation, but "forbids" potentially dangerous movements. This approach, which allows surgical gestures within a "safe space" has been successfully implemented in routine clinical use for cutting bone pieces in the implantation of total knee prostheses. As before, the "enhancement" of the surgeon leads to more reliable and precise gestures.

6.4.4   Active Systems

"Active" robots are capable of independently performing a part of the surgical strategy, and during this phase of the operation the surgeon retains only the ability to interrupt the action using a device such as the "dead man" switch. Here, the movement of the robot is interrupted as soon as the surgeon releases his pressure on the switch. The ROBODOC® system, developed by CUREXO Technology Corporation® (http://www.robodoc.com/) is the most successful example (Schulz et al. 2007; Hananouchi et al. 2008). Indeed this robot, which was the first surgical robot to obtain FDA clearance, is capable of maneuvering a bur used for milling the medullar shaft of the femur of the patient, so as to make it congruent with the end of the femoral prosthesis. ROBODOC ® is the only system capable of performing such a complex task. In this approach, the "enhancement" of the surgeon means that he can achieve a precision in his gestures that would be unthinkable without robotic assistance.

6.4.5   Tele-Controlled Systems

The surgeon is installed at a remote control console. The movements of the surgeon's hands are reproduced (scaled, filtered and translated) by the endoscopic instruments at the site of surgery in the patient. The system includes both a stereoscopic endoscopic camera and several articulated endoscopic instruments. The DaVinci® Intuitive® Surgical System is a typical example. It has been widely sold to assist surgeons in the field of radical prostatectomy. Here, the "enhancement" of the surgeon is linked to his removal from the immediate site of the operation and potentially enhanced perception of the site. From the console the

Fig. 6.2 The ViKY® robot
positioned above a patient

implementation of precise gestures are performed, with the system seamlessly
translating the surgeon's hand, wrist and finger movements into precise, real-time
movements of the surgical tools, increasing dexterity in particular (Freschi et al.
2012).

Enhancement of the Surgeon Using Robotic Assistance

Robotic assistance systems contribute to the "enhancement" of the operator
(usually a surgeon), allowing more precise, more rapid and more reproducible
targeted surgical gestures. The robot ViKY ® (see Fig. 6.2) developed by
Endocontrol® illustrates these concepts. This robotic system assists the
surgeon in maneuvering the endoscopic camera. Fixed at the bedside, it is
controlled by the voice of the surgeon to move the camera through six degrees
of freedom. The robotic assistance system showed better surgeon satisfaction
in terms of the stability of the view obtained.

6.5   Enhanced Medical Interventions (EMI)

Earlier, we presented two main types of approach to "enhance" the operator (or surgeon), either in terms of information and its use (see the intrinsic methodology of surgetics) or from the perspective of the surgical maneuver or gesture (with robotic assistance procedures), always with the aim of improving the quality of surgical operations. The goal of "EMI" is not only to provide the user with "low level" tools that will allow him/her to optimize their strategy with increased information, but to make use of "high level" technologies. These take advantage of highly sophisticated modeling techniques, as well as existing expertise in the field, giving the surgeon access to shared knowledge, to consider (or not) in his/her therapeutic strategy. In other words, it is more than just increasing 'technical' aspects but improving consideration of the pertinence of the medical procedure to be performed (in terms of medical benefit). The ripening of reflection in this area has recently been recognized by the creation of the Excellence Center for Computer Assisted Medical Interventions (http://www.eccami.com) dedicated to the improvement and added value (particularly in terms of medical benefit) provided by computer-assisted interventions.

6.5.1   EMI and Surgetics

What is the difference between "Surgetics" and "Enhanced Medical Interventions"? The scope of Enhanced Medical Interventions covers that of surgetics. Nevertheless, it is far more extensive, encompassing all radiological procedures, radiotherapeutic and endoscopic, the vectorization of therapeutic agents, and other procedures in which the operator is "augmented" in order to improve the benefit/risk associated with the intervention. In fact, the degree of maturity in surgetics achieved so far, gives a truly "enhanced" character to surgetic interventions. Taking the example of reconstructive ACL surgery, the challenges of this computer-assisted surgery is now to propose optimal insertion points for the ligament graft, taking into account not only the anisometric criteria during flexion/extension, but also the stability criteria for all possible joint movements (forced internal rotation, pivot shift, etc.) conventionally used to evaluate the stability of the joint.

6.5.2   EMI and Robotics

Here, the goal of EMI is to rely on robotic systems as "effectors" to better perform interventions designed to have high added medical value. As before, the inclusion of "high level" knowledge and shared expertise in the realization of surgical procedures could allow robots to be designed that are far more pertinent and useful in terms of the Medical Benefit provided. Taking this vision to the extreme, it could be the clinical "usefulness" which ultimately brings robots to be used in a truly "independent" way at the surgical site, this "autonomy" remaining nonetheless restricted, as any intervention must be supervised by a qualified medical expert.

Take the example of the robot endoscope holder VIKY® previously presented. As it stands, this robot is controlled by the surgeon using his voice or a foot pedal, to move the endoscopic camera through six degrees of freedom. The challenge is now to "extend" the attributes of this robot so that it anticipates the course of the operation, i.e. to invest the robot with a real "artificial intelligence" that, during the various steps of the operation, will permit it to recognize, analyze and adapt "automatically" the camera position, or to propose contextual information pertinent to the implementation of the surgical step. Obtaining such a degree of autonomy for the robot, that is robust, reliable and safe for the patient, naturally requires that we have extremely powerful tools, especially in terms of information processing (automatic image processing, knowledge bases etc.), with, as a result, real gains (in precision, efficiency, safety, duration of the intervention, hospital stay, etc.) for the surgical procedure.

6.5.3 EMI and Robots Implanted in the Body

This approach represents a real rupture with previous concepts of assistance (automated or not). This rupture is characterized by at least two main differences. The first difference is the operator, by definition it is not the surgeon or the doctor, but the patient who carries the robot. The second difference is in the type of improvement sought. As for the assistance systems presented above, "enhancing" the operator was primarily directed at ways to perform a more precise surgical procedure, or at the information made available to the surgeon to best achieve the intervention. Here, the operator (the patient) is assisted by automating mechanical tasks (see the definition presented earlier). These mechanical tasks are required for his/her disease management. While they were previously carried out by the patient, they are now performed partly or wholly by the robotic system. This improvement is only possible through advances in the field of mechatronics. Health applications of these technologies constitute a newly emergent field. We leave it to the reader to imagine all the possibilities offered by such systems in compliance with ethical and societal rules, and present below the first attempts to develop an implantable robot aimed at addressing the problem of urinary incontinence in men.

Men with severe urinary incontinence can nowadays benefit from a urinary prosthesis (American Medical Systems AMS800®) to palliate this handicap. Fully hydraulic, it consists of an occlusive cuff placed around the urethra which compresses the urethra thanks to a pressurization balloon. The pressure can be modulated using a pump placed in the scrotum. When the man feels the need to urinate, he repeatedly presses on the pump with consequent opening of the lumen of the urethra. The occlusion pressure is restored after a few minutes with a mechanical system incorporated into the pump. However, this medical device is far from perfect, with inefficient systems for some patients (20 % of implanted patients) or the need for regular revision (35 % of implanted patients) related to failure or urethral atrophy. An innovative solution, the active artificial urinary sphincter, is currently under development by Uromems® and aims to address this problem using an implantable mechatronics system. Eventually, a fully implantable system should automatically adjust and exert pressure on the urethra, not only to prevent leaks but also to prevent the deleterious consequences of excessive pressure exerted during closure of the urethra. We can, from this example, readily imagine all the incremental steps needed to eventually have truly autonomous implantable robots. In particular, a judicious compromise will need to be made between miniaturization, power consumption and the desired level of information processing or mechanical activation, a compromise naturally guided by the medical benefit that you want to actually provide to the patient.

6.6 Estimation of Expected Medical Benefit and Demonstration of Rendered Medical Benefit

6.6.1 Medical Benefit

As initially introduced, the ultimate goal of medical or surgical intervention is to improve and restore the health of the person undergoing the operation, in compliance with societal rules. It thus seems natural to identify the steps toward evaluating medical devices involved in the implementation of Enhanced Medical Interventions, and to prove their pertinence, especially in terms of Medical Benefit Rendered. However, such an assessment is difficult because of the complexity and multiplicity of factors that are involved in achieving the therapeutic act and which must naturally be taken into account when estimating the relevance of the therapeutic solutions implemented, as accurately as possible. Schematically, it is not only necessary to consider the factors related to the patient, (use of a functional score, a score of quality of life, or even survival rates, etc.), but also all aspects that are related indirectly to the patient, whether in the field of healthcare (surgeon, doctor, patient associations, competent authorities, the Health Authority, etc.), the industrial sector (Industrial technology developer, the target market, industrial property rights, etc.) and the area of Research (researchers, intellectual property rights etc.).

In this section, firstly we describe the broad principles that apply to the development cycle of a technological innovation in the field of healthcare. Principles from which it is possible to draw a methodological approach to assess the medical benefit associated with the development of an innovative medical device. In the second part, we highlight some specific features of EMI in terms of the demonstration of the Medical Benefit Rendered.

6.6.2   The Development Cycle in Innovative Technology (IT)

Several different stages can be identified throughout the development cycle of a Technological Innovation intended to be used in the healthcare domain (http://www.cic-it-grenoble.fr/presse.php). They are organized in a temporal manner around the three groups of actors usually involved in the development of any Technological Innovation in this field. They are: healthcare recipients and providers (patients, but also surgeons, patient associations, etc.), actors in the field of Research (Researcher, Research Laboratories, Research Agencies, etc.), and actors in the field of Industry (Industrial developers, ministry, etc.). These steps are schematically and briefly presented below and illustrated in Fig. 6.3.

- **Genesis of the project idea.** A first step is to identify a medical problem and an idea. The latter is the initial response to the medical problem, which may eventually lead to the creation of an innovative medical device.
- **From idea to product concept.** This is a stage of formalization and maturing of the idea, with initial functional specifications and consideration of the scientific and technical issues. At the end of this second step, the notion of product concept should have emerged.
- **From product concept to preclinical prototype.** In this third step, and from the results of work upstream, the first preclinical prototype is manufactured. The first tests are conducted to assess the adequacy of the technological innovation with the specifications previously drawn up.
- **From prototype preclinical to clinical prototype.** The objective of the fourth stage is to reach maturity. Iterations by the different actors involved in the innovation can make the preclinical prototype more compatible with clinical requirements.
- **Improvement of the technical innovation with use.** The first clinical prototype is manufactured and must be evaluated.
- **Commercialization.** The sixth and final stage of this development process is to put the fully developed medical device on the market, ideally accompanied by a full evaluation of its Medical Benefit.

Following this description, two phases can be defined that hinge around the intrinsically translational character of the development of a Medical Device. This "translational character" can of course refer to different steps in the development process of an innovation and will vary according to the degree of development of the innovation or depend on the benchmarks defined by the actor in question. In this context, by "translation", we mean the "conveyance" of the medical device and all information and knowledge associated with the final preclinical stage to a first clinical stage. Thus, the first period includes all the individual steps upstream of the

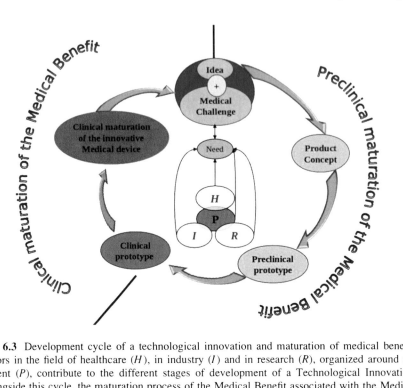

Fig. 6.3 Development cycle of a technological innovation and maturation of medical benefit. Actors in the field of healthcare (H), in industry (I) and in research (R), organized around the patient (P), contribute to the different stages of development of a Technological Innovation. Alongside this cycle, the maturation process of the Medical Benefit associated with the Medical Device can (must) be elaborated. This is organized into two main periods, respectively, pre-clinical (Expected Medical Benefit) and clinical (Rendered Medical Benefit)

first clinical evaluation and therefore comprises the first four stages of development. This can be considered to include a phase of "Preclinical evaluation of the Medical benefit." During this phase, all assessments of the use of the innovation can contribute toward an estimation of what the final "Medical Benefit" will be. It is therefore an estimate of the Expected Medical Benefit (EMB) of the medical device. The second period of "Clinical maturation of Medical Benefit", combines the last two steps of the Technological Innovation development cycle, that is, the clinical use of the Medical Device. In these last two steps the Rendered Medical Benefit of the newly developed Medical Device will be evaluated.

6.6.3 Enhanced Medical Interventions (EMI) and Medical Benefit (MB)

In this section our goal is to alert the reader to the concepts essential to this field and we describe three examples of the medical benefit that might be obtained by the use of EMI.

Unlike the field of pharmacology and drug development in which standardized methodologies have been developed to evaluate the clinical pertinence of new drugs, EMI are not completely covered by these approaches. Three major differences are worth noting. Firstly, and in most cases, an EMI relies on the use of an innovative medical device, for which the conventional methods developed for the clinical assessment of drugs cannot be simply transposed. As an example, the methodologies which provide a "high level of evidence", such as double-blind trials, are hardly applicable to assessments in the field of robotics.

In an evaluation of the medical benefit of an EMI it is necessary to consider not just one but two actors (the patient and the operator). Indeed, while it is possible to standardize the administration of a drug (dose, form, dosage, route of administration), it is much more difficult to standardize the use of a new medical device by the operator. Given that the clinical outcome of the intervention depends, certainly on the patient, but also on the way the device is used, then evaluations of such use must be considered when making clinical assessments so as to make them more pertinent. In fact, failure to consider this type of information can actually lead to completely erroneous conclusions, especially when introducing new technologies into daily clinical practice. Thus, it seems judicious to pose a different set of questions before considering an evaluation of the Medical Benefit: is the medical device being used for its intended purpose? Is it easy to use for its intended purpose? Is the operator trained to use it? Is training necessary? Should the notion of a learning curve be considered? Was the planning of the intervention done properly? etc. There are so many questions where the answers can help enrich the evaluation of the medical benefit associated with the use of the innovative medical device.

Finally, new approaches are being undertaken in the field of EMI to make the best use of Innovative Medical Devices (IMD) that are used during the intervention. Essentially there are two uses that may be attached to an IMD during an EMI. The first that could be considered as "explicit", is the "conventional" or "typical" use of a medical device. By "conventional use", we are referring to a use consistent with the purpose for which the device was designed, i.e. in order to achieve the objectives for which it was developed. Thus the robot is used to make a gesture more precisely without tremor. The second, more "implicit", has to be thought as "stealthy". It is the opportunity to enrich the functionality of the IMD involved in EMI, in order to contextualize the use to which it is put during the course of the EMI. Such an approach may then provide further insights into its contribution. Indeed, following the conventional approaches for evaluating an intervention, the assessment often takes an "overall" or comprehensive perspective and is sometimes based on the results of selected intermediate intraoperative acts, that are linked to the "explicit" use of the IMD. One of the particularities of EMI is that it can provide, thanks to the technologies "involved" in the MD (sensors, computers, etc.), valuable intraoperative "unconventional" information during the surgery. For instance, the complete set of "machine" data needed to run the IMD may be recorded, in an stealthy way, in order to identify the reason of a mechanical failure if such failures happened during the EMI. In fact, these data may also be processed to follow the use to which the medical device is put during the EMI. Based on these facts, new approaches are currently being developed to allow real contextualization of the intraoperative gestures which may well provide further insights into the Medical Benefit actually provided by these innovative approaches.

6.7   Conclusion

In this chapter, we have presented the different approaches, robotic or not, that have been developed in recent years to increase the quality of medical and surgical interventions, giving examples in current use. The state of the art of the approaches developed to date, as well as advances in technology, now allow us to envisage Enhanced Medical Interventions that target, not only technical improvement but also an increase in the pertinence of the medical procedure performed. To achieve this, we have stressed the importance of high quality information for the implementation of such procedures. Finally, we conclude this chapter by emphasizing the need to take an objective view and to demonstrate the Medical Benefit provided by Enhanced Medical interventions. Such a demonstration, that provides information pertinent to healthcare, is a crucial step in convincing society of the high benefit to be gained from these interventions for patients. This will facilitate their dissemination, which is both in the interest of patients, and more generally in the interest of society.

6.8   For More Information

The web sites listed throughout this chapter illustrate the concepts presented. For more details on the concepts of Augmented Interventions in Medicine, the reader should consult www.ncbi.nlm.nih.gov/pubmed/ and use the medical concepts introduced here in the MeSH ® thesaurus, together with "Computer-Assisted".

You could also consult reviews and conference proceedings in the field (French Medical Informatics workshops, European Medical Informatics Conference, Medical and Health Informatics), as well as more highly specialized journals or conference proceedings in the field of computer assisted surgery (International Journal of Computer Assisted Radiology and Surgery, The Journal of Bone and Joint Surgery, Computer Assisted Orthopaedic Surgery, IEEE Transactions on Medical Imaging, IEEE Transactions on Biomedical Engineering, etc.). On the subject of Technological Innovation and Medical Benefit, we recommend the special issue: CIC-IT-ITBM RBM published in 2010, which presents the main concepts to consider in the field of Technological Innovations with many references (and also research reports).

Exercises

Q1 You are a young fellow in Public Health specializing in the evaluation of innovative medical devices. You are asked by your colleague, a laparoscopic surgeon, to evaluate a new robotic endoscope-holder developed to support and manipulate the endoscope in response to voice commands. What would be your strategy? How would you acquire the appropriate information to implement your strategy?

Q2 True/False
Which statements seem relevant to the Medical Benefit (MB) associated with a navigation system to assist procedures in interventional radiology under a scanner?

1. As the interventional radiologist is satisfied with the new system available to him in his department, demonstrating the medical benefit associated with such a system is of no interest.
2. One of the components to be taken into account when appropriately assessing the Medical Benefit rendered by this system comes under the area of health economics
3. A knowledge of the use made of the medical device during the procedure may be essential to estimate the medical benefit rendered by the medical device in a meaningful way.
4. In general, no method is currently being considered in a systematic way to acquire information on the use of the navigation system interoperatively to assist in Interventional Radiology procedures.

R1 Possible answers:
Implement a biomedical research project with objectives that reflect the requirements and expectations of the various actors implicated with this MD. In practice, several different objectives could be considered:

- Demonstrate the equivalence of LER in laparoscopic surgery, comparing the two following surgical procedures: surgery with robot and the surgeon (LER group: new treatment) and intervention with an operating room assistant (OA) and the surgeon (OA group: standard treatment), the primary endpoint would be the number of useful hands, the robot being considered a useful mechanical hand.
- Assess the reliability and limitations of the LER, by counting the number of breakdowns and the number of failures (manipulation, failure of voice command, etc.).
- Compare the visual comfort (declarative score) of the surgeon between the two groups REL and OA;

(continued)

Exercises (continued)

- Count the intraoperative, immediate and late postoperative complications (with data collection during the monitoring visit at 1 month);
- Assess the opinions of the different people (medical and paramedical) who have used the device; (study of the learning curve and time needed to perform the procedure);
- Perform a medico-economic evaluation in the form of a cost benefit analysis, with in particular an assessment of potential human costs avoided.
- Others

The second part is an open question. The following lines of thought could be proposed: paper case report form, electronic case report form, patient medical records, machine data (parameters) from the REL, video made of surgical site etc.

R2

R21 False. See Sect. 6.6.1. To demonstrate the Medical Benefit, it is not only necessary to consider the factors related to the patient, but also the all the factors related indirectly to the patient, which are essential and subject to other actors, whether in the field of healthcare or in industry.

R22 True. See Sect. 6.6.1 and the above answer.

R23 True.

See Sect. 6.6.3. Because the clinical outcome of the intervention depends, of course on the patient but also on the use of the assistance system made by the interventional radiologist, the feedback of this use must be considered when making clinical evaluations focused on the estimation and demonstration of medical benefit associated with the navigation system.

R24 True. See Sect. 6.6.3 and the new "stealthy" use of technologies adopted by Innovative Medical Devices (IMD) and implicated in the performance of Enhanced Medical Interventions.

References

Auyang ED, Santos BF, Enter DH et al (2011) Natural orifice translumenal endoscopic surgery (NOTES(®)): a technical review. Surg Endosc 25(10):3135–3148

Baumann M, Mozer P, Daanen V et al (2012) Prostate biopsy tracking with deformation estimation. Med Image Anal 16(3):562–576

Cinquin P, Bainville E, Barbe C et al (1995) Computer assisted medical interventions. IEEE Eng Med Biol Mag 14(3):254–263

Clark MP, Qayed ES, Kooby D (2012) Natural orifice translumenal endoscopic surgery in humans: a review. Minim Invasive Surg 2012:189296

Freschi C, Ferrari V, Melfi F et al (2012) Technical review of the da Vinci surgical telemanipulator. Int J Med Robot. doi: 10.1002/rcs.1468

Hananouchi T, Nakamura N, Kakimoto A et al (2008) CT-based planning of a single-radius femoral component in total knee arthroplasty using the ROBODOC system. Comput Aided Surg 13(1): 23–29

Koulalis D, O'Loughlin PF, Plaskos C et al (2011) Sequential versus automated cutting guides in computer-assisted total knee arthroplasty. Knee 18(6):436–442

Long JA, Cinquin P, Troccaz J (2006) Preclinical development of the TIMC LER (light endoscope robot). Prog Urol 16(1):45–51, French

Long JA, Tostain J, Lanchon C et al (2013) First clinical experience in urologic surgery with a novel robotic lightweight laparoscope holder. J Endourol 27(1):58–63

Plaskos C, Cinquin P, Lavallée S et al (2005) Praxiteles: a miniature bone-mounted robot for minimal access total knee arthroplasty. Int J Med Robot 1(4):67–79

Schulz AP, Seide K, Queitsch C et al (2007) Results of total hip replacement using the Robodoc surgical assistant system: clinical outcome and evaluation of complications for 97 procedures. Int J Med Robot 3(4):301–306

Chapter 7
Medical Diagnostic Decision Support

B. Séroussi, P. Le Beux, and A. Venot

Abstract Medical diagnostic decision-making is a complex task that consists in finding the right diagnosis from the signs and symptoms presented by a patient. Computers have rapidly been considered as potential diagnostic aids in medical decision-making. This chapter first presents medical diagnostic modeling as a hypothetico-deductive reasoning process. Then, the different approaches developed to provide computerized medical diagnostic decision support are proposed. Initial numerical approaches, either statistical or probabilistic, are first presented. Examples of clinical scores, more recently developed, are given. Then, medical expert systems are described. The three components of an expert system, the knowledge base, the base of facts, and the inference engine, are introduced. A focus is given on knowledge representation formalisms with the description of production rules, decision trees, semantic networks, and frames. The subsection describing the inference engine starts with a presentation of the three types of inference (deduction, induction, abduction). The principles of formal logic are given and the main ways the inference engine may operate are described (forward and backward chainings). Finally, historical medical expert systems, such as Mycin and Internist, as well as systems currently available for medical diagnostic decision support (DXplain™) are described.

Keywords Decision support • Expert systems • Knowledge base • Production rules • Decision trees • Inference engine • Forward chaining • Backward chaining

B. Séroussi (✉)
Département de Santé Publique, UFR de Médecine, UPMC, Paris 6, Hôpital Tenon, 4 rue de la Chine, 75020 Paris, France
e-mail: brigitte.seroussi@tnn.aphp.fr

P. Le Beux
Laboratoire d'Informatique Médicale, Faculté de Médecine, INSERM U936, 2, av. Léon Bernard, Rennes 35043, France

A. Venot
LIM&BIO EA 3969, UFR SMBH, Université Paris 13, 74 rue Marcel Cachin, Bobigny Cedex 93017, France
e-mail: alain.venot@univ-paris13.fr

A. Venot et al. (eds.), *Medical Informatics, e-Health*, Health Informatics,
DOI 10.1007/978-2-8178-0478-1_7, © Springer-Verlag France 2014

> **After reading this chapter, you should:**
>
> - Know the main approaches developed for medical diagnostic decision support,
> - Be able to describe the different components of an expert system,
> - Know knowledge representation formalisms, especially production rules,
> - Understand the role of the inference engine and know how it operates,
> - Be familiar with historical medical expert systems

7.1 Difficulties of Medical Diagnostic Decision

Medicine was initially essentially a magico-religious practice. Diagnosis belonged to soothsayers, oracles and priests. With Hippocrates, medicine became more rational. More recently, the medical diagnostic decision process gained in formalization (Ledley and Lusted 1959). In agreement with the Webster dictionary, we define medical diagnostic decision process as "the art or act of identifying a disease from its signs and symptoms."

7.1.1 Medical Diagnostic Decision-Making: A Hypothetico-Deductive Reasoning Process

Medical diagnostic decision-making is complex (Elstein and Schwarz 2002). It is based on the confrontation of facts either coming from the patient interview, or the clinical examination, or the results of laboratory exams, with a combination of theoretical knowledge and empirical skills acquired from experience. One usually considers that setting a medical diagnosis follows a hypothetico-deductive reasoning process (see Fig. 7.1).

The physician forms hypotheses from clinical observations. (S)he usually considers only the diagnoses compatible with theses hypotheses. However, (s)he will start reasoning while considering only most probable diagnoses, for example, on the basis of a frequency criterion. Coronary disease, pulmonary embolism, or aortic dissection are only considered in case of chest pain, less frequent etiologies are not considered in the first place. For each considered diagnosis, the physician looks for signs either to confirm the hypothesis or to invalidate it. For example, in the case of a coronary disease, (s)he will seek for personal or family history, topography of pain, specific abnormalities of the ECG, etc.

At the end of this process, when it remains only one hypothesis, it corresponds to "the" diagnosis. However, if there remain several assumptions, the process should be repeated while widening the circle of signs to look for. Finally, if there is no remaining hypothesis, the process should be repeated while widening the circle of

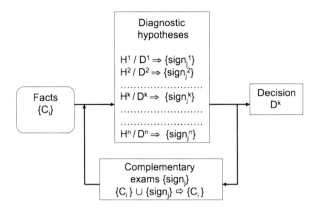

Fig. 7.1 Hypothetico-deductive process of medical diagnostic decision-making

hypotheses to be considered, and including rarest diagnoses, initially excluded. The expression hypothetico-deductive describes the loop made from the generation of hypotheses and the deduction of signs to be assessed. Thus, for each diagnostic hypothesis D_j, deduction "If diagnosis D_j then ($sign_j^1$ AND $sign_j^2$ AND .... AND $sign_j^d$)" is used to elicit signs associated with D_j which are missing and must therefore be sought. If a sign $sign_j^k$ is not validated, the hypothesis D_j may be ruled out.

Due to the continuous growth of medical knowledge, signs, symptoms and diseases are becoming more specific, and complementary investigations have multiplied. It is increasingly difficult for a physician to master all the knowledge necessary to recognize diseases by their symptoms and organize the hypothetico-deductive reasoning process necessary to distinguish diseases from the others. The qualities of computers (memory, speed, power) have emerged as good solutions to solve this problem and computerized systems have been developed to support medical decision, whether diagnostic or therapeutic (Schoolman and Bernstein 1978). While first systems were based on numerical approaches, their successors, also named expert systems, have been developed in a logico-symbolic approach (Miller 1994).

7.1.2 Errors of Diagnosis in Recent Medical Literature

Numerous studies (Gandhi et al. 2006; Singh et al. 2007; Schiff et al. 2009; Lucas et al. 2012) have been published which outline the frequency of diagnostic errors in medicine (see for instance the results of a query on PubMed with "diagnostic errors" as keyword). These studies have been carried out in various contexts (primary care, emergency units, hospital departments, etc.) including many medical specialties (family medicine, pediatrics, internal medicine, radiology, surgery, etc.). *To Err is Human: Building a Safer Health System* is a report issued in November 1999 by the U.S. Institute of Medicine that resulted in increased awareness of U.S. medical errors.

Among diagnoses subject to errors, the identification of drug side effects, pulmonary embolism, lung cancer, colorectal cancer, breast cancer, coronary syndrome, viral infections often treated as bacterial infections, and appendicitis are frequently reported.

These diagnostic errors can be related to systemic or cognitive factors. Communication problems within hospital teams and delays to get the results of lab tests are examples of systemic factors. Various cognitive factors are also reported:

- Improper interpretation of patient interview and clinical examination,
- Poor memorization of patient data,
- Bad choice of lab tests and imagery to reach the good diagnosis,
- Erroneous estimation of probabilities for the different suspected diagnoses,
- Shortcomings of medical knowledge in certain areas.

Using diagnostic decision support systems could improve the quality of physician decision and ultimately patient care:

- By providing physicians with the medical knowledge necessary to improve diagnostic reasoning process,
- By helping physicians identify differential diagnostic hypotheses and prescribe appropriate additional tests to progress in the diagnostic reasoning process.

7.2 Numerical Approaches for Medical Diagnostic Decision Support

7.2.1 Statistical Approaches

Statistical approaches for medical diagnostic decision-making are based on multivariate classification techniques, especially discriminant analysis. Schematically, we consider a set of N observations (patients), described by k variables (signs, symptoms and results of laboratory tests), divided into n classes (diagnoses).

Discriminant analysis aims at producing a new representation system made of the linear combinations of original variables, which improves the separation of classes. The goal is thus to construct a classification function able to predict the class of a patient on the basis of the values taken by the variables that characterize her/him. Thus, this technique is similar to supervised techniques of machine learning. Many methods have been proposed (Nakache 1976).

For example, if one seeks to discriminate between sick and healthy persons (here $n = 2$), we have to find the plan that best separates the points corresponding to sick persons and the points corresponding to healthy persons in the k-dimensional space of the variables describing the observations. This function is learnt from a learning sample, and tested on another sample of data to assess its validity. Then, it is calculated on any new patient for whom it is needed to diagnose the disease.

7.2.2 Probabilistic Approaches: Application of Bayes' Theorem

Similarly, we use a set of N observations, each one described by k variables X_j forming the vector X, and divided into n classes, the n diagnoses D_i considered. For every new patient, Bayes' theorem allows to calculate the posterior probabilities of different diagnostic hypotheses D_i:

$$\forall\, i = 1, \ldots, n \quad P(Di/X) = \frac{P(X/Di) \cdot P(Di)}{P(X)}$$

$P(D_i)$ are the "a priori" probabilities of the different diagnoses considered. They are estimated by the frequency of D_i. Assuming that diagnoses D_i are exhaustive (the set of D_i actually covers all possible diagnoses) and exclusive (you cannot have D_i and D_j at the same time), we have:

$$\forall\, i = 1, \ldots, n \quad P(Di/X) = \frac{P(X/Di) \cdot P(Di)}{\sum\limits_{i=1}^{n} P(X/Di) \cdot P(Di)}$$

The Bayesian approach has resulted in many applications. Results from De Dombal and colleagues in Leeds on the diagnosis of acute abdominal pain are particularly noticeable. Each patient is defined by about 40 variables, such as topography of pain, exacerbation factors and healing, the presence of nausea, vomiting, fever, etc. The assumption of independence of signs can be written as:

$$P(X/Di) = \prod\limits_{j=1}^{k} P(Xj/Di)$$

With eight diagnoses (De Dombal et al. 1972), appendicitis, acute cholecystitis, bowel obstruction, pancreatitis, perforated ulcer, acute diverticulitis, non-specific abdominal pain, and "other pain", the performance of the computer system was 91.8 %, significantly higher than the one of the human experts of the field (79.6 %).

Yet, despite their good performance, decision support systems based on numerical approaches have not been used in clinical routine. On the one hand, numerical approaches, whether statistical or probabilistic, are based on a database allowing, for example in the case of probabilistic approaches, to learn a priori probabilities P (D_i) and marginal probabilities P (X/D_i), and it has been shown that the database has to be local to the site where systems would be used, which limited their dissemination and sharing. In addition, poor interfaces, based on the mere display of the probabilities of diagnostic hypotheses, without explanation, did not convince physicians. Figure 7.2 shows an example of the interface in the Leeds' system (Horrocks et al. 1972).

```
POSSIBLE DIAGNOSES
 APPEND DIVERT PERFDU NONSAP CHOLEC SMBOBT PANCRE
PROBABILITIES ARE
      0·0     0·0      2·7      0·0      0·9      3·1      93·2

CLINICIANS DIAGNOSIS
      PRIMARY    -CHOLEC
      SECONDARY -SMBOBT

COMPUTERS DIAGNOSIS
      PRIMARY    -PANCRE  93·2
      SECONDARY -SMBOBT    3·1

NEITHER OF YOUR DIAGNOSES SEEM LIKELY. PROBABILITIES INDICATE
PANCRE AS PRIME POSSIBILITY

 ++ SUGGEST CHECKING THE FOLLOWING.......
AMYLASE
TENDERNESS....
SITE PRESENT
```

Fig. 7.2 Display of a posteriori probabilities with comments and recommendations

7.2.3 Clinical Scores

Numerous clinical scores have been proposed to help physicians in medical diagnostic decision-making. Some of them are very well known by physicians such as the Mini Mental Score (MMS), broadly used for detecting dementia in the elderly, or the Fagerström test which allows to quantify tobacco addiction.

The principle of such diagnostic scores is as follows: the physician is given a fixed number of standardized questions and responses that can be binary (true/false), ordinal or numeric, are recorded. The final score often takes the form of a sum (weighted or not) of the results of the answers to questions. Thresholds lead to the conclusion (e.g. intellectual deterioration: absent, moderate, high). In the case of MMS, the doctor asks the patient a series of standardized questions, 30 in total (see Fig. 7.3). He counts the number of right answers. Several cognitive abilities are successively explored:

- Orientation in time and space,
- Learning and transcribing information,
- Attention and mental arithmetic,
- Recall of information and memorization,
- Language,
- Ability to organize a series of movements for a specific purpose (constructive praxis).

The computerized implementation of the MMS makes it much easier to use in consultation. The doctor has just to click on the appropriate boxes. The total score is automatically calculated (see Fig. 7.4).

Maximum	Score	
		Orientation
5		• What is the (year) (season) (date) (day) (month)?
5		• Where are we (state) (country) (town) (hospital) (floor)?
		Registration
3		• Name 3 objects: 1 second to say each.Then ask the patient all 3 after you have said them.Give 1 point for each correct answer. Then repeat until he/she learns all 3.Count trials and record. Trials_____
		Attention and Calculation
5		• Serial 7's. 1 point for each correct answer.Stop after 5 answers. Alternatively spell "world" backward.
		Recall
3		• Ask for the 3 objects repeated above.Give 1 point for each correct answer.
		Language
2		• Name a pencil and watch.
1		• Repeat the following "No ifs,ands or buts."
3		• Follow a 3-stage command: "Take a paper in your hand,fold it in half and put it on the floor."
1		• Read and obey the following CLOSE YOUR EYES.
1		• Write a sentence.
1		• Copy the design shown.

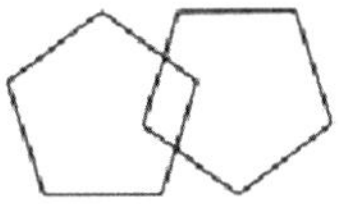

Fig. 7.3 The 30 standardized questions of the Mini Mental Score

7.3   Logical Approaches for Medical Diagnostic Decision Support

With the advent of Artificial Intelligence (AI), new computer-based approaches for medical diagnostic decision support have been proposed. AI was defined by its creators as "the design of computer programs to support tasks usually performed more efficiently by humans because they require high-level mental processes". These programs were called "expert systems" because they were able to simulate the reasoning of experts (from mathematicians to doctors). Based on a formalization of a given domain knowledge, they may derive conclusions or actions from data or facts.

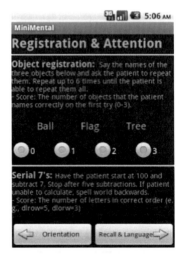

Fig. 7.4 The BuiPlus! App to interactively evaluate the Mini Mental Score (http:/buiplus. blogspot.fr/2010/06/minimental-application.html, last consultation on January 2013)

7.3.1 Expert Systems

7.3.1.1 Definition

The term "expert system", very common today, is used for any computerized system designed to simulate human expertise. *"An expert system is a program or a software which requires a fair amount of knowledge, and is able to reproduce all or part of the behavior of a human expert in a limited area"*. This definition calls for a number of observations. First, it does not restrict the function of the expert, neither in the field nor in the type of activity. It is not limited to diagnostic reasoning process or problem solving. Finally, this definition emphasizes the importance of knowledge. In fact, these systems are also referred to as "knowledge-based systems" or "cognitive systems".

7.3.1.2 Expert Behavior

An expert is a person whose knowledge allows him/her to understand and efficiently solve problems (s)he faces. In addition, this qualification of expert may be assigned only if a large number of his/her peers recognize him/her as such. An expert system must meet these two requirements, being able to solve problems by using large amounts of knowledge, and successfully pass validation testing, especially in the perspective of dissemination.

The medical expert engaged in a diagnostic reasoning process implements the knowledge (s)he has by confronting it with the situation (s)he has to handle. (S)he reasons, makes conjectures, comparisons, analogies. (S)he is able to redirect his/her search at any time. The complexity of the process may vary according to the characteristics of the domain and the problem. The reasoning process may be either expressed as a sequence of clearly identified operations, or the expression of a cognitive process more difficult to formalize because mostly unconscious. In all cases, these activities have a common denominator: they use knowledge, whether theoretical or practical, explicit or implicit. What characterizes expert behavior, and makes the difference between experts and novices, is thus essentially the ability of experts to mobilize at any time a piece of knowledge, usually not found in books, that is the result of years of experience, practice and analysis of large numbers of cases.

Knowledge refers primarily to the description of a field: the expert generally has a structured synthetic view of situations, important concepts are grouped together and notions are associated, often empirically. This comprehensive understanding does not preclude a more technical knowledge.

Expert knowledge is also related to the efficient exploitation of facts. When the expert has the complete information, his/her deductions provide undeniable results and the resolution process always leads to a solution. In this case, reasoning mechanisms can be reproduced and may usually be easily computerized. When "uncertainties outweigh certainties", finding the solution requires a specific type of knowledge, most of the time expressed as "heuristics". Heuristics allow to relate probable causes to a data set, to combine uncertain facts, and conduct an approximate reasoning. Unlike deductions in deterministic situations, heuristics do not guarantee to get to the result; they "only" guide the search towards promising directions.

As a conclusion, expert systems aim at *simulating the expert behavior,* which means that a software with man–machine interfaces allowing for appropriate communication may reproduce the tasks, attitudes, and reasoning processes described above. Indeed, computerized systems do no longer execute a sequence of operations elaborated in advance and encoded in a program, but they can extract the information needed from their direct environment, and reason, or at least solve problems for which there is generally no algorithmic solution or for which combinatorial complexity prohibits the use of conventional techniques (Shortliffe 1986).

7.3.1.3 General Structuration of Expert Systems

The idea of a computer that "quickly but blindly calculates" has shown its limitations and shortcomings. The new focus is thus to introduce knowledge. Two opposite solutions were first considered: encode knowledge in the body of the program, which was the approach adopted for the expert system Dendral that aimed at identifying the chemical components of a material from mass spectrometry and nuclear magnetic resonance (Lindsay et al. 1993), or separate the encoding of knowledge from other programs, in particular from the encoding of reasoning mechanisms. The second solution proved to be better than the first one since domain knowledge is a priori

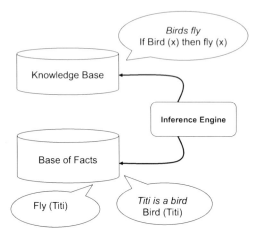

Fig. 7.5 The three modules of an expert system

specific to a given application and subject to evolutions, whereas information processing mechanisms of reasoning (inferences) are stable. Later, Dendral was subsequently modified to isolate from the knowledge base the module encoding reasoning mechanisms called Meta-Dendral. As a conclusion: *in an expert system, knowledge, whether expert, descriptive, or relational, must be represented in a different module than the one used for the encoding of processing mechanisms.*

An expert system is made out of three main modules (see Fig. 7.5):

- The knowledge base that contains the knowledge of the domain of expertise, in our case, it is the theoretical and empirical medical knowledge (heuristics) needed for the diagnostic reasoning process,
- The base of facts that represents the data characterizing the problem to solve, in our case, the description of the clinical case for which the physician seeks to establish the medical diagnosis,
- The inference engine, which is the module that carried on the reasoning process from both the knowledge base and the base of facts to produce the decision support. The reasoning process is based on logic principles.

Methods of knowledge representation and acquisition are currently widely investigated through the design of sophisticated formalisms by researchers of an active knowledge engineering scientific community.

7.3.2 The Knowledge Base

Expert knowledge can be decomposed into small homogeneous units. Each of them represents an elementary piece of expertise. This is called modularity. This feature is important because it provides a flexible use of knowledge and improves knowledge management. There are many knowledge representation formalisms. Each of them has specificities including a specific expressiveness.

7.3.2.1 Production Rules

Production rules (defined by the logician Post) are used to represent elementary pieces of knowledge. They are expressed as "**If** condition **Then** Action". Conditions and actions may be either atomic or complex formulas. The condition or left part of the rule is also called the premise of the rule:

If ((proteinuria> 5 g/l) **and** (oedema) **and** (albumin <30g/l)) **Then** nephrotic syndrome

This "rigid" formalization cannot handle uncertainty. Therefore, certainty factors have been associated to non-deterministic rules. Certainty factors are numbers, a priori between -1 and 1, that quantify the belief of the expert in a rule. Chaining production rules to handle the reasoning process involves the combination of certainty factors (e.g. choice of the largest coefficient on a disjunction, or the smallest in case of a conjunction).

Example (excerpt from Mycin):

If ((the stain of the organism is gramneg) **and** (the morphology of the organism is rod) **and** (the aerobicity of the organism is aerobic)) **Then** the class of organism **is enterobacteriaceae** with a **certainty factor** of 0.8.

7.3.2.2 Decision Trees

The medical knowledge of an expert system may be represented as a decision tree. In the deterministic version, the decision tree consists of a set of decision nodes. Each node corresponds to a condition. Each condition examines the value of a criterion, for instance the value of a sign or a symptom (clinical, laboratory test results, or imaging signs) in the case of medical diagnostic decision support. Values of signs correspond to the labels of the arcs coming out of the node. The leaves represent the diagnosis.

A decision tree is a graphical representation of a classification procedure. Indeed, each patient is associated with only one path, therefore a single diagnosis. This association is built when going through the decision tree from the root to the leaves, using the arcs, thus the criteria, that fit the patient condition. The diagnosis is associated with the leaf that matches the description. Figure 7.6 shows the decision tree for the diagnosis of galactosemia. Infants are routinely screened for galactosemia in the United States, and the diagnosis is made while the person is still an infant. Infants affected by galactosemia typically present with symptoms of lethargy, vomiting, diarrhea, failure to thrive, and jaundice. None of these symptoms is specific to galactosemia, often leading to diagnostic delays. Luckily, most infants are diagnosed on newborn screening.

When taking uncertainty into account (see Fig. 7.7), the nodes of the decision tree are of two types, the *decision* nodes (arcs D_1 and D_2) and the chance events

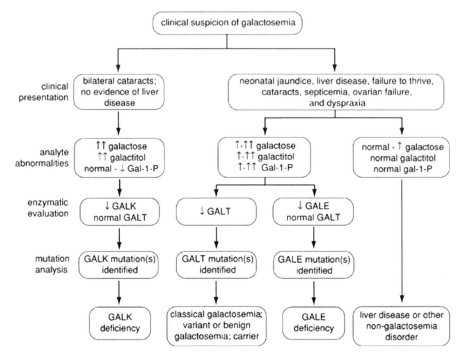

Fig. 7.6 Decision tree to support the diagnosis of galactosemia

(leading to patient states) denoted Ei. For each decision, we compute the utility function of the resulting patient state. The best decision is the one that maximizes the expected utility.

7.3.2.3 Semantic Networks

First used in cognitive psychology to model memory representation and "common sense" knowledge, semantic networks are ones of the oldest knowledge representation formalisms. They are used to represent sets of concepts linked by semantic relations as graphs (in principle acyclic graphs). Nodes represent concepts, and arcs represent the relationships between these concepts:

- Link "*is-a*": a human being *is-a* mammal, a mammal *is-a* vertebrate, a feline *is-a* mammal, a feline *is-a* carnivore, etc.
- Link "*part-of*": head is *part-of* body, trunk is *part-of* body, etc.
- Link "has": mammal *has* body, mammal *has* hair, etc.

Fig. 7.7 Decision tree under uncertainty (probability)

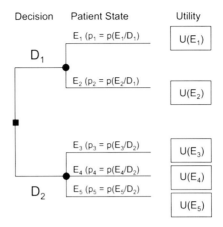

$$E(U/D_1) = U(D_1) = p_1.U(E_1) + p_2.U(E_2)$$

$$E(U/D_2) = U(D_2) = p_3.U(E_3) + p_4.U(E_4) + p_5.U(E_5)$$

Figure 7.8 displays the semantic network representing these relationships. Properties are inherited when following *is-a* or *part-of* relations. Other reasoning mechanisms can be performed when using pattern-matching methods.

7.3.2.4   Frames

When using production rules or semantic networks, knowledge is dispersed. Frames introduced by Minsky in the 70s, are a knowledge representation formalism that proposes to aggregate notions as concepts (Minsky 1975). An example of a frame is displayed in Fig. 7.9. Frames combine the description of a situation as a set of attributes along with the information on how it can be used.

7.3.3   *The Base of Facts*

Usually, the base of facts or data set corresponds to the description of the case to be solved. It is the working memory of the expert system, and evolves as the reasoning process progresses. At the beginning, the base of facts is made of what is known about the case to be solved, before any inference is made. From these initial facts, it is possible to infer new facts until reaching the final conclusion. In medicine, the base of facts may contain a list of symptoms at the beginning of the execution, and the diagnosis at the end. When the execution is completed, the base of facts is cleared.

Facts are usually qualitative data (color, quality, properties, etc.) or numeric values. The same fact can be represented as a linguistic expression, "the patient has

Fig. 7.8 Knowledge representation using the semantic network formalism

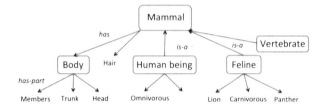

Fig. 7.9 Representation of the concept "nephrotic syndrome" as a frame

```
(Frame Nephrotic Syndrome
    Is-a-type-of                  Clinical State
    Signs                         Proteinuria > 5 g/l
                                  Massive oedema
                                  High plasma cholesterol
                                  Low plasma albumin
                                  Lipiduria
    Does not have                 No proteinuria
    Pathognomonic signs           Massive oedema and proteinuria
    May be caused by              Glomerulonephritis
                                  Nephrotoxic drugs
                                  Lupus
                                  Insect bites
    May be complicated by         Hypovolemia
                                  Cellulitis
    May cause                     Sodium retention
    Differential diagnosis
    If Ascites Then consider Cirrhosis
    If Pulmonary Embolism Then consider Renal Vein Thrombosis
    If Jugular Vein Distension Then consider Constrictive Pericarditis)
```

fever", or as a measure, "the patient's temperature is 39 °C". In the latter case we can represent the fact by pairs such as <Attribute,Value>. More generally, facts are represented as triples <Object, Attribute, Value>. For example, the size of a patient can be represented by <Patient Size 180>.

7.3.4   The Inference Engine

The inference engine is an algorithm that allows to produce (infer) new facts leading to terminal facts, conclusions, decisions or actions starting with initial facts (base of facts) and using the knowledge base.

7.3.4.1   Three Types of Inferences

1. Deduction. This is the mechanism for deriving a conclusion from a given set of axioms or theoretical knowledge, and facts or observations. This is the *modus ponens* or syllogism used for theorem proving. It is a mechanism that preserves "the truth". If the facts are true, the inferred conclusions are true too.

- All men are mortal (axiom)
- Socrates is a man (fact)
- Inference: Socrates is mortal (conclusion)

2. Induction. This is the mechanism for deriving a general rule or axiom, from domain knowledge, and facts or observations. It is used in machine learning to infer a general knowledge from examples. This mechanism does not preserve "the truth". The general inferred rule is true until it is questioned by a counter-example.

- Socrates is a man (domain knowledge)
- Socrates is mortal (observation/example)
- Inference: Men are mortal (generalization)

3. Abduction. This is the process for deriving a premise from a known axiom or theory, and facts or observations. For example:

- Men are mortal (theory)
- Socrates is mortal (observation)
- Inference: Socrates is a man (diagnosis)

Abduction is often used in expert systems for medical diagnostic decision support. This is a mechanism that proceeds on a principle of causality without preservation of "the truth".

- Flu $=>$ temperature $> 38\,°C$
- Patient X has a temperature $> 38\,°C$
- Inference: Patient X has got the flu.

7.3.4.2   Principles of Formal Logic

Today, we talk about digital technology to encompass everything related to information technology hardware and we forget that computers are first logic machines! There are many logical theories, and we only present classical logic studied and defined by philosophers and mathematicians since the Greeks (including Aristotle).

A logical language is defined by a syntax, i.e. a system of symbols and rules, allowing to combine symbols to form formulas called "well-formed formulas". Furthermore, semantics is associated with language. It allows to interpret the language, i.e. to attach a given meaning to formulas and symbols. A deduction system allows reasoning by constructing demonstrations. Logic includes classical propositional logic (also called propositional calculus), and predicate logic (predicate calculus).

Propositional calculus: A proposition is a statement that gives a quality to an object or a human being: *the sea is calm, John is a doctor, Titi is a bird*, etc. To complete the alphabet, we define logical operators (connectors): negation ($\neg$), disjunction ($\vee$), conjunction ($\wedge$), implication ($\Rightarrow$), and equivalence ($\Leftrightarrow$). Using the rules, it is possible to define well-formed formulas, containing no variables, as the association of *atomic propositions* through logical connectors.

Such formulas can be evaluated from the interpretation of atomic formulas and the truth tables associated with the connectors used in these formulas. It is possible, in particular, to demonstrate the following properties and relationships:

Double negation: $\neg\,(\neg\,P) = P$
Distributivity:

P1 $\wedge$ (P2 $\vee$ P3) = (P1 $\wedge$ P2) $\vee$ (P1 $\wedge$ P3)
P1 $\vee$ (P2 $\wedge$ P3) = (P1 $\vee$ P2) $\wedge$ (P1 $\vee$ P3)

Implication:

P1 => P2 $\Leftrightarrow$ $\neg$ P1 $\vee$ P2
P1 $\vee$ P2 $\Leftrightarrow$ $\neg$ P1 => P2
P1 => P2 $\Leftrightarrow$ $\neg$ P2 => $\neg$ P1

De Morgan's laws:

$\neg$ (P1 $\wedge$ P2) $\Leftrightarrow$ $\neg$ P1 $\vee$ $\neg$ P2
$\neg$ (P1 $\vee$ P2) $\Leftrightarrow$ $\neg$ P1 $\wedge$ $\neg$ P2

Predicate calculus: a predicate is a general statement about objects including variables that can be instantiated. The interpretation of formulas such as P (x) can be true for all values of x, or for at least one value of x. There are two quantifiers to represent such situations:

- The universal quantifier "for all" denoted "$\forall$". Thus, one can express the expression "all men are mortal" as: ($\forall$ x) (MAN (x) = > MORTAL (x)).
- The existential quantifier "there exists" denoted "$\exists$". Thus, one can express the expression "there are individuals who have brown hair" as: ($\exists$ x) (MAN (x) $\wedge$ HAIR (x, BROWN)).

The well-formed expressions of the predicate calculus can be universally or existentially quantified. Predicate calculus is said *first order* because only variables can be quantified, functions or predicates themselves cannot be.

As a conclusion, in expert systems, the principle is to use domain knowledge to reason from the facts and produce new facts, in order to reach the answer to the question set by the problem to be solved, following a correct reasoning. Most expert systems are thus based on existing mechanisms of formal logic. Simplest expert systems rely on propositional logic, and use only propositions that are true or false. Other systems are based on predicate logic (also known as "order 1 logic") that can easily be handled by algorithms.

The inference engine can operate in forward chaining or backward chaining. Forward chaining is a method of deduction that applies rules starting from the premises to deduce new conclusions. Findings produced by the reasoning process enrich the working memory and become the premises of new rules. On the contrary, the backward chaining starts from the conclusion and goes back to the premises to determine, when premises are true, that the conclusion is true too.

7.3.4.3   Forward Chaining

Rules which premises are true are triggered until the goal is reached or no rule can be triggered anymore. We analyze each fact and review all the rules where this fact appears as a premise. Conclusions of triggered rules are added to the base of facts. We say that facts are propagated. The new deduced facts are part of the final result. They themselves are again propagated until all the facts are produced. A simple use of a forward chaining reasoning process is given below. Consider the following knowledge base:

R1: If give-milk then mammal
R2: If has-hair then mammal
R3: If eat-meat then carnivorous
R4: If (claws and sharp teeth) then carnivorous
R5: If (carnivorous and mammal and fawn and black-points) then leopard
R6: If (carnivorous and mammal and fawn and black-stripes) then tiger
R7: If (carnivorous and mammal and digitigrade) then feline
R8: If (feline and fawn) then lion

Suppose we consider a base of facts made of known or observed facts on a given animal, for instance, (has-hair and eat-meat and fawn and black points), and we want to know what is the animal. Using a simple iteration over the set of rules, the inference engine triggers the rule R2 and gets a new fact *mammal*, then the rule R3 and gets the new fact *carnivorous,* and finally the rule R5 that gives the conclusion *leopard.*

An inference engine operating in forward chaining is easy to program as it is an iterative algorithm. It is called data-driven since it operates from the facts contained in the base of facts.

7.3.4.4   Backward Chaining

The mechanism of backward chaining starts from the goal that one wishes to establish. The principle consists in searching all the rules that conclude that goal, then listing the premises of these rules that have to be proven to trigger the rules. The same mechanism is applied recursively to the facts contained in these lists, which become the new goals to prove. We use the previous example to illustrate the use of a backward chaining reasoning process.

Consider a base of facts made out of the facts known for or observed on a given animal, for instance, (has-hair and eat-meat and fawn and digitigrade). It raises the question of whether the animal is a *lion.*

The reasoning process starts with the rule R8 which conclusion matches the goal to prove. To trigger R8, we need *fawn* and *feline* to be true. *Fawn* is true as it is in the base of facts. Thus, we only need to prove *feline.* We consider the rule R7 whose conclusion is *feline.* We must then prove the two sub goals *carnivorous and mammal* since *digitigrade* is in the base of facts. Thus, we trigger rules R2 since *has-hair* is true because it is in the base of facts, and R3 since *eat-meat* is true because it is also in the base of facts, which allows to conclude that *lion* is true.

If we set as goal *leopard* or *tiger*, we would not be able to conclude because neither the fact *black-points* nor the fact *black-stripes* is present in the base of facts and they cannot be proven from the facts available. Nevertheless, it is always possible to progress in the reasoning process by opening the possibility of questioning the user, for example by asking about new facts that may be missing from the base of facts. Thus the backward chaining mechanism is goal-driven and proceeds in a recursive way until proving the first goal.

7.3.5 Historical Medical Expert Systems

7.3.5.1 Mycin

Mycin is an expert system designed in the early 70s at Stanford University by E. Shortliffe (Shortliffe et al. 1975). The objective was to support the physician in the diagnosis of infectious diseases and the determination of the appropriate antibiotic treatment (which explains the name of the system since the majority of antibiotics had a name ending in *Mycin*, e.g. *erythromycin*). The knowledge base was made of about 500 production rules weighted by certainty factors. The inference engine was operating in backward chaining. Mycin has been extensively studied and discussed because of the importance of the medical specialty it was applied to (infectious diseases) and especially because of the high quality of its diagnostic and therapeutic options. Furthermore, it was very much attractive to physicians because based on an interface in natural language. Nevertheless, despite all these qualities, the system has never been used in clinical routine because of both unsolved liability reasons and ethical considerations.

7.3.5.2 Internist/QMR

Internist/QMR has been developed at the University of Pittsburgh (Miller et al. 1982). This is an expert system applied to internal medicine. The knowledge base includes profiles of 650 diseases and about 4,500 signs and symptoms. Each sign is connected to all the diseases in which it may occur. This connection is twice quantified to express the frequency of the association and the evocative power of the sign for the disease. From the elements of the clinical observation, the program proposes a list of diagnostics compatible with the hypotheses, thus reproducing the hypothetico-deductive reasoning of a doctor. A score is used to classify the resulting differential diagnoses.

Although experimental, this system was one of the most evaluated expert systems. Its performances are close to those of human experts. Nevertheless, it was not routinely used because the computation time of the system was not compatible with the duration of a medical consultation. An optimized version of Internist has been developed under the name Caduceus (Banks 1986). A further version available on PC was produced as the QMR (Quick Medical Reference) for training purposes (Miller and Masarie 1989). A recent article outlines the history of these systems (Miller 2009).

7.4 Examples of Current Medical Diagnostic Decision Support Systems

Today, some medical diagnostic decision support systems are available. They may be either "specialized" or "general". The first ones provide a diagnostic decision support in a given domain, e.g. MMS can confirm the diagnosis of intellectual deterioration. Similarly, we find on the site http://medcalc.com/ the calculation of a number of clinical scores, and extracts of decision trees (last consultation on January/2013). Following the development principles of systems such as Internist, general systems allow to input patient features and suggest possible diagnoses ordered according to their probabilities. They cover all the domains of medicine and are consequently very long and difficult to build.

The development of DXplain™ has been conducted by the Massachusetts General Hospital from 1984 (Barnett et al. 1987). In its current version, DXplain™ is based on a knowledge base that describes the signs encountered in 2,400 diseases (http://lcs.mgh.harvard.edu/projects/dxplain.html, last consultation on January 2013). Patient data concern nearly 5,000 clinical signs, symptoms, laboratory results, radiological and endoscopic examinations. Frequencies of signs in each disease are included in the knowledge base. The prevalence of diseases is quantified by qualitative values such as "very common", "common", "rare", and "very rare". When the user enters data about a patient, the system proposes a set of possible diagnoses listed in order of decreasing probability. The user can know why a particular diagnosis is discussed. (S)he may easily obtain additional information about the various signs that can be observed in a given disease and get information about the most discriminating features. There is a demo available at http://dxplain.org/demo2/frame.htm (last accessed on January 2013, see Fig 7.10).

Evaluations of DXplain™ have been performed. (Elkin et al 2010) showed that the use of DXplain™ was financially interesting and could allow hospital savings.

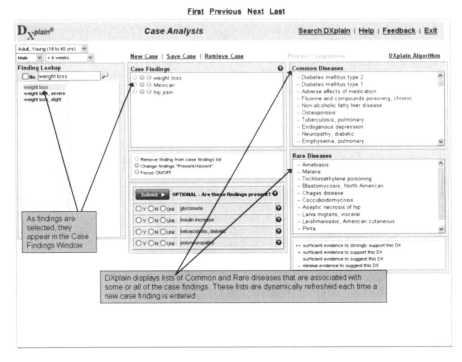

Fig. 7.10 Screenshot of the demo of DXplain™ with the proposition of diagnostic hypotheses making the difference between common and rare diseases

7.5   Conclusions and Perspectives

With the development of information technology, and because of the continuous inflation of medical knowledge, the idea of using the computing power and the memory capacity of computers has rapidly emerged and many computerized systems have been developed to support medical diagnostic decision. Numerical approaches were first developed and had good performances. However, the presentation of results as a list of diagnoses ordered by their probability was not convincing and these systems have not been used in clinical routine. With the advent of artificial intelligence, expert systems have been developed with the ambition to simulate human expertise. They first seduced doctors attracted by a futuristic technoscience: expert systems could explain their reasoning in quasi-natural language. Yet, despite these qualities, expert systems, have also not been used in clinical routine.

First, the performances of early versions of expert systems were inadequate as soon as they had to solve "real" problems out of the "micro-worlds" in which they were developed. Moreover, the systematic evaluation of these systems, essential prerequisite before effective use, has not been achieved, in particular due to a lack of methodology. From a technical perspective, the quality of a medical expert system is based on both the quality of its knowledge base and the quality of the inference engine, and it is difficult to assess separately the two modules. In addition, the construction of these two modules is complex. Thus, the construction of the knowledge base has long been considered as the limiting step in the development of expert systems, the experts used as sources of knowledge having indeed always difficulties to formalize their knowledge (by definition "compiled" by the experiment). So, many studies have been conducted in machine learning to learn knowledge bases from resolved clinical cases. Currently, knowledge sources are clinical practice guidelines. Guidelines are usually in text format and need to be translated and formalized, which is a complex task due to the structural incompleteness and ambiguity of these documents. Furthermore, the inference engine must process complex reasoning where forward chaining and backward chaining are interconnected (mixed chaining) and logics are multiple (valued logic, fuzzy logic, exception management, conflicts between rules, etc.). Finally, in order to process patient care, these systems have to be integrated with a base of facts made from data extracted from electronic patient records. It requires that the medical record is coded, which is still today, a goal to reach.

We must also question the real need of physicians of support systems for medical diagnosis, and if the need is confirmed, consider the modes of interaction favoring the "usability" of these systems. Currently, if we put aside active systems that are automatically triggered and can make decisions without human intervention (monitoring systems), there are two types of operating systems:

- "Passive" systems or "on demand" systems that doctors would consult if they wish to, in a voluntary quest for decision support. These systems may be used as consultant systems (the user provides information about the patient's condition and receives in return the diagnosis) or as critical systems (the user provides information about the patient's condition and the diagnosis which may be criticized by the system if necessary).
- "Semi-passive" systems that trigger automatically even if doctors do not want them to. These systems oversee the reasoning of doctors and remind them of important elements of the decision which are missing (recall systems), or warn them of errors that should be corrected (warning systems).

When developed in a passive mode, diagnostic support systems are rarely used. Developed in a semi-passive mode, they are unwelcome. In clinical practice, it seems that doctors want to keep the noble part of medical activity, i.e. reasoning, except in cases of rare diseases (http://www.orpha.net/consor/cgi-bin/Disease_DiagnosisAssistance.php?lng=EN last accessed on January 2013). Thus, expert systems which are existing today are primarily used for educational purposes by medical students. Another trend concerns the development of online medical diagnostic decision support systems for patients (http://www.bettermedicine.com/symptom-checker/, last accessed on January 2013).

7.6   For More Information

- http://www.openclinical.org/dm_cacddst.html
- About medical diagnostic decision support systems: search Pubmed with keywords such as "decision support system", "diagnosis", "computer-aided diagnosis"
- About diagnostic errors: search PubMed with keywords such as diagnostic Errors/statistics & numerical data, diagnostic Errors/classification

---

Exercises

Q1 Bayes' Theorem
In winter, doctors have difficulties to diagnose many patients that consult with a syndrome combining the two signs "fever" and "aches or myalgia".
We want to develop a computerized decision support system to help physicians better manage these patients. We consider that only three diseases can cause these symptoms: cold, angina and flu.
We have the following data on each of these diseases:

Cold: It is the consequence of a viral infection (rhinovirus, influenzae, adenovirus) that starts in 10 % of cases by a slight fever. Myalgia are observed in 10 % of patients. This disease is so common in winter that 50 % of patients who consult their GP actually have a cold.
Angina: It is a viral or bacterial infection of the throat that often starts with fever (90 % of cases). Aches (myalgia) are experienced in 10 % of cases. Then the patient feels pain in the throat. In winter, 20 % of patients who consult have angina.
Flu: This viral disease is often due to orthomyxovirus. It frequently associates fever (90 % of cases) and aches (80 % of cases). In winter, 30 % of patients who consult their GP have got the flu.

We want to develop a medical diagnostic decision support based on the Bayes' theorem to help discriminate between these three diseases.

1. Calculate the frequency of each disease and the frequency of each sign for each disease.
2. Calculate the posterior probability of each disease when the patient has "fever"
3. Calculate the probability of each disease if we observe both "fever" and "myalgia"
4. What is the most probable diagnosis when the patient has both signs?
5. Is-it interesting to consider the sign "tachycardia = increase of heart rate" that often comes with "fever"?

Exercises (continued)

Q2 Truth tables

1. *Compare the truth functions of* $(p \Rightarrow q) \wedge p$ *and* $q \wedge p$
2. *Given p and m are true, evaluate the following formula:* $[(p \Rightarrow q) \wedge p] \vee m$
   $\Leftrightarrow (q \Rightarrow m)$
3. *Is the following reasoning valid?*
   If we spent Christmas on a balcony, then we spend Easter at the embers.
   But this year Easter has been very cold,
   So, this year Christmas was mild.
4. *Are these two propositions equivalent?*
   If hens had teeth, we could put Paris in a bottle.
   We do not put Paris in a bottle and hens do not have teeth.
5. *Evaluate the following expression:* $\forall x, P(x) \Rightarrow \exists x, P(x)$

Q3 Logic brainteaser
Pures who always tell the truth and Worses who always lie, live on an island. Every inhabitant is either a Pure or a Worse. A stranger meets three people A, B and C. He asks A: "Are you a Pure or a Worse?". A answers mumbling but is incomprehensible. The stranger asks B: "What did he say?". B answers: "He said he is a Worse." C intervenes and says: "Do not believe B, he is lying." Determine if B and C are Pures or Worses.

Q4 Expert system
Here are the rules of an expert system knowledge base developed to represent recommendations on how to manage blood transfusion in case of surgery.

1. If platelets $\geq 50. \, 10^{+9}/l$ and normal fibrinogen then normal coagulation
2. If platelets $<50.10^{+9}/l$ or decreased fibrinogen then disturbed coagulation
3. If pulse ≥ 120 beats/min then tachycardia
4. If pulse <120 beats/min then normal pulse
5. If systolic blood pressure <100 mmHg then arterial hypotension
6. If systolic blood pressure ≥ 100 mmHg then normal arterial tension
7. If hypotension and tachycardia then important blood loss
8. If normal blood pressure and normal pulse then minimal blood loss
9. If minimal blood loss and disturbed coagulation then recheck coagulation function
10. If minor surgical risk and minimal blood loss and normal coagulation then there is nothing to do
11. If major risk surgery and disturbed coagulation then platelet transfusion
12. If major risk surgery and important blood loss then platelet transfusion

(continued)

Exercises (continued)

1. Draw the tree that represents production rules according to the usual formalism
2. The expert believes that in 70 % of the cases of hypotension with a normal pulse, there is a major blood loss. How is it possible to account for this observation?
3. The patient must undergo a surgery with major risk. The systolic pressure is 130 mmHg, his pulse is 70 beats/min. All these clinical data are 100 % certain. Platelets are 45.10^{-9}/l. Fibrinogen is decreased. All these biological data are 100 % certain. For each "evaluation-execution" cycle of a forward chaining reasoning, give the rules triggered and the corresponding base of facts. What is the action to be chosen?

R1

R1.1 Frequency of diseases: Cold (0.5), Angina (0.2), Flu (0.3)

R1.2 Applying Bayes' theorem, we get P(Cold/fever) = 0.10, P(Angina/Fever) = 0.36, P(Flu/Fever) = 0.54

R1.3 Applying again Bayes' theorem with the a priori probabilities computed in question 2, we get P(Cold/Fever and Myalgia) = 0.02, P(Angina/Fever and Myalgia) = 0.07 et P(Flu/Fever and Myalgia) = 0.90

R1.4 The most probable diagnosis is the flu.

R1.5 "Fever" and "Tachycardia" are correlated (fever comes almost always with tachycardia). It is not interesting to take tachycardia into account

R2

R2.1 When building the truth tables of the two expressions, we find that they are equivalent.

R2.2 For all q, the expression is always true.

R2.3 Let A be: Spend Christmas on a balcony = Christmas is mild
B be: Easter at the embers = Easter is cold
Then, A $\Rightarrow$ B is true, and B is true, thus A can be either true or false: we cannot conclude.

R2.4
Let A be: Hens have teeth
B be: Paris is in a bottle
Then, the first proposition is A $\Rightarrow$ B and the second proposition is $\neg$ A $\wedge \neg$ B. Thus, there are not equivalent.

Exercises (continued)

R2.5 Reductio ad absurdum

Consider that $\forall$ x, P(x) $\Rightarrow$ $\exists$x, P(x) is false

$\exists$ x, P(x) is false

$\forall$ x, P(x) is true

Let a be a value of x such as P(a) is false. When taking the previous value of a : P(a) is true. Contradiction.

R3

It is impossible that a Pure or a Worse says: "I am a Worse"

So A could not say he was a Worse, B lied and B is therefore a Worse.

Saying that B lied, C told the truth, and C is a Pure.

If A has said he was a Pure it can be true or false, we cannot conclude anything about A.

R4

R4.1 Representation by an and/or tree

R4.2 You must add a rule to the knowledge base:
13. If hypotension and normal pulse then major blood loss

R4.3 Triggered rules are (with updated base of facts)

Rule 2: disturbed coagulation

Rule 4: normal pulse

Rule 6: normal arterial tension

Rule 8: minimal blood loss

Rule 11: platelet transfusion

It is not necessary to trigger Rule 9 (recheck coagulation) because biological data are 100 % certain. The recommended action is platelet transfusion.

References

Banks G (1986) Artificial intelligence in medical diagnosis: the Internist/Caduceus approach. Crit Rev Med Inform 1(1):23–54

Barnett GO, Cimino JJ et al (1987) DXplain. An evolving diagnostic decision-support system. JAMA 258(1):67–74, 3

De Dombal FT, Leaper DJ et al (1972) Computer-aided diagnosis of acute abdominal pain. Br Med J 2(5804):9–13

Elkin PL, Liebow M et al (2010) The introduction of a diagnostic decision support system (DXplain™) into the workflow of a teaching hospital service can decrease the cost of service for diagnostically challenging Diagnostic Related Groups (DRGs). Int J Med Inform 79(11): 772–777

Elstein AS, Schwarz A (2002) Clinical problem solving and diagnostic decision-making: selective review of the cognitive literature. Br Med J 324(7339):729–732

Gandhi TK, Kachalia A et al (2006) Missed and delayed diagnoses in the ambulatory setting: a study of closed malpractice claims. Ann Intern Med 145(7):488–496

Horrocks JC, McCann AP et al (1972) Computer-aided diagnosis: description of year adaptable system, and operational experience with 2.034 cases. Br Med J 2(5804):5–9

Ledley RS, Lusted LB (1959) Reasoning foundations of medical diagnosis. Science 130:9–21

Lindsay RK, Buchanan BG et al (1993) DENDRAL: a case study of the first expert system for scientific hypothesis formation. Artif Intell 61(2):209–261

Lucas AE, Smeenk FJ et al (2012) Diagnostic accuracy of primary care asthma/COPD working hypotheses, a real life study. Respir Med 106(8):1158–1163

Miller RA (1994) Medical diagnostic decision support systems- past, present, and future: a threaded bibliography and brief commentary. J Am Med Inform Assoc 1:8–27

Miller RA (2009) Computer-assisted diagnostic decision support: history, challenges, and possible paths forward. Adv Health Sci Educ 14:89–106

Miller RA, Masarie FE (1989) Use of the Quick Medical Reference (QMR) program as a tool for medical education. Methods Inf Med 28(4):340–345

Miller RA, Pople HE Jr et al (1982) Internist-1, an experimental computer-based diagnostic consultant for general internal medicine. N Engl J Med 307(8):468–476

Minsky M (1975) A framework for representing knowledge. In: Winston P (ed) The psychology of computer vision. Mc Graw Hill, New York, pp 211–277

Nakache JP (1976) Multidimensional data analysis in medical decision. In: De Dombal FT, Grémy F (eds) Decision making and medical care. Amsterdam, North Holland

Schiff GD, Hasan O et al (2009) Diagnostic error in medicine: analysis of 583 physician-reported errors. Arch Intern Med 169(20):1881–1887

Schoolman HM, Bernstein LM (1978) Computer use in diagnosis, prognosis, and therapy. Science 200(4344):926–931

Shortliffe EH (1986) Medical expert systems–knowledge tools for physicians. West J Med 6: 830–839

Shortliffe EH, Davis R et al (1975) Computer-based consultations in clinical therapeutics: explanation and rule acquisition capabilities of the Mycin system. Comput Biomed Res 4: 303–320

Singh H, Sethi S et al (2007) Errors in cancer diagnosis: current understanding and future directions. J Clin Oncol 25(31):5009–5018

Chapter 8
Computerized Drug Prescription Decision Support

B. Séroussi, J. Bouaud, C. Duclos, J.C. Dufour, and A. Venot

Abstract Drug prescription has to satisfy three quality criteria. Orders have to be adapted to the patient state, be compatible with all the other drugs of the prescription, and in compliance with the recommendations described in clinical practice guidelines (CPGs). Computer provider order entry systems (CPOEs) have been developed to secure drug orders and they address the first two criteria. Clinical decision support systems (CDSSs) have been developed to improve the implementation of CPGs and promote evidence-based medicine. This chapter first introduces the different medication errors. Then, the general architecture of CPOEs (user interface, drug database, interface with electronic medical records (EMRs) and inference engine) is presented. The main modalities of entering drug orders are described. Alert generation for contra-indications, or drug-drug interactions, are detailed. CDSSs are tools to provide patient-specific recommended treatments. They rely on a knowledge base embedding CPGs. The translation process of CPGs from their original narrative format to a structured formalized representation is described. The difficulty of text translation is emphasized and documentary tools such as GEM that help formalize guideline content are described. The main guideline representation formalisms, Arden Syntax, decision trees, EON and GLIF, are presented. Then, ways of operating CDSSs are described, from the totally

B. Séroussi (✉)
Département de Santé Publique, UFR de Médecine, UPMC, Paris 6, Hôpital Tenon, 4 rue de la Chine, 75020, Paris, France
e-mail: brigitte.seroussi@tnn.aphp.fr

J. Bouaud
INSERM, UMR_S 872, eq. 20, CRC, 15 rue de l'école de médecine, 75006, Paris, France

C. Duclos • A. Venot
LIM&BIO EA 3969, UFR SMBH, Université Paris 13, 74 rue Marcel Cachin, 93017 Bobigny Cedex, France

J.C. Dufour
SESSTIM - UMR 912, INSERM/IRD/Aix-Marseille Université, Faculté de Médecine, 27 boulevard Jean Moulin, 13385, Marseille CEDEX 5, France

A. Venot et al. (eds.), *Medical Informatics, e-Health*, Health Informatics,
DOI 10.1007/978-2-8178-0478-1_8, © Springer-Verlag France 2014

automated alert-based mode, to various documentary approaches where the user navigates through a structured knowledge base. Finally, examples of clinical decision support systems currently routinely used are given.

Keywords Medication errors • Computer provider order entry systems • Drug contra-indications • Drug-drug interactions • Alert generation • Clinical decision support systems • Clinical practice guidelines • Evidence-based medicine • Guideline representation formalism • Documentary approaches

After reading this chapter, you should be able to:

- Explain the main issues set by the therapeutic management of patients
- Make the difference between computerized physician order entry systems and clinical decision support systems
- List the different services offered by computerized physician order entry systems as well as the main alerts they may generate
- Know the problems that the use of computerized physician order entry systems may generate
- Explain the principles underlying the certification of computerized physician order entry systems
- Describe the main formalisms used to represent clinical practice guidelines in clinical decision support systems
- Know the different methods for processing knowledge bases in clinical decision support systems

8.1   Introduction

8.1.1   Challenges of Drug Prescription: Patients, Drugs, Orders

History of Drug Prescription: The French Example

In Europe, until the tenth century, those who practiced medicine had also to cultivate, harvest, prepare and manage the medicinal plants that composed the materia medicina. They were called "apothecaries" (from the Greek word *apotec* meaning shop).

With the diversification of the substances used, and the increase of the knowledge required to master the art to cure and the art to prepare drugs, the separation between these two activities became necessary as one person

(continued)

> **History of Drug Prescription: The French Example (continued)**
>
> could not perform both functions. Hence, the "medicus" and the "pigmentarius" (or spice merchant) that appeared in France, quickly divided into two different fields: "grocers" only allowed to handle alimentary drugs and "apothecaries" in charge of medicinal drugs. In 1271, the French king forbade apothecaries to provide drugs apart from the presence of a doctor except for common treatments. At that time, the written prescription did not exist yet and doctors gave their prescriptions in the pharmacy, indicating orally which drugs should compose their remedy.
>
> In France, the official birth of the written prescription dates back in 1322 when a new edict forbade apothecaries to sell or give laxatives, toxics or abortives without a doctor's prescription that they could not renew.

Prescriptions are currently made out of a set of orders that physicians consider as appropriate to solve medical problems of patients. Orders must verify two types of criteria to be appropriate. At first, and at least, they should not harm patients in a way that could have been anticipated: a drug prescribed should not be contra-indicated (allergies, renal failure, pregnancy or breastfeeding), should be prescribed at the "right" dosage, and be compatible with all the other drugs of the prescription (drug interactions). All these controls are related to issues about "securing" orders. Computerized provider order entry systems (CPOEs) are tools that may help physicians securing their prescriptions. However, these systems are not expected to provide any decision support and cannot help physicians to find what should be the recommended treatment for a given patient. For example, a physician using a CPOE could prescribe an antibiotic to treat insomnia. The antibiotic will be ordered at the right dosage, and interaction with the other drugs of the prescription would be checked. However, antibiotics are not the recommended treatment for insomnia!

Clinical practice guidelines (CPGs) have been elaborated in order to optimize the therapeutic management of patients. They are usually provided as narrative documents, structured as a set of clinical conditions for which evidence-based best therapeutic recommendations are provided. In this case, we consider the "quality" of orders and the aim is to evaluate if the drugs prescribed are those recommended to solve the medical problem of a patient. However, CPGs proved to have a limited impact on physician compliance when disseminated as texts. By providing the best patient-specific guideline-based recommendations, clinical decision support systems (CDSSs) are expected to be tools able to impact physician behavior and help promote the use of CPGs and improve the quality of care.

8.1.2 Medication Errors

The problem of medical errors has been confidential for a long time. It became however a hot topic in the late 90s because improving patient safety and quality of care, as well as reducing health costs, especially the additional costs generated by a priori preventable errors, became of major concern in all developed countries. In 1999, a publication of the Institute of Medicine entitled "To Err Is Human" estimated that between 44,000 and 98,000 deaths each year were related to medical errors in U.S. hospitals. In Europe, the situation was similar.

Medical errors can be of several types, either related to diagnosis (misdiagnosis of an illness, failure to diagnose, delay of a diagnosis, use of non-recommended or obsolete tests, etc.), to treatments (errors in the implementation of an intervention or a procedure, giving the wrong drug, wrong patient, wrong chemical, wrong dose, wrong time, or wrong route), to prophylaxis (forgotten screening or immunization, etc.) or to other reasons (non-running equipment, poor organization, etc.).

Medication errors are the first cause of medical errors. Medications errors are defined as "any preventable event that may cause or lead to inappropriate medication use or patient harm while the medication is in the control of the health care professional, patient, or consumer. Such events may be related to professional practice, health care products, procedures, and systems, including prescribing, order communication, product labeling, packaging, and nomenclature, compounding, dispensing, distribution, administration, education, monitoring and use" (US Coordinating Council for Medication Error Reporting and Prevention). Medication errors may concern either therapeutic indications or modalities of prescription, which are under the responsibility of physicians, or transcription, which is under the responsibility of nurses or pharmacists, or dispensation, or drug administration, which are under the responsibility of nurses. Medication errors, especially those related to orders, are equally encountered in both children and adults, but the risk of harmful consequences is three times higher for children, particularly because of the greater frequency of calculation errors in pediatrics and neonatology (errors in cross multiplication, errors by a factor of ten in computations, etc). Emergency care services, intensive care units, and neonatal units are considered at high risk of medication errors.

Two interventions are considered to prevent these potentially dangerous errors. The first one is the use of a computerized prescription system enhanced by decision support services, and the second is to improve hospital organization, including the systematic supervision of prescriptions by a pharmacist, and the definition of clear procedures for nurses. In the following paragraphs of this chapter, we will describe computerized physician order entry systems which are used *to secure drug orders* (CPOEs) and clinical decision support systems which are used *to improve the quality* of a therapeutic strategy (CDSSs).

8.2 Computerized Physician Order Entry Systems

The quality of drug orders has often been studied (Coste and Venot 1999; Aronson 2009). The most frequently reported errors of drug orders are incorrect dosages, drug interactions, and non-compliance with therapeutic indications. Thus, computer-

Fig. 8.1 Architecture of a computerized drug prescription system

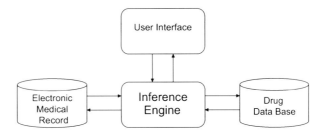

assisted drug prescription systems have been developed. The main objective is to secure the prescription of each drug (Kaushal et al. 2003).

8.2.1 Architecture of Physician Order Entry Systems

CPOEs do not exist as standalone software. These systems are indeed built as modules, integrated within software used by practitioners, either for the management of their patients in ambulatory care, or within hospital information systems. CPOEs share a common architecture (Fig. 8.1). They are usually connected with electronic patient records, and possibly to the pharmacy.

- User interface: the user interface is used by physicians to enter the prescribed drugs and dosages, as well as to read the alerts that may be generated by the inference engine in case of errors.
- Drug database: for each drug, whether commercially sold or available at a medical institution, the drug database provides all the information about the recommended dosage, therapeutic indications, drug interactions, contra-indications, especially allergies, precautions of use and warnings. Data characterizing drugs must be structured and encoded in order to be processed.
- Interface with electronic medical records (EMRs): when the physician prescribes drugs, (s)he has to take into account some patient data to fit the treatment to patient conditions. Thus, age, sex, height, weight must be available as well as information about renal failure, characterized by the serum creatinine, liver failure, intolerances, hypersensitivity to active substances, and excipients with known effects. It is also important to indicate when the patient is a woman, if she is pregnant and the expected date of the term, or if she is breastfeeding. Drugs already included in the long-term treatment of the patient should be indicated in the EMR, so that the drug prescription system could take into account the whole picture of administered treatments and operate efficiently.
- Inference engine: the inference engine is the central component of the drug prescription system. Access to the drug database allows the generation of lists of drugs (commercial names), along with the dosages appropriate for the indication

chosen by the physician, and various types of alerts. Interactions with the EMR provide the patient information needed for controlling prescriptions and generate alerts, as well as the record of drug orders. After being analyzed by the system, thanks to the inference engine, prescriptions may possibly be transmitted directly to the pharmacy.

8.2.2 Entering the Prescription

Entering the name of the therapeutic specialty to be prescribed would be laborious and error-prone. Instead, the prescribing physician identifies a list of specialties relevant to her/his prescription, and (s)he selects the desired product in the list of drugs. The list of specialties can be generated according to different modalities:

Main Modalities Provided by CPOEs to Enter Drug Orders

- Search by commercial name. The prescriber enters the first letters of the commercial name of the drug to prescribe. The system generates the list of drugs which name begins with the entered letters. The physician selects in this list the drug (s)he wishes to prescribe, defined by a commercial name, a dosage (e.g. 500 mg), a form (e.g. tablets), as well as a presentation (e.g. box of 28 tablets).
- Search by international non-proprietary name (INN). The prescriber enters the first letters of the INN of the active molecule (e.g. MET for METFOR-MIN). Once the INN is selected, the system displays all commercial products that contain this specific INN.
- Search through the navigation within a therapeutic classification. The system proposes the first level of a pharmaco-therapeutic classification, e.g. ATC (Anatomical Therapeutic Chemical). Then, the physician selects the level of the drug (s)he wants to prescribe (e.g. Dermatology). The system proposes all the possible values of the corresponding second level. The physician then chooses the desired second level (e.g. Anti-psoriasis), etc. The process is repeated until the leaves of the classification are reached where the prescribing physician got the list of expected relevant commercial products.

Some systems are able to generate lists of therapeutic specialties that share the same indications (e.g. all the products that have hypertension or prophylaxis of angina in their indications), or the products sold by a given pharmaceutical company.

Depending on the country, prescriptions may or must use INN. The pharmacist has then to give the patient the specialties which INN is specified in the order (e.g. the physician enters Amoxicillin 500 mg tablets, and the pharmacist dispenses Clamoxyl © 500 mg tablets). In this case, the system no longer generates lists of pharmaceutical products but lists of "virtual drugs" involving INN, dosage and form.

Once the drug is selected, the system proposes a list of usual dosages associated with the therapeutic indications. The dosage to be used generally depends on parameters such as age, weight, height, body mass index, body surface area, etc. Dosage calculations are automatically performed by the CPOE, which saves valuable time to the prescribing physician and reduces the risk of errors. The prescriber selects the suitable dosage and enters the duration of treatment in accordance with the laws of her/his country.

To help decrease the cost of useless prescriptions, some CPOEs automatically provide the prescribing physician with information on the cost of each order, reimbursement from health insurance, and remaining costs for patients.

The physician is expected to assign a reason for prescription (indication) for each drug order. This information is not visible on the prescription but is stored in the EMR.

In some cases, the prescribing physician may build her/his own base of pre-orders as a set of commercial products, with usual dosage and duration of treatment. This can help save time, especially during epidemics (viral gastroenteritis, viral nasopharyngitis, etc.) since the physician has not to enter separately each specialty with the dosage.

8.2.3 Alert Generation

Once the prescriber has entered the name of the specialty (s)he wants to prescribe along with the dosage, the CPOE performs real time checks: verification of the absence of known allergy to the drug prescribed, verification that the prescribed dosage is consistent with patient characteristics (weight, height, body surface area, renal or liver function, etc.) (Kuperman et al. 2007). These checks may lead to the generation of alerts to reduce medication error (Radley et al. 2013). To avoid overloading the prescriber with information, or interfere in her/his exercise and interrupt her/his workflow, alerts are usually not displayed as blocking windows but are often characterized by color codes (red, orange, etc.) used for features of the user interface (buttons, icons, orders). The doctor can then click on these features to get more information on the nature of the alert. In this case, a comment is displayed (e.g. "The drug is contraindicated in asthma") (Horsky et al. 2012).

Some alerts may indicate that the drug prescribed is not adapted to the patient condition. This is the case of checks performed on dosage, pregnancy or lactation, presence of allergies, intolerances, or contraindications. These kinds of alerts may be generated only if the interface between the CPOE and the EMR allows for importing coded data.

Alert Generation and Coding Diseases in EMRs

Two conditions must be met for controls on drug contraindications to operate properly:

1. The physician must have coded patient diseases in the EMR using a terminology similar to those detailed in Chap. 2.
2. Contraindications of each therapeutic specialty must be encoded in the drug database using the same terminology as the one used for coding diseases in the EMR. The selected coding system should then allow to represent, in the same way, diseases in the medical record, and indications and contraindications in the drug database (e.g. we should find the concepts of type 1 diabetes and type 2 diabetes, but not the concepts of insulin-dependent and non insulin-dependent diabetes mellitus which are used, for instance, in the CISP2).

Some alerts do not depend upon the patient condition but depend on the nature of the other drugs concurrently prescribed. These alerts are generated by the search for drug-drug interactions, or physical and chemical incompatibilities. Other alerts are generated when a redundancy of active molecules present in prescribed drugs is detected.

On the basis of these alerts and their severity, the physician may need to modify her/his order. This is of course not compulsory and the physician always remains the ultimate owner of her/his prescription contents.

Alert Generation for Drug-Drug Interactions: An Example

A 70-year-old patient suffering from an effort angina consults for a secondary lung infection resulting from a chronic bronchitis. He also complains of migraine attacks. His doctor has prescribed a macrolide (Josamycin 500 mg tablets) for his infection and Ergotamine (ergotamine tartrate with caffeine tablets) for migraine. Fortunately, the doctor has used a computer-assisted drug prescription system which has identified the presence of drug interactions and serious contraindications:

- Interactions between Josamycin and Ergotamine: Josamycin potentiates the effect of Ergotamine (risk of extremities necrosis due to decreased hepatic elimination of Ergotamine)
- Contraindications: Ergotamine should never be prescribed in cases of diseases predisposing to vasospastic reactions such as coronary artery disease, severe sepsis, shock, vascular occlusive disease, peripheral vascular disease such as Raynaud's syndrome, history of transient ischemic attack or brain damage, or poorly controlled hypertension.

8.2.4 Order Archiving

Once input in the CPOE, the drug prescription is archived in the EMR. According to the policies of the country, it can be either printed and delivered to the patient, or directly transmitted to a drug prescription server or to the pharmacy where the patient purchases her/his medications when it is known.

Compared with handwritten orders, printed orders have many advantages for the patient and the pharmacist who dispenses the medications:

- It is easy to read,
- Advices about how to use each drug can be automatically printed, e.g. "To be taken with some water during meals"
- The names of the specialties are not ambiguous as it is often the case with handwritten prescriptions where dosage and form are often forgotten.

8.2.5 Certification of CPOEs

Some countries have organized a certification process for CPOEs. The aim is to push vendors to offer the functions that help providers improve their practices, but also to control the neutrality of the information provided (especially with respect to the comparative choices of different commercial products), and possibly contribute to lower medication cost while keeping the same quality of care (Classen et al. 2007; Wright et al. 2009).

In France, the certification is a voluntary process and non-certified systems may still be sold. However, the commercial promotion of a CPOE is easier when the system is certified. Certification is carried out by an independent organization on the basis of a set of generic tests randomly chosen.

8.3 Clinical Decision Support Systems

Like CPOE systems, clinical decision support systems (CDSSs) are integrated in healthcare professional software. Similarly, their architecture relies on basic CDSS components including (i) an API with an EMR to access patient data that is relevant to the current problem, (ii) specific knowledge bases, and (iii) a dedicated inference engine which role is to match the appropriate knowledge to a patient situation in order to provide the recommended care plan.

8.3.1 Clinical Practice Guidelines and Recommendations

In the 1990s, medical practice evolved to follow evidence-based medicine principles (EBM) and to integrate scientific knowledge or "evidence" in the medical reasoning. EBM was defined by Sackett et al. (1996) as "the conscientious, explicit, and judicious use of current best evidence in making decisions about the care of individual patients [. . .] integrating individual clinical expertise with the best available external clinical evidence from systematic research". Practicing EBM means to take into account, in a rigorous manner and for every medical decision, the scientific results of clinical research. These results should be issued from methodologically well-conducted studies and validated through their publication in scientific journals (Lohr and Carey 1999). They represent the current "state of the art". Implementing EBM is nowadays a major challenge in developed countries to promote best healthcare practices and provide care quality at optimal costs. This trend is well demonstrated by the numerous developments of certification, accreditation, or professional-practice evaluation procedures for healthcare settings and individuals. However, EBM remains an idealized view, difficult to implement in routine. It is indeed well documented that clinicians can hardly and not systematically appraise and synthesize the state of the art for each patient (Bates et al. 2003).

"Clinical practice guidelines" (CPGs) are currently elaborated as a synthesis of current scientific productions on a given medical problem to provide guidance to healthcare professionals. CPGs, "best practices", or "clinical pathways" are similar terms. CPGs are often developed in the case of population-wide, public-health-related problems, like for instance, the therapeutic management of types 2 diabetes or arterial hypertension, or for specific medical problems, for instance Alzheimer's disease diagnosis. CPGs are published by professional societies (e.g. the Canadian Medical Association) or National health agencies (e.g. the AHRQ in the USA, the NHS in the UK, etc.). They are mainly textual, narrative, documents structured as a set of peculiar identified clinical situations to which recommended management plans or therapeutic prescriptions are associated.

The EPOC group (Effective Practice and Organization of Care, http://epoc. cochrane.org) of the Cochrane Collaboration is an international research group focusing on interventions that are designed to improve the delivery, practice, and organization of health care services, and to measure their impact on the health of individuals or populations. It is acknowledged that the simple dissemination of textual CPGs has a very low impact on clinician behavior, whatever their format, on paper or through electronic publication on promoters' web sites. The relevant use of such material for individual care is indeed difficult. Let's consider the case of a 62-year-old patient with a history of myocardial infarction, and suffering from diabetes and angina. The physician has primarily to identify the major health problems of the patient. Then, (s)he should search for the CPGs related to the identified problems, get them and read them entirely, decide which parts are relevant for the current clinical situation, evaluate the proposed recommendations and their applicability with respect to available resources, and take the patient's preferences into account. Finally, (s)he has to decide what is the best sequence of actions and register her/his decision in the patient medical record. During the next patient's visit, (s)he would check whether new recommendations or updates have been published and should be integrated in the care plan. The workload of this procedure can be largely reduced when CPGs are computerized. Computerizing CPGs is a prerequisite that enhances the availability and the usability of recommendations for it allows to integrate them in routine care, delivering only what is relevant to the specific context, at the time and place decisions are made. Clinical decision support systems (CDSSs) rely on a formalized knowledge embedding CPGs and are able to deliver patient-specific recommendations. Using CDSSs has the potential to improve the implementation of CPGs by health care professionals, and promote quality.

8.3.2 *Translation from Textual CPGs to Knowledge Bases*

CPGs knowledge is originally expressed with words and sentences, using a common vocabulary that people (practitioners, patients) to whom CPGs are intended can grasp without difficulty. The human intellect can indeed easily understand implicit or non-formalized information. It can deal with ambiguity, interpret, select, prioritize and qualify the information expressed as narrative text. But, a CDSS is a software which can only handle formal knowledge, i.e. knowledge expressed as logical and structured objects, where each piece of information must be explicit and codified.

Therefore, the formalization of the medical knowledge embedded within CPGs is a compulsory step prior to the computerization of CPGs. This is a complex translation process that requires informatics engineering tasks and medical skills. The difficulty is twofold. First, there is not one unique and consensual CPG representation model but several models depending on multiple factors. Depending on the analysis made about the goals to achieve (desired functionalities, granularity of information to provide, etc.), the model for computerized CPGs and its implementation within a CDSS are different. This variety of CPG models is a first obstacle to the standardization of knowledge representation, and consequently to the sharing of decision support materials. Fortunately, there are similarities between existing models, which allows to consider the setup of a consensual model, or at least can facilitate interoperability between different systems.

The second difficulty comes from the fact that CPGs are initially elaborated as narrative documents. Computerization of textual CPGs is a procedure that usually follows different steps:

- Select the portions of the text that should be formalized and encoded,
- Identify the medical concepts and decompose complex concepts into atomic concepts. Express atomic concepts as primitives or as codes that the model can support,

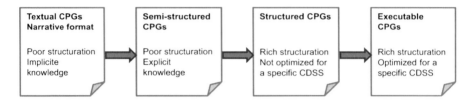

Fig. 8.2 From text to formal CPGs (Extract from http://www.partners.org/cird/cdsc/images%5CCDSC_ONC.pdf)

- Identify ambiguous or abstract notions which cannot be represented univocally (e.g. in order to determine if the concept "high cardiovascular risk" is present or not for a given patient, it is necessary to explicit all criteria defining the concept),
- Detect when clinical situations are incompletely described or recommendations are missing ("gaps of knowledge" that may be incidental or exist because evidence is lacking) and decide whether to supplement or not these gaps with expert opinions,
- Encode CPGs using the target formalism.

The sequence of the steps carried on to translate narrative CPGs into executable formalisms is shown in Fig. 8.2.

The information contained in CPGs is of two kinds, either focused on documentary notions (title, keywords, version of CPGs, editor, etc.) or focused on decision-making. Decision-making information must be evaluated and used directly or indirectly by the CDSS in order to identify the appropriate recommendations or actions for a specific clinical context. Generally, decision-making information is made of variables which values reflect the clinical context (patient data, available resources, emergency situations, etc.). Formalizing decision-making information is crucial in order to provide decision support able to automatically infer patient-specific recommendations from a given clinical profile.

Thus, "documentary tools" have been developed to facilitate the cutting of the initial monolithic document into structured elements. For example, the Guideline Elements Model (GEM) model (Shiffman et al. 2000) (http://www.gem.med.yale.edu, accessed on February 2013) offers over 100 elements hierarchically organized in order to label (i.e. characterize) and individualize the information contained in CPGs (see Fig. 8.3). GEM is not a CDSS. It is a model used to structure textual CPGs. GEM instances of CPGs can be used as a pivotal representation between the initial text and an executable format adapted to computer interpretation.

8.3.3 Knowledge Representation Formalisms

Since the production rules used in the first medical expert systems, knowledge formalisms have evolved to represent the complexity of guideline-specific features (Fig. 8.4). The first developed formalisms were the Arden Syntax, Eon and GLIF.

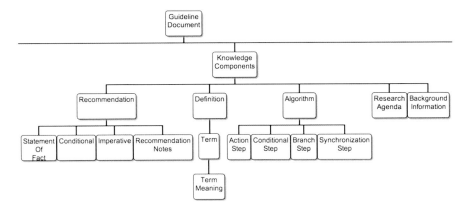

Fig. 8.3 Hierarchical structure of the "Knowledge Components" element of GEM (Extract from http://gem.med.yale.edu/Hierarchy/hierarchy.htm, accessed on February 2013)

8.3.3.1   Arden Syntax

The Arden syntax (Hripcsak 1991) is a formalism where knowledge is represented as a set of independent procedural rules named Medical Logic Modules, or MLMs, specialized to handle a specific task (e.g. display an alert telling that there is a risk of arrhythmias if serum potassium is low and patient is taking digoxin) and triggered by the events (event-driven) recorded in a computerized system (e.g. when a new serum digoxine is recorded in the EMR). MLMs can be thus considered as extensions of production rules. When patient criteria (possibly extracted from the EMR) map with a clinical situation described by a MLM (in the "If" part), the inference engine (event monitor) triggers the MLM and provides the patient management recommended in the "Then" part. Some CPGs can be represented by a sequence of MLMs.

8.3.3.2   Decision Trees

A decision tree is a formalism used to represent CPGs as a tree of all possible clinical situations encountered in a given pathology (Fig. 8.5). Nodes correspond to the variables describing the patient state, and arcs are the modalities of these variables. Paths are sequences of instantiated variables (i.e. variables with a given value, e.g. "Grade = 2"), and represent clinical profiles to which recommendations are associated at the level of the leaves. When patient criteria (possibly extracted from the EMR) correspond to a clinical situation described as a path of the decision tree, the inference engine provides the treatment recommended by the leaf.

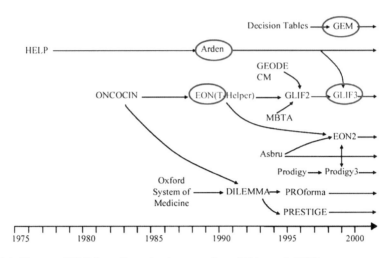

Fig. 8.4 History of CPG formalisms development; from (Elkin et al. 2000)

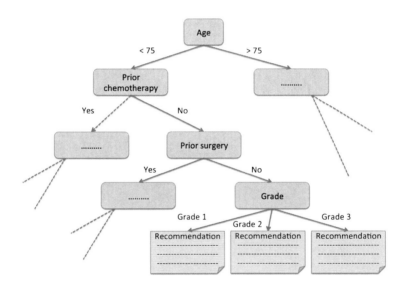

Fig. 8.5 A simplified example of the decision tree used in the OncoDoc system (Séroussi et al. 2001)

8.3.3.3 EON

EON is a guideline-based representation formalism centered on patient states (Musen et al. 1996). The knowledge base is structured as a graph whose nodes are patient states organized in "scenarios" characterizing stereotypical clinical situations (Fig. 8.6). Outgoing arcs represent the decisions or actions that are

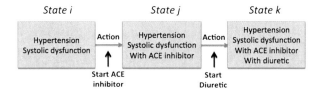

Fig. 8.6 A simplified example of EON to represent the management of arterial hypertension combined with systolic dysfunction (*ACE* is used to denote angiotensin converting inhibitor)

recommended in a given patient state and allow going from one state to the other (prescribe, conduct a review, send to a specialist, etc.).

8.3.3.4   GLIF

GLIF provides a representation of CPGs as a flowchart of structured steps that correspond either to actions, i.e. "action step" (clinical actions to perform, e.g. "Initiate therapeutic education program") or decisions, i.e. "decision step" (Ohno-Machado et al. 1998; Boxwala et al. 2004). Decisions represent patient criteria that should be verified while traversing the flowchart. Decisions may be "deterministic" (case step) and may therefore be automated (e.g. "Diagnosis of hypercholesterolemia" based on the evaluation of the cholesterol level from data extracted from the patient medical record as shown in Fig. 8.7). They may also require the intervention of physicians (choice step), either when they involve the safety of the patient, or when the needed information is not available, or when a human verification is required.

8.3.4   Inference Engine

In a systematic review of the literature analyzing interventions on health professional practices using computer systems, Hunt et al. (1998) defined a CDSS as "any software in which characteristics of a patient are matched with a computerized structured knowledge base in order to generate patient-specific recommendations or evaluations."

The classical interpretation of this definition corresponds to the automatic execution of CDSSs under the control of an inference engine. Once the clinical situation of a patient is automatically recognized on the basis of structured coded data extracted from the EMR, the inference engine triggers the adequate knowledge structure to generate the recommended patient-specific treatment. Two modes of operation are available. The critic mode issues an alert when the treatment prescribed by the physician does not match the treatment recommended. With the guiding mode, the physician does not have to enter a prescription. The system helps

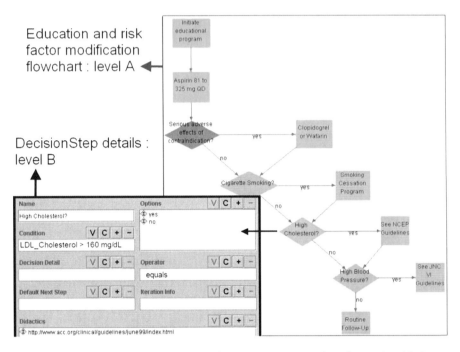

Fig. 8.7 GLIF representation of guidelines for the management of cardiovascular risk factors (From http://www.openclinical.org/gmm_glif.html)

her/him formalize the clinical situation of the patient and finally get to the recommended treatment.

"Documentary" approaches are at the opposite side of the spectrum. In this case, CDSSs do not automatically match patient data and knowledge. On the contrary, the matching is performed by physicians. In this case, the inference engine is reduced to its simplest form, the computerized system is just used to support knowledge bases through which the user navigates. Thus, in documentary approaches, the physician operates an hypertextual navigation through the knowledge base: during the traversal, (s)he plays the role of a mediator providing data characterizing the patient, with no need to have these data coded, and interpreting the information given in return, according to the clinical context of the patient.

There is a continuum between completely automatic and fully documentary approaches. Some authors (Shahar et al. 2004) have proposed intermediate stages prior to knowledge formalization in order to help practitioners find patient-specific recommendations more easily than in original texts, while keeping the flexibility of the interpretation of medical notions. Thus, even if the knowledge base is semi-structured, structured or totally formalized, it is not automatically executed. The scale proposed in Fig. 8.8 illustrates different guideline-based knowledge bases from textual CPGs (left), such as those developed by national agencies, to completely

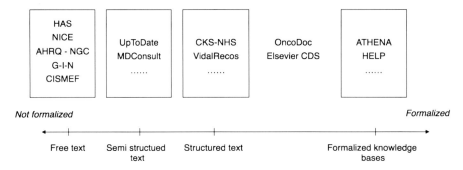

Fig. 8.8 Running CDSSs, from the reading of free text to the execution of knowledge bases

structured knowledge bases automatically executed (right). In the middle, approaches are mixed: knowledge bases may either be browsed by physicians or automatically executed. This is the case of the OncoDoc system applied to the management of breast cancer (Séroussi et al. 2001, Séroussi and Bouaud 2003).

In alert-based approaches, CDSS act as safeguards. They are very simple to "use" since CDSSs operate automatically when it is necessary (Lamy et al. 2010). However, it is difficult in practice to get the data quality and the right calibration to make CDSSs working in a totally satisfactory manner. "Silence" due to missing data, or poorly coded data, or "noise" resulting in redundant alerts or obvious alerts are frequently observed. Moreover, the lack of flexibility in the interpretation of the patient condition is often criticized: if the reduction of the patient singularity into a set of coded data is a necessary step to document the EMR, the medical reasoning involved in the decision-making process sometimes requires a flexible and contextual interpretation of much broader medical notions. This difficulty is described as a " formalization bias". Patient reduction in a set of data, although necessary for electronic medical records, does not completely reflect reality, thereby showing that the practice of medicine is not a science. Instead, documentary approaches allow physicians to use non-coded patient data and interpret information according to specific patient contexts.

These two different approaches probably meet distinct but complementary needs (Bouaud et al. 2006). In the management of "simple clinical cases" for which doctors think they a priori don't need any help, medical decision usually relies on a limited number of criteria, and alert-based decision support (critical mode) is technically feasible and potentially efficient. Besides, in simple cases, doctors are sure to know the proper solution and will not engage in an information retrieval process to get the "right" one or even to check their own. Thus alert-based decision support is compulsory to avoid errors. The situation is different for complex clinical cases since physicians are aware of the difficulty and they may want to get some support either because they do not know how to solve the problem or because they assume their choice could be suboptimal. Complex cases often involve a lot of past and current criteria and documentary approaches are more appropriate to guide them in the search for the best treatment.

8.3.5   Examples of Clinical Decision Support Systems

First CDSSs were developed in the 1970s, essentially by major hospitals involved in research and development of clinical information systems. The HELP system developed by and implemented at LDS Hospital in 1967 (Salt Lake City, Utah, USA) is emblematic of projects that have simultaneously integrated the use of a patient medical record and the use of decision support tools (Pryor et al. 1983; Gardner et al. 1992). Initially developed to support physicians in the diagnosis and evaluation of patients with cardiac disorders, HELP has been extended to other clinical areas (infectious diseases, prevention of adverse drug reactions, etc.), to other hospitals, and is currently open to different types of professionals (pharmacists, nurses, etc.). The Arden syntax has been initially developed for the HELP system.

Some research projects carried on to promote the implementation of CPGs rely on CDSSs using the EON formalism. It is the case of the PRODIGY project developed in UK for general practitioners in collaboration with major vendors of medical software (Johnson et al. 2000). Computerized CPGs cover the therapeutic management of various acute and chronic pathologies. The first two phases of the project resulted in the development of an executable knowledge base interfaced with the computerized medical record of partner vendors. Evaluation of the system showed that it was satisfactory for the management of acute diseases, but not adapted to patient with chronic diseases, numerous comorbidities, and who require to be followed over time, by different health professionals. A third phase of the project was then started. The use of production rules was abandoned in favor of EON to represent CPGs, with better results. Subsequently, Clinical Knowledge Summaries or CKS, available at http://www.cks.nhs.uk/home (accessed on March 2013), and developed by the NHS, continued the PRODIGY project with a similar approach but with a focus on using a documentary navigation through structured recommendations rather than the integration with the electronic patient record. Similarly, the ATHENA system specialized in the management of hypertension has been developed using EON. It is integrated within the VistA system (information system of the VA, Department of Veterans Affairs) and operates automatically (critical and guiding).

In France, the OncoDoc system, applied to the management of breast cancer, is routinely used in multi-disciplinary staff meetings of Tenon Hospital (Assistance Publique – Hôpitaux de Paris) (Séroussi et al. 2007). The system is developed in the documentary paradigm of decision support that provides for any patient, a personalized therapeutic care plan, with all stages of treatment (surgery, chemotherapy, radiotherapy, hormone therapy), as shown in Fig. 8.9. The knowledge base is structured as a decision tree through which the user interactively navigates to characterize her/his patient.

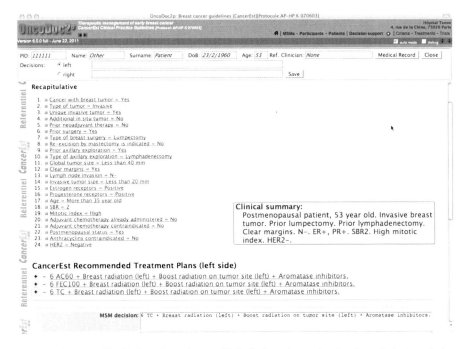

Fig. 8.9 OncoDoc2's display of a patient profile built from the navigation through the knowledge base along with the therapeutic recommendations

8.4 Conclusions and Perspectives

CPOEs are now widely disseminated and actually used in medical practice. They contribute to the production of quality drug prescriptions. However, several disadvantages should be noted. First, manually writing is faster. Moreover, the multiplication of alerts, especially concerning drug interactions, may irritate doctors. According to some authors, only a small half of alerts are appropriate, which lead practitioners to either not read them or shut down the CPOE in 49–96 % of the cases (Hsieh et al. 2004). Besides, with the development and the deployment of computerized patient records and technically improved CPOEs, unintended consequences are now observed. The use of health information technology could indeed introduce new errors (click on the wrong patient name, the wrong drug, etc.). This is known as e-iatrogenesis (Weiner et al. 2007). More work to better calibrate the triggering of alerts and to improve the usability of such systems should help correct these problems.

On the contrary, CDSSs that implement CPGs are not currently widespread. The complexity of textual CPGs, the multiplicity of agencies that produce textual CPGs, the diversity of possible formalisms used to represent CPGs (Peleg et al. 2003), the various decision variables and recommendations to generate, the constraints related to the update of knowledge bases, the small number of patient records that contain structured and coded data, are probably the factors that explain why these systems are neither widely developed nor available and used by doctors. However, things have changed from the 2000s, and the promulgation of the American Recovery and Reinvestment Act in 2009, has generated a strong pressure from the Obama administration for the widespread adoption of health information technologies. Through these initiatives, many public and private hospitals have started to use computerized patient records and CPOEs. Meanwhile, the Institute of Medicine encourages the use of health information technology, especially CDSSs, to improve the quality of care. Because of financial incentives tied to the "Meaningful Use" of new technologies, CDSSs become very attractive.

8.5 For More Information

- CPOE: http://www.ahrq.gov/research/findings/evidence-based-reports/chapter6.html
- Arden Syntax: http://www.openclinical.org/gmm_ardensyntax.html
- Eon:
  - http://www.openclinical.org/gmm_eon.html
  - http://bmir.stanford.edu/projects/view.php/eon
- GLIF: http://www.openclinical.org/gmm_glif.html
- More generally: http://www.openclinical.org/

Exercises

Q1 Several pharmaco-therapeutic classes are indicated for hypertension. Physicians may have difficulties to choose the class best suited to their patients. Do you think a computerized prescription support system, able to generate the list of all specialties with the indication "hypertension" may help? If this is the case explain why and how. Is-it easy to use the drug knowledge contained in monographs in order to compare the efficiency of different specialties?

Q2 A patient takes 20 mg of atorvastatin per day. He consults his dermatologist for a pityriasis versicolor. The treatment received is oral itraconazole.

1. By accessing drug databases, could you tell if this prescription is safe for the patient? What information has to be encoded in the knowledge base for the detection of drug interactions to be automated?
2. The patient has a disturbed liver function. How a system able to secure the prescription may warn the prescriber that atorvastatin is contra-indicated?

(continued)

Exercises (continued)

3. Is the CPOE able to assess how adequate is the prescription of atorvastatin for this patient?

Q3 Study the Canadian recommendations for the management of hypertension, edition of 2011 (http://download.journals.elsevierhealth.com/pdfs/journals/0828-282X/PIIS0828282X1100256X.pdf, last accessed 16/01/2012) and separate the chapters dedicated to the diagnosis of hypertension from those dedicated to therapy. Using production rules and a decision tree, represent the knowledge that covers the therapeutic treatment of the combination of hypertension, ischemic heart failure and cardiac insufficiency. What do you conclude? It may help to use guidelines given when increasing drug therapy is needed, in particular how to choose synergistic combinations (see Section III "Choice of treatment in adults with hypertension without drugs constraint" for the transition to bi or even triple therapy). Manually run the rules built to identify the medical management of the following clinical case:

1. Consider the case of a 68 year-old-patient, with stable angina. She has a moderate hypertension, normalized with beta-blockers. She consults for increased blood pressure. This increase is confirmed during two successive consultations. You decide to change her treatment. What treatment do you decide?
2. After 3 years of treatment by a combination of two therapeutic specialties, the patient had a myocardial infarction for which she was hospitalized. Due to ischemia, she has now a cardiac failure with systolic dysfunction. What would be her new treatment?

R1 Visit the site DailyMed at http://dailymed.nlm.nih.gov/ (last accessed 16/01/2012) and study the indications of several antihypertensive drugs belonging to different classes.

R2

1. There is an interaction between the two drugs. Their combined prescription is thus contra-indicated. The required information embeds a structured drug repository, the list of currently administered drugs, drug composition, and dangerous combinations.
2. The contra-indication should be coded in the patient record using the same terminology system as the one used for coding contra-indications in the drug database.
3. It is not the role of the CPOE but that of a clinical decision support system applied to therapeutic management.

R3 Knowledge about the management of clinical conditions combining arterial hypertension with ischemic heart disease and heart failure is given

(continued)

Exercises (continued)

in Chapter VI. Rules such as "IF hypertension and coronary heart disease THEN prescribe ACE inhibitors or angiotensin II receptors", or "IF stable angina THEN prescribe beta-blockers", or "IF stable angina THEN prescribe calcium channel blockers" can be formalized.

1. When using the rules formalized in the knowledge base, it is possible to add a calcium channel blocker to the beta-blocker already administered.
2. You have to use the rules that could be formalized from the section about the therapeutic management after myocardial infarction and those described in the section dedicated to the management of heart failure. One option could be to move from the combination beta-blocker + calcium channel blocker to the combination beta-blocker + ACE inhibitor, but this latter rule does not exist as such in the Canadian guidelines and has to be interpreted from the text.

References

Aronson JK (2009) Medication errors: definitions and classification. Br J Clin Pharmacol 67(6): 599–604

Bates DW, Kuperman GJ et al (2003) Ten commandments for effective clinical decision support: making the practice of evidence-based medicine a reality. J Am Med Inform Assoc 10(6): 523–530

Bouaud J, Séroussi B et al (2006) Design factors for success or failure of guideline-based decision support systems: an hypothesis involving case complexity. AMIA Annu Symp Proc. 71–75

Boxwala AA, Peleg M et al (2004) GLIF3: a representation format for sharable computer-interpretable clinical practice guidelines. J Biomed Inform 37(3):147–161

Classen DC, Avery AJ, Bates DW (2007) Evaluation and certification of computerized provider order entry systems. J Am Med Inform Assoc 14(1):48–55

Coste J, Venot A (1999) An epidemiologic approach to drug prescribing quality assessment: a study in primary care practice in France. Med Care 37(12):1294–1307

Elkin PL et al (2000) Toward the standardization of electronic guidelines. MD Comp 17(6):39–44

Gardner RM, Maack BB et al (1992) Computerized medical care: the HELP system at LDS Hospital. J AHIMA 63(6):68–78

Horsky J, Schiff GD et al (2012) Interface design principles for usable decision support: a targeted review of best practices for clinical prescribing interventions. J Biomed Inform 45(6): 1202–1216

Hsieh TC, Kuperman GJ et al (2004) Characteristics and consequences of drug allergy alert overrides in a computerized physician order entry system. J Am Med Inform Assoc 11(6):482–491

Hripcsak G (1991) Arden syntax for medical logic modules. MD Comput 8(2):76–78

Hunt DL, Haynes RB et al (1998) Effects of computer-based clinical decision support systems on physician performance and patient outcomes: a systematic review. JAMA 280(15):1339–1346

Johnson PD, Tu S et al (2000) Using scenarios in chronic disease management guidelines for primary care. Proc AMIA Symp. 389–393

Kaushal R, Shojania KG, Bates DW (2003) Effects of computerized physician order entry and clinical decision support systems on medication safety: a systematic review. Arch Intern Med 163(12):1409–1416

Kuperman GJ, Bobb A et al (2007) Medication-related clinical decision support in computerized provider order entry systems: a review. J Am Med Inform Assoc 14(1):29–40

Lamy JB, Ebrahiminia V et al (2010) How to translate therapeutic recommendations in clinical practice guidelines into rules for critiquing physician prescriptions? methods and application to five guidelines. BMC Med Inform Decis Mak 10:31

Lohr K, Carey T (1999) Assessing "best evidence": issues in grading the quality of studies for systematic reviews. Jt Comm J Qual Improv 25(9):470–479

Musen MA, Tu SW et al (1996) EON: a component-based approach to automation of protocol-directed therapy. J Am Med Inform Assoc 3(6):367–388

Ohno-Machado L, Gennari JH et al (1998) The guideline interchange format: a model for representing guidelines. J Am Med Inform Assoc 5(4):357–372

Peleg M, Tu S et al (2003) Comparing computer-interpretable guideline models: a case-study approach. J Am Med Inform Assoc 10(1):52–68

Pryor TA, Gardner RM et al (1983) The HELP system. J Med Syst 7(2):87–102

Radley DC, Wasserman MR et al (2013) Reduction in medication errors in hospitals due to adoption of computerized provider order entry systems. J Am Med Inform Assoc 20(3): 470–476

Sackett D, Rosenberg WMC et al (1996) Evidence based medicine: what it is and what it isn't. BMJ 312(7023):71–72

Séroussi B, Bouaud J, Antoine EC (2001) OncoDoc: a successful experiment of computer-supported guideline development and implementation in the treatment of breast cancer. Artif Intell Med 22(1):43–64

Séroussi B, Bouaud J (2003) Using OncoDoc as a computer-based eligibility screening system to improve accrual onto breast cancer clinical trials. Artif Intell Med 29(1–2):153–167

Séroussi B, Bouaud J et al (2007) Supporting multidisciplinary staff meetings for guideline-based breast cancer management: a study with OncoDoc2. AMIA Annu Symp Proc. 656–660

Shiffman RN, Karras BT, Agrawal A et al (2000) GEM: a proposal for a more comprehensive guideline document model using XML. J Am Med Inform Assoc 7(5):488–498

Shahar Y, Young O et al (2004) The Digital electronic Guideline Library (DeGeL): a hybrid framework for representation and use of clinical guidelines. Stud Health Technol Inform 101:147–151

Weiner JP, Kfuri T et al (2007) "e-Iatrogenesis": the most critical unintended consequence of CPOE and other HIT. J Am Med Inform Assoc 14(3):387–388

Wright A, Sittig DF et al (2009) Clinical decision support capabilities of commercially-available clinical information systems. J Am Med Inform Assoc 16(5):637–644

Chapter 9
Computerized Medico-Economic Decision Making: An International Comparison

P. Landais, T. Boudemaghe, C. Suehs, G. Dedet, and C. Lebihan-Benjamin

Abstract A large part of public health care costs come from hospital expenditures. Activity-based payment has gradually become the most common system for hospital reimbursement in high income countries. It was variously implemented over the last decade in order to achieve shared goals such as improving overall efficiency, quality or transparency, and to help in hospital management. Activity-based funding is also supposed to help targeting where and how money is being spent, and thus orient policy and decisions on the behalf of patients. The system also provides for payment adjustments and promotes high quality of care via reward payments.

Different schemes have been adapted to each country according to their individual developmental contexts, or to their conception of a welfare state. Variations relate to differences in the health system models used, the relationships between providers and funders, the degree of centralization, the separation between purchasing and provision, the structure of the hospital market, the type of facilities, or the level of competition between public and private structures.

European countries have adopted rather different regulation and monitoring of health expenditure. To better understand the international complexity of the framework we recalled the basic principles of activity-based payment. We then explored

P. Landais (✉)
Montpellier 1 University & CHU de Nîmes, Département de Santé Publique, BESPIM, CHU de Nîmes, Hôpital Carémeau, place Robert Debré, 30029 Nîmes, France
e-mail: paul.landais@chu-nimes.fr

T. Boudemaghe • C. Suehs
CHU de Nîmes, Département de Santé Publique, BESPIM, CHU de Nîmes, Hôpital Carémeau, place Robert Debré, 30029 Nîmes, France

G. Dedet
London School of Hygiene and Tropical Medicine, London School of Hygiene and Tropical Medicine, Keppel St, London WC1E 7HT, UK

C. Lebihan-Benjamin
Institut National du Cancer, Pôle Santé Publique et Soins, Département Observation, Veille, Évaluation, Institut National du Cancer, 52 av Morizet, 92513 Boulogne Billancourt, France

A. Venot et al. (eds.), *Medical Informatics, e-Health*, Health Informatics,
DOI 10.1007/978-2-8178-0478-1_9, © Springer-Verlag France 2014

the context of development of activity based payment in five countries: United States, Australia, the United Kingdom, France and Germany. We also described the diagnosis related groups they are based upon, the basis of setting up costs and tariffs, the fields of application, and the regulatory mechanisms.

Keywords Activity-based payment • Health planning • Hospital funding • Hospital expenditures • Hospital performance • Diagnosis related groups • Pricing activity • International policies • Decision making

> **After reading this chapter, you should be able to:**
>
> • Identify and explain the key principles of health planning and hospital funding
> • Understand the principles of activity-based payment
> • Explain the origin of hospital stay rates
> • Understand the principle of DRGs and related pricings
> • Understand draft estimates of revenue and expenditure
> • Describe the basis for pricing activity and its assessment
> • Describe how hospital expenditures are regulated
> • Describe the similarities and differences in hospital funding among certain nations.

9.1 Activity-Based Payment: General Principles

9.1.1 Context

Hospital expenditures represent a large part of public health care costs, and for this reason, several related regulatory mechanisms exist Paris et al. (2010). Activity-based payment has gradually become the most common system for hospital funding in high income countries. Initially implemented as an indicator of relative hospital performance (Wiley 2011), today it mainly represents a hospital payment system. This was variously implemented over the last decade in order to achieve shared goals such as improving overall efficiency, quality, transparency, and to help in hospital management (Scheller-Kreinsen et al. 2009). Activity-based funding is also supposed to help target where and how money is being spent, and thus orient policy and decisions on the behalf of patients. The system also provides for payment adjustments and promotes high quality of care via reward payments. An efficient price can be used to maximize the volume of high-quality care through cost efficiency.

9.1.2 Motives

The motives underlying the introduction and the development of DRG systems vary greatly from country to country. The retained schemes have been adapted to each country according to their individual developmental contexts, or to their conception of a welfare state (Beveridge, Bismarck or mixed models). Variations relate to differences in the health system models used, the relationships between providers and funders, the degree of centralization, the separation between purchasing and provision, the structure of the hospital market, the type of facilities (e.g. profit versus non-profit), or the level of competition between public and private structures (Dexia and HOPE (2008), Ettelt et al. (2009)).

9.1.3 Implementation

The implementation of activity-based payment has also varied among countries. Hospital payment mechanisms include global budgets, service fees, daily rebates, as well as case payments; each of the latter differentially influences provider behaviour and efficiency through various incentives and disincentives (Langenbrunner et al. 2009). The models developed incentivize providers to improve coding quality. Activity-based payment was therefore beneficial in some aspects but also generated unexpected side-effects such as upcoding. Upcoding is the practice of maximizing activity-based reimbursement by exaggerating case severities when coding. Another unwanted effect is selecting patient populations in order to have the most profitable casemix. The casemix, which represents the distribution of DRGs for a facility, also illustrates the relative complexity and intensity of services required to treat patients in a hospital due to diagnosis, disease severity, and personal characteristics such as age. Thus, contrasting hospital policies, such as responding to a public health need versus maximizing profits, can greatly affect their activity and casemix.

9.1.4 Assessment

Evaluation of the hospital activity-based payment program points towards unwanted shifts away from its intended purpose. The reasons for performing such evaluations include but are not limited to:

- Identifying those incentives which generate unwanted effects, such as a risk for decreased quality of care;
- Assessing incentives for productivity, effectiveness of the activity-based reform, and implementing additional funding to overcome the imperfections in the model;
- Ensuring facility sustainability without generating annuities;
- Maintaining the economic model (classification, rates, etc.);
- Ensuring system consistency via control procedures.

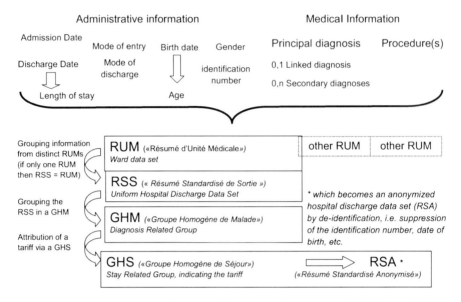

Fig. 9.1 The coupling administrative and medical information to generate DRGs': example of the French system

9.1.5   The Uniform Hospital Discharge Data Set

Activity-based payment is based on Diagnosis Related Groups (DRGs). Diagnosis Related Groups are assigned to stays for acute inpatient care. Information is collected for hospital stays in the fields of medicine, surgery and obstetrics, including ICU stays. Activity analysis can be carried out via the automated treatment of minimum and standard administrative information as declared in the Uniform Hospital Discharge Data Set (UHDDS; see Fig. 9.1).

The medical information contained within an UHDDS varies from country to country. The administrative data generally describe the main characteristics of stays, for example, facility identification, patient age, gender, postal code, stay identifier, the ward involved, date/mode of entry and discharge from the ward, or type of transfer if any. The medical data include the diagnoses and procedures carried out during the stay, and other country-specific data such as a severity index for patients in ICUs, intensive care procedures or continuous monitoring, or radiotherapy-specific information (dosimetry standards, type of device).

9.1.6   Diagnoses and Procedures

Diagnoses are prioritized. As specifically related to the French system, the principal diagnosis is the diagnosis which motivated the hospital stay. With some exceptions, all principal diagnoses are dispatched into Major Diagnostic Categories (MDC) that generally correspond to a single organ system. For instance, Diseases and Disorders of the Nervous System, Diseases and Disorders of the Eye, Diseases and Disorders of the Ear, Nose, Mouth and Throat, and so forth. A major diagnostic category represents a category of diagnoses generally based on a single body system or disease aetiology that is associated with a particular medical specialty.

Some diagnoses have specifically been identified as associated with higher resource consumption, and are consequently termed "complications" in the coding lexicon. Depending on the intensity of this effect, several categories of complications have been defined, eventually splitting DRGs into different severity levels, priced accordingly.

Therefore the presence of a "complication" will result in a higher priced DRG (this is the primary mechanism via which upcoding can occur).

Diagnoses are coded according to the international classification of diseases (ICD), whereas procedures are coded according to country-specific classifications. This information is then submitted to a grouping algorithm which associates each standardized discharge summary with the corresponding DRG.

9.1.7   DRGs

Diagnosis related groups (DRGs) have been created in such a way as to result in homogeneity in terms of both medical characteristics and resource consumption. In consequence cases that share the same DRG are medically and economically similar (Cylus and Irwin 2010).

The DRGs are assigned to inpatient stays via an algorithm (DRG grouper) using the principal diagnosis and additional diagnoses, the principal procedure and additional procedures, as well as gender and discharge status. One and only one DRG is assigned to each stay, and each DRG has a tariff.

Payment rates are expressed as a base rate (e.g. observed past cost) multiplied by a cost weight (adjustment for contextual evolutions).

9.1.8   Budget Allocation

While differing from country to country, budget allocation has a general pattern that mixes one or several of the following elements.

- General principles of cost studies
  The way tariffs are determined is based on two main principles. *Top-down costing* (or full absorption model) tries to ensure that the full cost of the stay is allowed for in the calculated cost. This requires simple data and fairly basic technical knowledge. Furthermore, it is cost-effective. Requirements include

identifying the total, hospital-wide costs that are expected to be incurred in the year. This is calculated on the basis of resource inputs and utilization, expected levels of outputs, and any agreed surplus/deficits generated in a year. Then costs are classified in a standardized form such that they can be allocated to the service specialties that form the basis of the global budget. The whole process takes the total cost down to the level of the specialty or subspecialty of the budget. Conversely, *bottom-up costing* is a patient-based micro-costing model. It determines the unit cost of a medical service by summing the cost of all inputs used to provide said service in the most recent year, and then dividing the sum by the annual total number of provided services. This approach is precise but more delicate and demanding to carry out.

- DRG base rate

    The hospital base rates are determined as a function of average stay costs per hospital across a group of hospitals. It has an attached monetary value even though cost-weights usually are a relative measure determined according to the relative resource use within a given DRG, where the average cost-weight is taken as '1'. Approximately, the hospital base rate identifies the reimbursement that a hospital would receive for treating the average patient.

- DRG weights

    Each DRG may be assigned a weight. Weights are updated annually to reflect changes in medical practice patterns, use of hospital resources, diagnostic and procedural definitions and DRG assignment criteria. To estimate the reimbursement that a hospital will receive for a particular DRG, the hospital's base rate is multiplied by the DRG weight. Software which makes this calculation is called a "pricer". For convenience, providers can bundle the DRG grouper and the pricer within their software. However, one must not confuse grouping DRGs and pricing, the latter being based on tariffs.

- Fixed payment

    Even though fixed payments may be accompanied by supplementary money that takes into account particular situations (for instance, a stay in an ICU), hospitals are paid a set fee for treating patients in a given DRG, regardless of the actual cost for that case. This is why accurate and complete coding by professionals is essential for appropriate DRG assignment and subsequent reimbursement.

9.2   International Policies

Management of Medicare expenditures is multiple. A transposition from one country to another is not possible. The control of expenditure of health insurance holds to several factors, all of which are not reducible to a pattern of governance or to preformatted rules. Practices, involvement of stakeholders, public financial constraints, the facility considered or the period involved, differ enough to limit the comparability of national health systems. Moreover, detailed rules for the administration of the health budget are very dependent on the modality of budget allocation and periodicity.

Funding the health system is operated by compulsory deductions in several European countries. The prediction of the periodic national health expenditure is region driven in Denmark or Germany, or performed according to a particular type of activity or a specific medical act such as in Belgium. The procedure is most often annual or multi-year as in Denmark, or on a longer period, for instance triennial in the UK. There is most of the time a pretty good match between the authorized or voted health expenditures, and the money finally spent in Germany, Denmark, Belgium and the UK. The UK which has the most restrictive mechanism of regulation, also presented the largest increase of its expenditures.

Germany, Denmark and Belgium have frequent projections of future health expenditures. Thus, the Danish regions publish quarterly reports. German follow-up is based on frequent projections, but all of them are not made public. A group of experts in health expenditure predictions meets quarterly and publishes aggregated projections twice a year. The detailed projections are provided only to members of the group.

Compared to the framework of regulation defined over 1 year, the levers of interventions to correct possible overruns during the year are seldom clarified. However, local manager put a persistent pressure on professionals either health insurance funds in Germany, Primary Care Trustees in the United Kingdom, or regions and municipalities in Denmark. In Belgium, the management of health insurance expenditures is performed through the annual setting of an overall health insurance expenditure target. The target of the overall budget, determined by the Belgian "National Institute for Health Insurance and Invalidity", is divided into 40 partial budget targets for each care sub-sector.

Fixing the overall budget target implies fixing partial budget targets. A description of the total expenditure by sector is made monthly. A statement on the expenditures is done quarterly by sector, with details about classification codes subject to refund by the health insurance. Each of these steps can be a source of possible savings measures. Quarterly, in case of significant overrunning, a warning mechanism is provided for information purposes only. Finally, every 6 months audit reports for each sector are provided.

The control of expenditures passed from an accounting approach to an analytical approach. These reports determine pre-emptively the potential causes of exceeding a target budget. This intra-year follow-up is only valid if decisions necessary to correct trajectories are controlled and observed. So far, this mechanism has never resulted in recommendations for new measures of money saving outside the annual process.

Thus, European countries have adopted rather different regulation and monitoring of health expenditure. Thus any attempt to evaluating the impact of activity-based payment in a given country or between countries was seldom performed or focused on specific aspects (Commission des Comptes de la Sécurité Sociale 2009). Thus for instance, it has been shown (O'Reilly et al. 2012) that activity-based funding has been associated with an increased activity, a reduced length of stay and/ or a diminished rate of growth in hospital expenditure.

Assessing and comparing activity-based funding is seldom available in the literature (Street et al. 2011; Paris et al. 2010; Busse et al. 2011; O'Reilly et al. 2012) as well as the impact of DRG payment systems on cost-containment, efficiency and quality of care (Cylus and Irwin 2010; Scheller-Kreinsen et al. 2009).

To better understand the complexity of the framework it is however possible to describe the context of development, the diagnosis related groups they are based upon, the basis of setting up costs and tariffs, the fields of application and the regulatory mechanisms. As an example, five systems will be described, in the United States, Australia, the United Kingdom, France and Germany.

9.3 United States

9.3.1 Context

Activity-based payment was introduced in the United States in 1983 for the payment of elderly patients stays covered by Medicare, a health insurance programme set up in 1966. At that time, a large number of elderly patients could not afford their hospital care. The hospital funding mechanism was based on the principle of "1 US$ refunded for 1 US$ spent". This system encouraged hospital expenditures since increased expenditures results in increased hospital income (Miraldo et al. 2006).

In this context, Congress set up the "Prospective Payment System" (PSS) in 1983. The aim was to control hospital expenditures, which had almost doubled between 1970 and 1982 (Rosenthal et al. 2006). Presently, three disbursing organisms are involved in the funding of American hospitals, each one contributing for one third of the whole, namely (1) Medicare, (2) private insurance companies that negotiate their tariffs directly with hospitals, and finally (3) Medicaid, focussing on uninsured patients and those with poor resources. Hospitals can negotiate with the various funders and can be paid at different price levels for the same service.

9.3.2 Diagnostic Related Groups

The classification system presently used is called the "Medicare Severity – Diagnostic Related Groups" (MS-DRG). It takes into account the principal diagnosis, the secondary diagnoses and an index of gravity. The total number of DRGs reported was 470 in 1983 and 557 in 2006. In 2007 *Medicare* set up a new system in order to better target payments. DRGs were adjusted according to severity, as related to the presence of major or common complications. The total number of the MS-DRGs rose to 999. Patients who are not covered by Medicare are classified according to another system, the « all-patients-refined DRGs ». It includes 1100 categories that integrate disease severity, as well as the risk of mortality assessed on four levels (minor, moderate, major and extreme).

9.3.3   Costs and Tariffs

Costs are based on record keeping performed by 3,500 health facilities founded by Medicare, representing 60 % of all American health facilities. It defines the average cost of health care delivery.

Costs are distinct from tariffs. A cost calculation is estimated on the average resources consumed by all Medicare patients belonging to the same DRG, and then it is standardized to eliminate the differences that might remain between structures and regions. This becomes the national baseline tariff, based on a single model used by both public and private health facilities. Several factors are taken into consideration:

– Exogenous factors, such as health worker salaries, hospital supplies, or a geographical adjustment factor;
– Indirect costs such as medical teaching. In this case an index is estimated which reflects the ratio between the number of medical residents and the number of beds;
– Additional costs in order to support specific populations, such as low-income or vulnerable groups;
– The type of health facility (rural hospital, reference hospital).

Resources are allocated on the basis of these references. However, a part of these resources is also negotiated directly by hospitals with the "Health Maintenance Organizations". HMOs are health assistance organizations that manage health care by grouping the health care coverage provided by hospitals, doctors or by other suppliers under specific contracts. Finally, Medicare allows tariff increases for severe and expensive cases whose stay costs are more than three times the standard deviation of the mean.

9.3.4   Fields of Application

Activity-based payment is used for all medical, surgical and obstetric (MSO) in-patient facilities, and has been applied to full time hospitalization since 1985, and to ambulatory care since 2000. It does not apply to practitioner fees, to expensive drugs, rehabilitation, oncology, psychiatry, or to the direct costs related to education and research. It does not apply to small institutions, which are still funded according to their actual costs. The system provides additional funding not related to activity. Expensive drugs and prostheses are provided for by an additional lump sum payment. Considering health institutions catering for a high proportion of indigent patients, a bonus index is applied to their tariffs, as well as for university hospitals concerning their missions of education for medical students. For research, scholarships are offered.

9.3.5 Regulatory Mechanisms

Health institutions are not submitted to a price/volume regulation of expenditures: unlike other countries, a given facility collects different tariffs for a given stay according to the insurance company implicated, which curtails regulatory measures at the local or national level (Cash et al. 2003). Facilities are responsible for their deficits; on the other hand, profits are retained.

In order to prevent facilities from orienting their treatment choices or selecting patient groups based on financial considerations rather than medical needs, DRG payments have been adjusted. However, DRG classification only explained 30 % of the variation among stay costs. In 1995, the Commission of evaluation of activity-based payment, known as "ProPAC", recommended that specific secondary diagnoses that frequently affect health care costs, as well as differences in medical practices between establishments, be taken into consideration.

In 2005, MedPAC highlighted important variations in profitability between-DRGs and intra-DRGs. Despite similar clinical situations, DRGs related to surgery were more advantageous than those related to medical treatments. This cost variation was systematically related to the patient's secondary diagnosis. A recommendation was implemented in order to improve the adjustment of recruitment from one hospital to another, and to adjust prices in order to better reflect differences in severity between cases.

The DRG system generated a reduction in the length of stay in MSO facilities, an increase of ambulatory care and a shift to follow-up and rehabilitation care. These changes promoted the development of structures such as centres for ambulatory surgery. Case-based or day-based prospective payment increased efficiency payment systems in the fields of ambulatory, follow-up and rehabilitation care.

9.3.6 Evaluation

This funding mechanism made it possible to fund therapeutic innovations such as coronary angioplasty. Quality improvement programmes were also encouraged via the creation of quality of care indicators and penalties in case of non-quality, such as the occurrence of hospital acquired infections.

This tarification system does not have a durable impact on cost growth or variation. Because DRGs tariffs are negotiated, a given DRG tariff can considerably vary from one health facility to another. Private institutions specialized in specific types of programmed and ambulatory surgeries offer rates well below those of conventional facilities. Thus, they become even more competitive.

There is also heterogeneity in costs of various stays despite the fact that DRG payment is adjusted for daily costs. Successive modifications of DRGs did not lower this heterogeneity, which explains only 30 % of cost variability.

DRG refinements make classification more complex, requiring a very detailed information base. For facilities, the risk of not being fully paid increases if their data are not appropriately collected. A perverse effect can occur, resulting in better payment of large teaching institutions and poorer payment of smaller rural facilities that are already in trouble.

It should finally be emphasized that in a prospective payment system, the power of funders decreases when the number of payment categories increases. Indeed, bargaining power is lost when these categories are numerous and very specific.

9.4 Australia

9.4.1 Context

Initiated in 1984, the Australian Medicare plan is funded by taxes, under the aegis of the Australian federal government and by the six member states. However public hospitals are controlled solely by regional governments (Australian government 2010). Before the introduction of activity-based payment, hospitals received their global budgets according to their activity together with the number of patients in their waiting lists. Hospitals used these queues to negotiate an increase of their financings. This mechanism was criticized because it encouraged health facilities keeping patients in their waiting list, particularly for surgery. This led to an overhaul of the funding system. The main goal of the introduction of activity-based payment was to reduce these waiting lists. The State of Victoria, representing one quarter of the Australian population, initiated this process in 1993.

9.4.2 AN-DRG Classification

Classification of hospital stays incorporates a hierarchy of benefits and medical situations. Stays are distributed in one of 23 classes and 667 AN-DRGs created in the 1990s. AN-DRGs classification has been adopted by most Australian states. The State of Victoria has gradually changed the AN-DRGs to finally create its own groups, the VIC-DRGs, for a total of 760 groups. They adjusted the payments of stays of less than 24 h duration on the basis of the technical difficulty of treatment. An annual review is performed by a panel of clinicians.

9.4.3   Costs and Tariffs

Costs of stays were calculated from patient data by a bottom-up method. This type of model aggregates all available information at the patient level by medical specialties in dedicated cost centres. This type of modelling is supposed to improve the accuracy of costs measurement for each patient compared to other methods the costs of which are allocated to patients solely on the basis of their length of stay.

Uniform tariffs for each AN-DRG are applied to hospitals that are similar regarding their patients' case-mix. They are used to negotiate price-volume contracts. Prices are also adjusted on several criteria, including the socio-demographic conditions of the population represented by the health facility. For each hospital a fixed tariff is generally applied to each AN-DRG for a given target level of activity determined by its historical activity.

To encourage health facilities to exceed their target level of activity, additional payments mechanisms were introduced. Regarding the management of patients requiring urgent care or waiting for a long time, a classification according to the rate of emergency was introduced: urgent, semi-urgent or non-urgent. Moreover, a ceiling on payments for additional activity was established.

A pre-defined budget is shared between health facilities. It is correlated with each establishment's own contribution to the total additional activity. Finally, access to additional funding has been conditioned on the achievement of specific care goals related to the waiting list of patients. This pricing policy has proven effective as shown by the drastic reduction in the number of "urgent" patients waiting for more than 1 month.

9.4.4   Field of Application

The Australian Health System is a mixture of both private and public sector health services providers, and a range of funding and regulatory systems. They involve:

- The Australian government. Its primary role is to develop broad national policies, regulation and funding. Funding includes three major national subsidy schemes; Medicare, the Pharmaceutical Benefits Scheme, as well as the 30 % Private Health Insurance Rebate.
- State and Territory, and Local governments. They are primarily responsible for the delivery and management of public health services. They are also in charge of maintaining direct relationships with most health care providers, including regulation of health professionals and private hospitals. Under Medicare, the

Australian and State Governments jointly fund public hospital services, providing free services to Australians who choose to receive treatment as public patients.
- Private practitioners including general practitioners, specialists and consultant physicians.
- Profit and non-profit organizations and voluntary agencies.

The Pharmaceutical Benefits Scheme and Medicare cover every Australian, subsidizing their payments for private medical services, as well as for a large portion of their prescribed medications. While these options are available to Australians, government funding of the 30 % rebate as well as additional key incentives support citizens choice to pursue and retain private health insurance.

Individuals make their contribution to the health care system through taxes and the Medicare levy based on their income, and through private financing such as private health insurance.

9.4.5   Regulatory Mechanisms

In 2009, the final report of the National Health and Hospitals Commission made 123 recommendations, which included the following: hospital funding tied to performance; introduction of activity based funding for all hospital services, with the Commonwealth paying a fixed percentage of the efficient price; promoting a personal electronic health record and national eHealth system; public reporting of public and private hospital performance; a range of measures related to choice in care for the elderly; a permanent, independent and expanded Australian Commission on Safety and Quality in Health Care; National Access Targets for timeliness of care.

To ensure that the nationally efficient price is determined on a fair and equitable basis, an independent expert will set the nationally efficient price and advise the Government on appropriate timelines and transition processes for all hospital services. In setting the nationally efficient price, the expert will be required to strike an appropriate balance between reasonable access, clinical safety, efficiency and fiscal considerations.

A main objective is eliminating the unnecessary use of resources in the production and delivery of services. In hospital systems, operational efficiency can be achieved by reducing length of stay, increasing quality, and looking closely at the reasons for significant variations in clinical practice.

The increased operational efficiency from activity based funding will provide savings for taxpayers or help fund additional services. Although the precise efficiencies from activity based funding are difficult to estimate, the NHHRC estimates that the introduction of activity based funding will lead to savings of between $0.5 billion and $1.3 billion each year.

9.4.6 Evaluation

The use of activity-based funding is supposed to drive increased operational efficiency across the hospitals system as it explicitly links the funds allocated to the services provided. It also allows for easy identification of underperforming providers so that the cause of underperformance can be remedied, while lessons from high performance can be disseminated.

In 2012, a new report, the "national Health and hospitals network for Australia's future" proposed several adjustments to the model. The Commonwealth Government will become the majority funder of the Australian public hospitals system. The Government will fund 60 % of:

– The efficient price of every public hospital service provided to public patients;
– Recurrent expenditure on research and training functions undertaken in public hospitals;
– Capital expenditure, both operating capital and planned new capital investment, to maintain and improve public hospital infrastructure;
– And over time, up to 100 % of the efficient price of 'primary health care equivalent' outpatient services provided to public hospital patients.

In return for providing a secure funding base for public hospitals in the future, the Commonwealth will require the states to commit to system wide reform to improve public hospital governance, performance and accountability.

In establishing Local Hospital Networks, states will be asked to create hospitals groups that can ensure geographic linkages, management quality, economies of scale, an appropriate service mix, and referral pathways within the network.

9.5 United Kingdom

9.5.1 Context

In Great Britain, the National Health System (NHS) is managed by the Ministry of Health. It is a decentralized system (Ellis and Vidal 2007). The NHS is organized around more than 304 local instances, called Primary Care Trusts (PCT), which provide primary and secondary care to citizens. PCTs are responsible for purchasing medical care for the whole population. Health care accessibility is "free" for everyone. The major specificity of the British system is that access to hospital care or specialists is controlled by general practitioners called "gatekeepers" (Mason and Smith 2006).

Before 2004, British hospital budgets were global contracts, called "blocks contracts". The predictable volume of care was used to determine the total amount a health facility might expect. At that time, British health expenditures were the lowest in Europe, but the reported quality of care was poor, including long waiting times for scheduled surgery.

To improve the health system, "Payment by Results" (PbR), similar to activity-based payment, has been introduced progressively since 2004. The main objective of this reform was to increase efficiency and productivity of care, while ensuring quality.

PbR was introduced at a time of relatively low levels of funding and high waiting lists. The explicit objective was to increase acute activity. In particular, surgical activity was required to reduce waiting lists and therefore the system was incentivized essentially by "piecework".

The English system is a mixed system. Health facility budgets are prospective, with 30 % of the budget financed by an overall envelope and 70 % funded by activity-based payment. UK hospitals are funded by a single-payer: PCTs. From taxes, the Ministry of Health allocates an annual budget to the NHS. Its amount is based on changes in care production's costs, previously allocated amounts and health indicators such as mortality rates.

The NHS allocates 80 % of its budget to PCTs who manage primary and secondary care. PCTs allocate secondary care to public, private or non-profit providers. These caregivers may be hospitals or groups of hospitals under joint management, "NHS Trust", "Foundation Trusts" meeting performance criteria, independent treatment centres, private hospitals or ambulatory specialists.

9.5.2 HRG Classification

Activity based payment is based on Healthcare Resource Groups (HRG), a classification which is close to the American model. This classification takes into consideration diagnosis and treatment complexity. It included 550 HRGs in its first version, and 1,400 in the 2009 version.

9.5.3 Costs and Tariffs

The accounting policy is similar to the French system. However, the national cost scale is based on all the hospitals involved in activity based payment. The price of each HRG is national, calculated from a baseline cost, using a top-down costing method. A reference cost results from a mean cost for a given HRG across all hospitals plus 350 Hospital Trusties, created in 1991 to introduce market rules. These Trusties are legal entities that often manage several health institutions.

Tariffs were designed to increase the level of activity beyond a pre-negotiated level; this is called the "Service level agreement". They reflect the average costs of care procedures, adjusted for several characteristics such as: scheduled stay or not, type of surgery, health procedure, ambulatory or not, income variations and consumer price index (using an index called the "Market Forces factor"), a geographic coefficient, costs generated by implementing the recommendations of the National Institute for Health and Clinical Excellence (NICE) on treatments and drugs considered as cost-effective. The payment unit is based on HRGs.

9.5.4 Field of Application

Activity based payment applies to hospital departments, including day care hospitalization in medicine, surgery and obstetrics. It also applies to emergency departments. It is not used for practitioner's fees, expensive drugs, chemotherapy, radiotherapy, dialysis units, cardiology intensive care, burn units or transplant units. Medical research and teaching is funded by the Ministry of Health and the Higher Education Funding Councils.

9.5.5 Regulatory Mechanisms

The English system is not very flexible (Street and Maynard 2007). Deficits are at the expense of the institutions. Any increase or decrease in activity will be charged according to the national scale, without any local negotiation. In addition, in order to regulate access to emergency departments, tariffs for emergency care services are reduced by 50 %, as soon as the volume of activity exceeds that of the previous year by more than 3 %. Conversely, there is no regulation of the volume of surgical activity.

Since the mid-1980s, the budgetary mechanism for health expenditure in the United Kingdom is multiannual and limiting. The health budget is voted by the Parliament as part of the State budget. Health expenditures included in the State budget are thus subject to the general evolution of the State revenue. These revenues do not depend, as in France, on the trends in total wages. There is therefore, by design, no possible deficit concerning social security insurance. The programming of expenditure is triennial. The preparation of multi-annual programming, or a "spending review", is held every 2 years. The third year is for the revision of the objectives. The amounts of expenditure are adopted in accordance with the objectives and priorities of the government, and available through multiannual envelopes. A deficit or a surplus compared to the forecast of expenditures, gives the following year an amputation or an increase in the amount of the excess or surplus. As noted above, the consequences of this very strict and effective discipline were attenuated by the authorized increase of health spending – identified as health priorities – for a decade, despite a slowdown over the last 2 years.

9.5.6   *Evaluation*

Though the original goal of payment by results was to incentivize activity, it has with some modifications been subsequently used for additional policy objectives: enabling choice, transferring funding so that money follows the patient, enabling benchmarking, creating efficiency or incentivizing quality (Sussex et al. 2008; Vogl 2013).

In hospitals where activity based payment has been introduced, the cost of care at a unit level fell faster than in other facilities, and length of stay decreased significantly, falling from 8.20 days in 2001–2002 to 5.98 days in 2005–2006. Nevertheless, the reduction in the length of stay cannot be formally attributed to the reform, since in parallel significant government pressures was exerted on hospitals to reduce waiting times. Result-based payment was associated with an increase in volume of care, but had no impact on the quality of care (Farrar et al. 2009). The latter was measured on the basis of hospital's global mortality rate, mortality rate 30 days after surgery, and emergency readmission rates after hip fracture. The introduction of activity based payment caused heavy financial losses for certain hospitals.

9.6   France

9.6.1   *Context*

The French health system is decentralized and based on Bismarck's model. It has many particularities, such as: a free path for patients, an important for-profit private sector, and the coexistence of several hospitalization sectors. Activity-based payment was introduced in France in 2004. It aims at encouraging better response to health care supply and improving efficiency. This system is based on a competitive approach between health facilities and encourages them to optimize a cost-effective health care supply, as well as the quality of care and patient satisfaction. It is a powerful incentive towards productivity. The challenge lies in the ability of the system to combine cost containment with maintaining the quality of care and innovation (IRDES 2009).

In keeping with the peculiarities of the French health system, a mixed model has been set up. On the one hand it covers the financing of activities related to diagnosis, treatment and care, and on the other hand it compensates expenses related to missions of general interest.

In order to equitably allocate resources to institutions, taking into account their specific activities and respecting the notion of equal access to care, activity-based payment defined according to four additional compartments was implemented, in agreement with health professionals:

- Stay-based prospective payment is the basis of this system. It consists of prospective pricing with a refund based on real and all-in costs. Stays are grouped into diagnosis related groups ("Groupes Homogènes de Malades", French GHM) according to medical criteria and applied similarly to all institutions of a given healthcare sector;
- The other three components are funded differently; they take into consideration the missions, activities and products that cannot be included either temporarily or permanently in the costing of the diagnosis related groups (DRGs'). These three components are:

  - Annual all-in packages (emergency care units, organ procurement and transplantation activities);
  - Extra funding for expensive drugs and medical devices included on a specific list and fully reimbursed;
  - Finally, Missions of General Interest and Contracting Assistance ("Missions d'Intérêt Général et d'Aide à la Contractualisation", French MIGAC) are funded, providing opportunity for financing specific services such as palliative care, consultation for drug users, halting tobacco use, care for prisoners,.. Some health care services are subject to specific invoices (intensive care supplement, dialysis package, or package for small equipment).

Since 2004, this reform is applied to both public and private sectors, but following different schemes, in particular with regard to costs and prices.

9.6.2 Classification System

The French classification system is based on GHM ("Groupes Homogènes de Malades") which corresponds to the American's Diagnosis Related Groups (DRGs). A GHM represent inpatient classifications on the basis of several items comprising diagnosis, procedure, age, gender, or discharge disposition. These groups were constructed to evaluate length-of-stay, which is related to illness severity and resource consumption.

Thus, DRGs provide a tool to analyse hospital stays. DRGs were designed to allow hospitals to operate on a more systematic basis, associated with resource allocation and cost analysis. DRGs are a tool used for the prediction of resource consumption for any given hospital stay, allowing comparisons between facilities.

This classification, which is the basis of the French payment system, was published for the first time in 1986. It is regularly updated. Its most important feature, compared to other classifications, is that it integrates day care hospital's activities.

The original version included 784 groups, the current one (version 11) consists of 2,296 GHMs. This last version is composed of 611 roots, each divided into up to four levels of severity to better take into account the complexity of care given patients.

9.6.3 Costs and Tariffs

Based on GHMs, Stay Related Groups ("Groupes Homogènes de Séjours" or GHS) have been created. Almost all GHM correspond to only one GHS (only three exceptions). GHSs are tariffs associated with GHMs. Thus GHMs are a basis for measuring resource consumption in the form of a normalized weight. The construction of a tariff is critical and complex. It has to meet several objectives:

- A tariffs' hierarchy conform to production costs in order to avoid under or over paid health care deliveries.
- An observance of the ONDAM (French national target for health insurance expenditures). The constructed hierarchy is thus subject to several constraints. The volume of all GHS can't generate higher expenditure than the amount provided by the ONDAM. Therefore, the ONDAM determines the tariff level, which is the first rate scale. This raw rate scale allows funds distribution (according to what is planned by the ONDAM) to health facilities while respecting the hierarchy of the classification;
- A stimulation of the tariff policy in order to achieve public health objectives or encourage the evolution of medical practice. These incentives are called 'marked tariffs';
- Taking into account the adaptive capacity of health facilities and the need for stability while limiting the effects of tarification scales on their revenues.

Tarification is centralized and generated at a national level using a top-down approach. Tariff weighting is performed on the basis of average costs observed in a 122 voluntary sample of facilities. A statistical adjustment is realised in order to take into consideration the selection bias related to the voluntary participation of each health facility. Two scales are defined:

- One for the public sector, institutions previously under global budget allocation;
- One for the private sector (where medical fees are not included in prices).

Tariffs are set up at a national level, determined on the basis of standards in order to meet targeted spending levels. Pricing is adjusted over regions, using a geographical factor which takes into consideration only geographical characteristics; therefore characteristics such as size of the facility or status are not included in this adjustment. In 2009, activity based payment accounted for 90 % of medicine, surgery and obstetrics expenditures and concerned 50 % of all French health care facilities.

9.6.4 Field of Application

Activity based payment is used for Medicine Surgery Obstetrics (MSO), and full time hospitalization. However, expensive drugs and medical devices are not included in GHS' tariffs, nor medical fees for private sector GHSs.

Another important element is that activity based payment is not used in Psychiatric hospitals follow-up and rehabilitation care.

Other remarks:

- Expensive drugs and prostheses have funding that can be added to those reported in the list of drugs and devices charged in addition to each GHS.
- MIG ("Mission d'Intérêt Général") funding is provided by an allocation, as a compensation of public service and the fulfilment of general interest missions.
- A budget called "Mission d'Enseignement de Recherche, de Référence et d'Innovation (MERRI)" provides funding for research and teaching. This additional funding is adjusted according to the declared activity in this field.

MIG budgets can better fund certain additional costs incurred by health facilities. Activities grouped in this field as well as their corresponding resources, are subject to negotiation between the Regional Health Agency (ARS) and the facility, embodied by a multi-year agreement of objectives and funding ("Contrat Pluriannuel d'Objectifs et de Moyens").

9.6.5 Regulatory Mechanisms

In France, the HPST law (N°2009–879) relative to hospital reform was dedicated to better articulating patients, health and territories. The creation of regional health agencies (ARS) aims at implementing, at the regional level, a coordinated set of programs and actions that put into practice the objectives of the national policy of health, the social and medico-social actions and the fundamental principles of the social security code. Thus, regional health agencies expand their regional expertise across the health field and create regional health projects. They were assigned the task of implementing health policy at the regional level, and in consultation with health professionals, regulate, direct and organize the provision of health services. The creation, conversion, and the grouping of care activities, as well as heavy equipment are subject to authorization. ARSs establish a multi-year contract with each institution, which defines objectives and means responding to the guidelines of the regional health project, of the regional organization of care (SROS) scheme, or interregional schema.

The law anticipated that if in a given year the budget is exceeded, adjustments can be made. Therefore, when MSO activity expenses are exceeded, a regulatory mechanism lowers the levels of the national tariff for all facilities. This operation is carried out annually by adjusting the amount of rates of hospital stays to the projected business of the following year. Surplus gains and deficits are retained by the institutions.

In 1996, the public and private hospital system was reformed and established regulation through costs based on two new items: annual funding of social security laws and the National Insurance (ONDAM) spending target.

- The annual funding of social security laws: these funding laws determine the general conditions of financial equilibrium and, given the forecast of revenue (contributions), set its spending targets, under the conditions and subject to the reservations provided by an organic law. The Act of funding for a given year includes four parts relating to (1) the period ended, (2) the current fiscal year, (3) revenues, and (4) the projected equilibrium and spending for the coming year. Three types of data are presented by the law of funding: revenues, objectives of projected expenditure and cash advance limits.
- The National Insurance spending target (ONDAM). This is a budgetary control passed under the control of the Parliament which votes the Act of financing social security, which sets the ONDAM.

9.6.6 Evaluation

Regarding efficacy and productivity, the effect of activity based payment remains unclear (Or 2010). In 2005, public and private institutions increased their activity in different ways. Public hospitals have increased the number of inpatients by 1.5 % and the number of outpatients by 5 %. During the same period, private clinics have reduced the number of inpatients by 3 % and increased their outpatient number by 9.5 %. In addition, a study undertaken in 800 hospitals reported that once facilities had understood and integrated the logic of activity based payment, most of the time they decided to change their activity rather than seeking to improve their efficiency. Thus, private clinics tended to over-specialize their activity (IRDES 2009). For instance, some of them report that 80 % of their global activity is invested into ambulatory surgery. Changes in the management of medical human resources were also introduced (Ministère de la Santé et des Sports 2010).

To date, it is not possible to measure the impact of activity based payment on quality of care. For this reason, indicators have been developed by the health authorities in order to measure the impact this reform has had on the quality of care in French hospitals.

Unlike the private sector, the public sector expenses have progressed faster than incomes between 2003 and 2007. Implementation of activity-based payment has been harmful to public hospitals, and their profitability has declined (DREES 2009). Public facilities face difficulties in reducing their costs despite increases in activity. Conversely, the private sector, which has seen its overall turnover and profitability rising by 8 % in 2005 and 4 % in 2006, seems to have benefited from this reform.

9.7 Germany

9.7.1 Context

The German health system is a federal system. Federal states, called "Länder", control public hospitals together with counties and municipalities. Hospital charging system borders were reformed in 2000 (Busse and Riesberg 2004). At this time, pricing was based on a day rate. Hospital expenditures accounted for 35 % of health spending. The main objective of the tariff reform was to optimize acute care by increasing efficiency, quality and transparency, while reducing the growth of hospital expenditures. The process was achieved step by step, based on three main principles: solidarity, benefits in kind and self-administration.

The classification system was created in 2000–2002 with the implementation of costing, which became stable between 2003 and 2004. Activity based payment became mandatory in 2004 for the MSO sector. The charging system converged between 2005 and 2010, and was provided by two funders. On the one hand, Länders together with municipalities, manage capital expenditures and infrastructures, and develop hospital planning. On the other hand, public health insurance funds manage current spending on the basis of a budget prepared according to the nature and volume of care in a given hospital. Finally, part of the funding comes from insured people who have to pay a fixed daily hospital charge.

Health facility budgets are prospective, based on annual estimates of activity and expenditures (Neubauer and Pfister 2008). Such budgets are applied to public and private institutions according to similar modalities. The budget is mixed, with 20 % determined by a fixed envelope, and 80 % determined by activity based payment.

9.7.2 Classification System: G-DRGs

Activity based payment is based on a classification system called "G-DRG" for German Diagnosis Related Groups (Hensen et al. 2008). It is modelled on the Australian AR-DRG. Principal and secondary diagnoses are coded using the International Classification of Diseases adapted to Germany (ICD-10-GM). The number of DRGs increased from 664 in 2002 to 1200 in 2010.

9.7.3 Costs and Prices

The cost for each DRG is self-administered, established annually by the Federation of Health Facilities, the Deutsche Krankenhausgesellschaft and by the Institut für das Entgelt system Krankenhaus (Fürstenberg et al. 2010), which groups the federal union of health insurance funds and the federation of private health insurers. This Institute is the property of the hospitals and of the health insurance funds. Objectives include classification improvements, estimating the relative weights of the DRGs and the creation of additional payments. The national costs scale is established by a sample of 250 voluntary hospitals (12 % of all facilities). The average costs for each DRG are published annually. Tariffs are negotiated between each hospital and the health insurance service for a 1 year period. Some geographic disparities are taken into consideration for certain tariffs, and may result in the creation of special funds. Regional tariffs are supposed to converge by 2014.

9.7.4 Fields of Application

Activity based payment applies to acute care medical departments, including emergency units, surgery and obstetrics. It is supposed to extend to psychiatry departments by 2013. It does not apply to private practitioner's fees or to expensive drugs or devices. For missions of general interest such as teaching, research or doctors' training, DRG rates are increased and special funds are allocated. Innovative treatments prices are increased. Tariffs are supposed to converge by 2014.

9.7.5 Regulatory Mechanisms

The federal government sets the regulatory framework for the health care system, including the rules of provision for social services and their financing (Vogl 2013). In case of a budget deficit, control is provided by the *Länders*, who are the key regulators of the hospital sector. When the hospital budget is exceeded, a *Land* can formulate a plan for restructuring activities over a group of hospitals. It can suspend certain activities, close facilities, or privatize a hospital. It can also offer financial assistance. Moreover, health insurance funds may minimize tariffs, increase tariffs of stays, and/or reduce the repayment of the funding the following year according to a principle known as "partial repayment".

9.7.6 Evaluation

The G-DRG system has made hospitals more organized and operational, through greater inter-disciplinary cooperation (Busse and Riesberg 2004). Significant restructuring has reduced hospital numbers by 4 % and the number of beds by 5 %. The length of stay has decreased by 5 %. Between 2004 and 2007, the bed occupancy rate increased by 2 %. Some sectors, such as obstetrics and gynaecology, internal medicine or surgery, have decreased, and other specialties such as neurosurgery, plastic surgery or paediatrics surgery increased. One hundred public hospitals have been fully or partially privatized. Between 2000 and 2007, the proportion of public institutions has declined from 38 % to 32 %, and private non-profit from 41 % to 38 %. Conversely, the proportion of the fully private sector rose from 22 % to 30 %.

In 2007, 53.5 % of public hospitals were mixed, public-private institutions. Nevertheless, activity did not increase as expected. Indeed, the number of hospitalized patients, adjusted on age, rose only by 0.4 % in 2006. A transfer of activity has been observed, mainly towards ambulatory surgery.

In the majority of medical departments, quality remained stable or improved, which is not the case in some disciplines such as cardiac surgery. However, it is difficult to attribute this improvement to activity based payment, since simultaneous measures to improve quality were developed, such as accreditation procedures. Moreover, the working conditions of medical and paramedical workers deteriorated, due to significantly increased workload. Usually, economic efficacy is better, despite additional costs such as expenses related to creation and maintenance of hardware. Between 2003 and 2006, the hospitals' total adjusted costs increased by 1.4 % annually. This increase was less pronounced than during the 1991–2006 period (+3.7 % every year). During the 2005–2008 period, health insurance spending increased by 2.6 % every year. In 2007, 30 % of hospitals were in deficit, 20 % balanced, and 50 % beneficiary. Large facilities and university hospitals were favoured to the detriment of small hospitals. The Act of March 17, 2009 thus proposed a 3-year aid program (Herwartz and Strumann 2012).

The G-DRG system is notable for its increased transparency and traceability of hospitals stays since it requires collection and publication of standardized and detailed data. The implementation of activity based payment has made the German system more effective and more organized, while maintaining quality of care. The strength of this system is due to the analysis and pricing of the G-DRG by the InEK. Nevertheless, it is difficult to attribute changes to just activity based payment implementation due to interactions with other economic parameters, such as rising wages, high inflation of energy costs, and the implementation of quality procedures.

9.8   Conclusion

Street et al. (2011) have considered a range of studies evaluating the introduction of activity based funding. Like others, they point out the lack of rigorous evaluation and the difficulty of conducting any controlled studies. They noted increases in volume of care, increasing day case activity and reductions in lengths of stay. However, they appeared unable to find any evidence that the quality of patient care has improved Or and Häkkinen (2012); conversely, there was no evidence of deterioration. Interestingly, their overview of DRG system goals for 12 European countries did not list quality improvement.

Further adaptations of activity-based funding are necessary. For instance, emphasis on quality of care is essential, as well as value-based purchasing. Delivering integrated care through episode-based payment is also a field to explore. These new adaptations to the model will require extending the reimbursement mechanism beyond the hospital sector. Recent developments indicate that progress is underway towards adapting activity-based funding in order to meet future challenges, among which improving quality of care is a key issue.

9.9 For More Information

Health policies and data: http://www.oecd.org/health/health-systems/healthworking-papers.htm

Investing in hospital of the future: http://www.euro.who.int/__data/assets/pdf_file/0009/98406/E92354.pdf

Designing and Implementing Health Care Provider Payment Systems: http://siteresources.worldbank.org/healthnutritionandpopulation/Resources/Peer-Reviewed-Publications/ProviderPaymentHowTo.pdf

From the origins of DRGs to their implementation in Europe: http://www.euro.who.int/__data/assets/pdf_file/0004/162265/e96538.pdf

The incentive effect of payment by results: http://www.york.ac.uk/media/che/documents/papers/researchpapers/rp19_the_incentive_effects_of_payment_by_results.pdf

USA: http://www.cms.gov/Medicare/Medicare-Fee-for-Service-Payment/HospitalAcqCond/HAC-Regulations-and-Notices.html

https://www.federalregister.gov/articles/2011/11/30/2011-28612/medicare-and-medicaid-programs-hospital-outpatient-prospective-payment-ambulatory-surgical-center

A National Health and Hospitals Network for Australia's Future: http://www.health.gov.au/internet/yourhealth/publishing.nsf/Content/nhhn-report-toc#.UU8OdFdrdBA

Europe, Health Care Systems in transition: http://www.euro.who.int/en/who-we-are/partners/observatory/health-systems-in-transition-hit-series/countries-and-subregions

For all references last access on March 24, 2013.

Exercise

A 18 year old patient presenting with an Alport syndrome (hereditary nephropathy associated with a collagen abnormality), responsible for end-stage renal disease, was hospitalized for a kidney transplantation. Transplantation was performed on March 1, 2011. The lack of diuresis in the post-operative recovery led to a transfer to a resuscitation unit where acute renal failure necessitated two haemodialysis sessions. At day 3, the resumption of diuresis permitted the transfer to a continuous monitoring care unit.

(continued)

Exercise (continued)
The postoperative was marked by:

- High blood pressure (HBP) requiring an anti-hypertensive treatment.
- Discovery by ultrasonography of a perirenal urinoma infected by enterobacteriaceae requiring surgical drainage. A retrograde pyelogram showed a calyceal injury due to the JJ catheter. The catheter was then removed.
- Digestive hemorrhage of the lower digestive tract. A complete coloscopy was then performed which proved to be normal and an upper gastrointestinal endoscopy which showed haemorrhagic duodenal ulcer.
- A 15 kg loss of weight which required an enteral nutrition for 4 days, then a parenteral nutrition for 30 days.

On day 50, the patient left the continuous monitoring care unit for the transplantation ward. The discharge to home occurred on day 60.

Questions

Q1 Propose a coding for diagnoses, while specifying the principal diagnosis and procedures which occurred in the three wards: transplantation ward, intensive care unit and continuous monitoring care unit, respectively.

Q2 Find the ICD-10 diagnostic codes for the stay. You can consult the ICD10 site on internet, for example: http://www.icd10.ch/ebook/FR_OMS_FR/FRS_Index.asp

Q3 Describe all the procedures used during the stay.

Q4 Below is the activity of the transplantation ward (French DRG root 27C06) of a given facility and the corresponding national data. Comment the results of the facility compared to national data.

DRGs	N	Facility		National data	
		Mean length of stay	Mean age	Mean length of stay	Mean age
27C061	6	11,8	50,6	13,0	47,7
27C062	42	17,3	52,4	16,7	50,1
27C063	15	31,3	44,2	22,9	53,3
27C064	7	15,3	39,4	29,0	51,1

N number of stays, *DRGs* according to the French classification which gives four levels of severity (increasing order from 1 to 4) for these DRGs dedicated to renal transplantation

Responses

R1

Identify diagnoses and procedures related to the stays in the different clinical units.

- Nephrology/Transplant ward:
    Principal diagnosis (PD): end-stage renal disease; *Associated diagnosis (AD)*: Alport syndrome; *Procedure*: renal transplantation.
- Intensive care unit:
    Principal diagnosis: Acute renal failure due with acute tubular necrosis; *Procedure*: two sessions of hemodialysis for acute renal failure.
- For the continuous monitoring care unit:
    PD: postoperative surveillance; *AD*: HBP due to a nephropathy, infected perirenal urinoma, infection due to enterobacteriaceae, calyceal injury due to a JJ catheter, hemorrhagic duodenal ulcer, malnutrition.
    Procedures: renal ultrasonography, surgical drainage, retrograde pyelogram, upper gastrointestinal endoscopy, total colposcopy, enteral nutrition then parenteral nutrition.
- Nephrology/transplantation ward:
    PD: postoperative surveillance; *AD*: HBP due to a nephropathy, duodenal ulcer.

R2

Diagnosis	CIM 10 code
End-stage renal disease	N18.5
Alport's syndrome	Q87.8
Acute renal failure with acute tubular necrosis	N17.0
Postoperative surveillance	Z48.9
HBP linked to a nephropathy	I12.0
Infected perirenal urinoma	N15.1
Enterobacteriaceae infection	B96.8
Calyceal injury caused by a JJ catheter	S37.00 + T83.8
Hemorrhagic duodenal ulcer	K26.0
Duodenal ulcer	K26.3
Severe denutrition	E43

(continued)

Responses (continued)

R3

Procedure
Kidney transplantation
Haemodialysis session for acute renal failure
Renal ultrasonography
Surgical drainage of a perirenal collection
Retrograde pyelogram
Upper gastrointestinal endoscopy
Complete coloscopy (crossing the ileo-colic orifice)
Enteral nutrition
Parenteral nutrition

R4

For the facility data, mean age is lower for stays of severity levels 3 and 4 (27C063 and 27C064 DRGs). Mean length of stay is higher for severity level 3, and lower for severity level 4.

When comparing facility data to national data for each French DRG, we noted that:

- For severity levels 1 and 2, the facility has results similar to the national data for both duration and age.
- For severity level 3, stays are longer and mean age is lower for the facility than for national data. One may wonder whether there is a relationship between age and length of stay.
- Severity level 4: for the facility, the length of stay is lower than for severity level 3 (but pay attention to the restricted number of stays), while for the national data the length of stay lengthens with the level of severity. It would be interesting to have additional information such as the percentage of deaths or early discharge by transfer to another facility.

This picture raises the question of comparability to a reference activity when the numbers are low and in the absence of additional information.

References

Australian Government (eds) (2010) A national health and hospitals network for Australia's future. Publications number: P3-6430. Commonwealth of Australia pp 1–74

Busse R, Riesberg A (2004) Health care systems in transition: Germany. Copenhagen, WHO Regional Office for Europe on behalf of the European Observatory on Health Systems and Policies Vol 6, No 9

Busse R, Geissler A, Quentin W, Wiley MM (eds) (2011) Diagnosis related groups in Europe: moving towards transparency, efficiency and quality in hospitals. Open University Press/ McGraw-Hill, Maidenhead, p 458

Cash R, Grignon M, Polton D (2003) L'expérience américaine et la réforme de la tarification hospitalière en France – commentaire de l'article de J Newhouse. Economie publique 13:35–47

Commission des comptes de la sécurité sociale (eds) (2009) Comparaison internationale des dispositifs de fixation des tarifs d'activité des établissements de santé pp 142–145

Cylus J, Irwin R (2010) The challenges of hospital payment systems. Eur Obs 12(Autumn):1–3

Dexia and HOPE (2008) Hospitals in the 27 member states of the European union. Dexia Editions, France

DREES (eds) (2009) Second rapport d'activité du Comité d'évaluation de la T2A. Document de travail No 94

Ellis R, Vidal-Fernandez M (2007). Activity-based payments and reforms of the English hospital payment system. Health Econ Policy Law 0:1–10

Ettelt S, McKee M et al (2009) Investing in hospitals of the future In: Rechel B, Wright S, Edwards N, Dowdeswell B, McKee M (eds) European Observatory on health systems and policies. Observatory studies series No 16

Farrar S, Sutton M, Chalkley M et al (2009) Has payment by results affected the way that English hospitals provide care? Difference-in-differences analysis. BMJ 339:1–8

Fürstenberg T, Zich K et al (2010) G-DRG-Impact evaluation according to sec. 17b para.8 Hospital Finance Act. Final report of the first research cycle (2004–2006) German Institute for the Hospital Remuneration System (InEK) (eds) pp 1–13

Hensen P, Beissert S et al (2008) Introduction of diagnosis-related groups in Germany: evaluation of impact on in-patient care in a dermatological setting. Eur J Publ Health 18:85–91

Herwartz H, Strumann C (2012) On the effect of prospective payment on local hospital competition in Germany. Health Care Manag Sci 15(1):48–62

IRDES (eds) (2009) Principes et enjeux de la tarification à l'activité à l'hôpital (T2A). Enseignements de la théorie économique et des expériences étrangères. Document de travail. DT No 23, pp 1–29

Langenbrunner JC, Cashin C, O'Dougherty S (eds) (2009) Designing and implementing provider payment systems. How-to manuals. World Bank, Washington, DC

Mason A, Smith PC (2006) Regulation and relations between the different participants in the English health care system. Revue Française des Affaires Sociales 2:265–284

Ministère de la santé et des sports (ed) (2010) Rapport 2010 au Parlement sur la tarification à l'activité (T2A) pp 1–56

Miraldo M, Goddard M, Smith P (2006) The incentive effects of payment by results, CHE research paper 19. Centre for Health Economics, University of York

Neubauer G, Pfister F (2008) DRGs in Germany: introduction of a comprehensive prospective DRG payment system by 2009. In: Kimberly JR, de Pouvourville G, D'Aunno T (eds) The globalization of managerial innovation in health care. Cambridge University Press, New-York, pp 153–175

O'Reilly J, Busse R et al (2012) Paying for hospital care: the experience with implementing activity-based funding in five European countries. Health Econ Policy Law 7(1):73–101

Or Z (2010) Activity based payment in hospitals: evaluation. Health Policy Monitor. Available at: http://www.hpm.org/survey/fr/a15/3

Or Z, Häkkinen U (2012) Qualité des soins et T2A : pour le meilleur ou pour le pire ? IRDES Paris. DT No 53, pp 1–20

Paris V, Devaux M, Wei L (2010) Health systems institutional characteristics: a survey of 29 OECD countries. OECD health working papers No 50, OECD Publ

Rosenthal MB, Landon BE et al (2006) Pay for performance in commercial HMOs. N Engl J Med 355(18):1895–1902

Scheller-Kreinsen D, Geissler A, Busse R (2009) The ABC of DRGs. Eur Obs 11(4):1–5

Street A, Maynard A (2007) Activity based financing in England: the need for continual refinement of payment by results. Health Economics, Policy and Law 2(Pt 4):419–427

Street A, O'Reilly J, Ward P, Mason A, Street A, O'Reilly J, Ward P, Mason A (2011) DRG-based hospital payment and efficiency: theory, evidence, and challenges. In: Busse R, Geissler A, Quentin W, Wiley M (eds) Diagnosis related groups in Europe. Moving towards transparency, efficiency and quality in hospitals. Open University Press, Maidenhead, pp 93–114

Sussex J, Farrar S, PbR team (2008) Payment by results. Evaluation newsletter 3. Office of Health Economics, London

Vogl M (2013) Improving patient-level costing in the English and the German 'DRG' system. Health Policy 109:290–300

Wiley M (2011) From the origins of DRGs to their implementation in Europe Diagnosis Related Groups in Europe – moving towards transparency, efficiency and quality in hospitals. In: Busse R, Geissler A, Quentin W, Wiley M (eds) Diagnosis-Related Groups in Europe: moving towards transparency, efficiency and quality in hospitals. Mc Graw Hill, Copenhagen, pp 3–8

Chapter 10
Public Health Decision Support

C. Jacquelinet, I. Belhadj, F. Bayer, E. Sauleau, P. Lévy, and H. Chaudet

Abstract This chapter presents potential uses of information and communication technologies (ICTs) for public health, with a special focus on decision-making and decisional support systems. An overview of public health sub-domains is given as potential functional areas for ICTs and business intelligence. The specifics and principles of decision-making for public health policy are explained. Main methods and observational tools used in epidemiology are overviewed. Relevant decisional support architectures are presented, ranging from data warehouse techniques, geographical information systems and simulation tools. The relevance and reusability of medico-administrative data sources is discussed, and balanced with data from registries and cohort studies in the perspective of evidence-based decisional support systems for public health. Last, public health decisional support systems are discussed as tools for translational research in public health.

Keywords Public health • Decision support system • Screening • Prevention • Health promotion • Environmental health • Occupational health • Nutrition • Health safety • Drug administration • Pharmacovigilance • Transplantation • Transfusion • Evidence-based decision • Uncertainty • Epidemiology • Registry • Cohort • Decisional support systems • Data warehouse • Simulation • Geographical information • Translational research

C. Jacquelinet (✉)
Agence de la biomédecine, Direction Médicale et Scientifique, 1, avenue du Stade de France, Saint Denis la Plaine 93212, France
e-mail: christian.jacquelinet@biomedecine.fr

I. Belhadj • F. Bayer
Agence de la biomédecine, 1, avenue du Stade de France, Saint Denis la Plaine 93212, France

E. Sauleau • P. Lévy
Département de Santé Publique des HUS, Hôpital Civil, 67091 Strasbourg, France

H. Chaudet
Faculte de Medecine, UMR 912 - SESSTIM - INSERM / IRD/ Aix-Marseille Université, 27, Bd Jean Moulin, 13385 Marseille, France
e-mail: herve.chaudet@univ-amu.fr

A. Venot et al. (eds.), *Medical Informatics, e-Health*, Health Informatics,
DOI 10.1007/978-2-8178-0478-1_10, © Springer-Verlag France 2014

After reading this chapter you should be able to:

- Define health and list factors influencing individual and population health status;
- Define health promotion, environmental and occupational health; explain what you can expect from the development of environmental geographical information systems;
- Define the concepts of health security, surveillance, vigilance and alert; specify main ICTs functional requirements for these domains;
- Describe main types of epidemiological studies and observational instruments; cite potential uses of ICTs for epidemiology;
- Define incidence and prevalence of disease;
- Know main steps of drug development and uses of ICTs in this context;
- Be aware of the complexity and variety of situations in public health decision-making; know about principles driving decision making under uncertainty conditions;
- Know main characteristics of a decisional support system and points that matter for health decisional support systems;
- Be aware of interest and relevance of simulation models;
- Explain relevance of decisional support system for translational research.

10.1   Introduction

The use of information and communication technologies (ICTs) to support public health decision making is likely to gain importance for the conception, implementation, monitoring and evaluation of public health policies.

Section 10.2 of this chapter presents the multiple aspects of public health. They appear as possible functional areas for ICTs and business intelligence. They can be considered as "business processes" governing the choice of data to collect, the range of possible decisions, and the functionalities expected from the information systems to be built.

Section 10.3 exposes the specifics of decision making in public health. Section 10.4 describes the methods and observation tools coming from epidemiology since they affect how to organize and aggregate data, build relevant indicators, analyze data, and use results to support decision making. Section 10.5 examines some relevant of ICTs for public Health Decision Support. The final section 10.6 discusses the relationship between Public Health decision support and translational research.

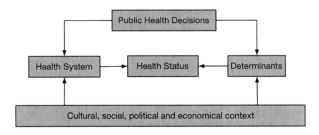

Fig. 10.1 Factors influencing health status

10.2 Public Health Components as Functional Application Domains for ICTs

Health is defined by the World Health Organisation (WHO) as "a state of complete physical, mental and social well-being and not merely the absence of disease or infirmity". Schematically (Fig. 10.1), we can consider that the health status of a population and its individuals is influenced by:

- A set of determinants comprising innate or acquired individual risk factors interacting with environmental, social or cultural risk factors;
- The organization of the health system, including curative and palliative care, as well as screening, prevention and health promotion;
- The cultural, social, political and general economic situation; and
- Public health decisions.

Public Health definitions vary according to which aspect is emphasized. It was initially defined as the "science and art of preventing disease, prolonging life and promoting health through the organized efforts of society" (Graham 2010). Others defined it as a "combination of knowledge, know-how, practices and legal rules that aim to understand, explain, preserve, protect and promote the health status of individuals and populations" (Bourdillon et al. 2007).

To account for the diversity of fields, disciplines, actors, occupations and institutions involved, Public Health could also be defined as an application of humanities and social sciences to human health. Public Health becomes then a domain with multiple aspects (Winslow 1926) related to sociology, anthropology, ethics, economics, politics, law, administration, demography, history and geography.

The American Medical Informatics Association defines Public Health Informatics as "the application of informatics in areas of public health, including surveillance, prevention, preparedness, and health promotion. Public health informatics and the related population informatics, work on information and technology issues from the perspective of groups of individuals. Public health is extremely broad and can even touch on the environment, work and living places and more" (AMIA 2013).

10.2.1 Health System and Healthcare Organization

Health systems comprise all actions, resources and organizations that societies implement to maintain the health status of their populations (WHO 2007a). Health systems perform four essential functions: service delivery, resource creation, financing, and administrative management.

Depending on the level of development and the economic model in place, the respective roles of states, national and regional authorities, private and public economic actors, and individuals will be highly variable. In underdeveloped countries, maternal and child health, malnutrition and infectious diseases are usually the main issues, in addition to the problems of safety of persons and property in the event of political instability. In emerging countries, morbidity and mortality from chronic diseases gradually supplant maternal, infant and infectious disease morbidity and mortality. This phenomenon is a corollary to the so-called "health transition" (Frenk et al. 1991). In developed countries, problems related to nutrition (obesity), aging of the population and control of health expenditures are often the major concerns. Health systems try to achieve three main objectives: (1) control of the medical and non-medical determinants of population health status, through screening, prevention, promotion of health, occupational medicine and environmental policy; (2) management of health problems (functional disorders, diseases, disabilities) in providing and organizing non medical and medical, diagnostic and therapeutic healthcare, as well as curative, replacement and palliative treatment; (3) control of access to health care, including underlying logistical, geographical, social, and economic dimensions.

Healthcare systems vary considerably from one country to another. The manner in which the creditworthiness of the health care consumer is managed is also a fundamental in health care systems, as it affects access to care. 'Socialized medicine' systems rely on broad or universal health insurance (Hussey and Anderson 2003; Danchin et al. 2011). The importance of ICTs to support public health decision must be appropriate to the sanitary, economic and technological situation, and to the expected added-value. The ability to measure and predict temporal and spatial variations of populations health needs is a key point in the healthcare organization.

10.2.2 Screening, Prevention, Health Promotion

Screening aims to identify, by means of simple tests, people with latent health problems (disease or individual risk factors). Tests used for screening usually prioritize the sensitivity, to minimize false negative rate and misdiagnosis, over specificity, to minimize false positive rates and over-diagnosis (Altman and Bland 1994). Screening can be individual or part of a program targeting a population or subgroup. The WHO has proposed criteria to determine the appropriateness of a screening program, recently revised in (Andermann 2008). Disease must be detectable at an early stage, when it is still latent. It must be an important health problem. Screening tests must be available and acceptable to the population. Diagnosis confirmation methods and facilities must be available. Screening cost must be balanced in relation to healthcare expenditure. Case-finding should be a continuing process. A treatment must be available for recognized disease. Screening (case-finding) is also used in the context of infectious diseases control and may lead to measures of isolation. Screening may raise ethical issues in human genetics, such as genetic counseling, or prenatal and pre-implantation diagnoses in reproductive medicine (Walters 2012).

Prevention includes all medical, environmental, financial, behavioral, legislative and political measures aimed at preventing the occurrence of diseases or health problems, by controlling the cause of disease (primary prevention), its progress (secondary prevention), or by limiting its consequences (tertiary prevention). Next to this categorization, based on the timing of the intervention to the occurrence of the risk, health promotion rather considers: "universal prevention" that applies to an entire population regardless of its health status; "oriented prevention" that applies to people with risk factors; and "targeted prevention" that applies to already sick-patients (Gordon 1983).

Health Promotion has been defined by WHO as "the process of enabling people to increase control over their health and its determinants, and thereby improve their health" (Minkler 1989). It aims to improve individual's and population's ability to control lifestyle, social, economic and environmental determinants of their health status. Health Education informs the population about social, economic and environmental determinants affecting health; about risk factors and high-risk behavior; and about health care system uses.

Therapeutic patient education (WHO 1998) aims more specifically to help patients and their family to understand and manage their disease and their treatment, to cooperate with healthcare providers and to improve their quality of life. Therapeutic patient education includes awareness and information actions, education, and psychological or social assistance related to disease, treatments, healthcare organization and behavior.

In these areas, ICTs and the internet are primarily used to provide appropriate information to either targeted populations or decision makers. They can, for example (www.healthcare.gov/compare/, www.platines.sante.gouv.fr), enable healthcare users to choose among healthcare providers according to relevant clinical performance indicators.

10.2.3   Environmental Health, Occupational Health, Nutrition

Environmental health comprises the study, control, and monitoring of environmental determinants of health. It focuses primarily on the physical, chemical and microbiological quality of the external milieu: soil, water and air, quality of food, exposure to toxins, radiation or noise pollution, and climatic conditions (WHO 1999). In its broadest sense, environmental health also includes social, economic, cultural, and behavioral factors: conditions of housing, stress, fatigue, risky behavior, which fall within the scope of the promotion of health (see above). Some risks related to physical, chemical or microbiological attackers are likely to alter the external environment on a large scale, to involve a large number of individuals and to generate real natural, artificial or industrial disasters. Environmental health then makes the headlines: Tsunami, earthquakes, heat waves or industrial accidents such as Bhopal or Seveso. Environmental health has often to do with media controversy: e.g. low frequency electromagnetic waves, adulterated oil, pesticides, urban air pollution, asbestos or mad cow disease. This can be related to the difficulty to quantify some health risks, or to relate them formally and quickly to potential environmental factors. This can also be due to the responsibility of public or private authorities concerning the exposition of populations to known environmental threats.

The management of environmental risks, in or outside a health crisis, and decision making in environmental health are anyway difficult. The case of hemolytic uremic syndrome due to verotoxin-producing *Escherichia Coli* (Corogeanu et al. 2012) or the case of mad cow disease show that environmental public health decisions are indeed frequently taken in a context of health crisis, strained by multiple, national and international, political and economical contradictory interests.

Environmental health is a domain on the move, which will benefit from the contribution of ICTs and from progress in spatial modeling and health geography. The creation of databases to record spatial and temporal variations of certain indicators related to pollution, industrial or natural environmental risks and socio-economic conditions of populations is an important step. It is a prerequisite for establishing correlations with health indicators from epidemiological registries or cohorts and for identifying, confirming or denying a relationship between a health event and a potential environmental factor. Such an approach will gradually make decision in environmental health more reliant on facts and evidence.

Occupational Health is part of the Health Promotion process. It aims to improve healthy working conditions by controlling risks related to the occupational environment: workplace injuries, work-related illness, musculoskeletal disorders, stress, depression or suicide. It is a comprehensive, multidisciplinary and integrated approach where prevention, information, hygiene, ergonomics, and safety of working conditions are important topics. Adapting workstation to the capabilities and health of workers, and treatment of occupational diseases are also part of the mission of occupational medicine. Occupational Health therefore has political, social and managerial implications (Vanhoorne et al. 2006).

Occupational health information systems in use do support primarily monitoring of employees by occupational medicine, survey of the risks to which employees are potentially exposed, monitoring of their compliance with measures of prevention and safety, and reporting of work-related injuries. These IT systems can produce reports required by the authorities, statistics on work-related injuries and absenteeism due to health reasons.

National or regional occupational health information systems should provide a more appropriate support for public health decision making. They can integrate data on occupational risks by workstation and achieve a first level of health surveillance about work-related injuries and illnesses, for those occurring at short or medium terms.

Indeed occupational diseases occurring after a long latency period, even after the cessation of activity, present more complex challenges for surveillance and remain an issue. In this situation, the integration of data on the professional history of individuals within diseases registries or within the information system of the health care system could become support for the surveillance of occupational risks. Pre-existing information sources can also be used for retrospective cohort studies or case–control studies to confirm or refute the responsibility of an occupational risk factor for the occurrence of a health event.

Nutrition acts either as protective or risk factor in many diseases: obesity, diabetes, lipid disorders, cancer, cardiovascular disease, high blood pressure. To assess the relative influence of nutritional, genetic, biological and environmental risk factors on the onset and progression of a disease, is a key issue to set up nutritional recommendations for prevention and to improve the health state of a population. The recent use of the internet for the Nutrinet study in France is an interesting example of the contribution of ICT to achieve a large cohort study in the general population with direct involvement of volunteers, referred to as "nutrinautes" (www.etude-nutrinet-sante.fr).

10.2.4 Health Safety, Surveillance, Vigilance and Alert

Health Safety aims at protecting people from health threats related to food, infectious diseases and environment (WHO 2007b).

Health Surveillance has become a pillar for decision-making and strategy in the field of Health Safety. Initially centered on the epidemiological surveillance of communicable diseases, health surveillance is now defined as a continuous observation of the health status of the population and its determinants (Bernstein and Sweeney 2012).

Health Surveillance focuses on diseases, on their distributions and variations in time and space, innate or acquired, acute or chronic. It includes sick people and vulnerable or at-risk populations as well. It is also interested in the quality of the environment and food. The practical organization of monitoring depends on the topic of interest and the type of health phenomenon to watch. It must be set up as to meet specified information needs. Surveillance information systems must provide functionality for: the registration of data relevant to the targeted objectives; management of data quality; analysis and interpretation of data; and the decision-making support.

Main national and international health statistics has focused for a long time on the medical causes of death, which are the historical pillar of health surveillance.

Infectious diseases surveillance depends on the evolutive characteristics of the disease, acute or chronic, sporadic or epidemic. This requires defining the cases and how they can be notified and confirmed. For certain diseases and in certain location, a notification to public health authorities is mandatory; some other diseases are monitored through the establishment of a specific sentinel network based on GPs (influenza, measles, chicken pox) or specialists, or on microbiological reference laboratories network. In France for example, except child lead poisoning, all mandatory notifiable diseases are infectious diseases. Surveillance of emerging diseases is critical in order to be able to identify the occurrence of a new disease in the population (e.g. the case of AIDS in the 1980s), a change in the phenotype of a known pathogen (resistance to antibiotics, acquisition of an epidemic severity factor), or a change in the biotope or in the communication vector. Specific monitoring is undertaken in many developed countries for hospital-acquired infections and rare diseases. A register is an ideal instrument for the surveillance of chronic diseases. Certain vulnerable groups may also be under surveillance for specific risks: pregnancy, newborns, children or the elderly. Occupational health and environmental surveillance have been addressed in Sect. 10.3.

Health Vigilance is a component of health surveillance devoted to incidents and adverse events related to health products. It is referred as pharmaco-vigilance for drugs, material vigilance for medical devices, bio-vigilance for human body elements (Pruett et al. 2012) and reacto-vigilance for in vitro diagnostic devices. It is also interested in adverse reactions to cosmetics and toxic effects of non pharmaceutical chemicals.

Health Alert is another component of health surveillance, enabling early detection of any abnormal health event presenting a threat to public health, thus allowing measures appropriate for the protection of the population. A Health Alert system comprises the collection, verification and analysis of data of varied nature. It must enable: the quick implementation of measures to control the risk; realization of further investigations to confirm an abnormal health event and its causes; and transmission of the alert at national or international levels if needed.

10.2.5   Drug Policy, Administration and Pharmacovigilance

A drug is a substance claimed to have curative or preventive properties related to some human or animal disease; or be available for medical diagnosis; or modify certain physiological functions by exerting a pharmacological, immunological or metabolic action. Among with human body products, medical and diagnostic devices, drugs are very important health products in modern healthcare systems. The design of drugs, their manufacture and their life on the market are highly controlled in developed countries which generally impose an approval procedure and pharmacovigilance controls. Drug policy comprises all actions and organizations to promote the availability of safe and effective medicines, including economic and financial aspects and constraints, especially in socialized health care systems.

The development of a new drug goes through several phases (NIH 2013). The preclinical phase includes toxicology, pharmacokinetics and pharmacodynamics studies in the animal. It permits to establish the potential efficacy and safety conditions of a new drug. Preclinical phase is completed by pharmaceutical formulation and marketing studies before considering possible use in humans. This step is crucial for rare diseases which often became orphans of a potential drug whose development costs exceeded its marketing possibilities.

Clinical research on drugs permits establishment and verification of data related to pharmacokinetic (absorption, distribution, metabolism and excretion of the drug), pharmacodynamic (mechanism of action of the drug) or therapeutic (efficacy and tolerance) properties of a new drug or a new use of a known medicament in humans. A clinical trial can be realized in sick or healthy volunteers. In the majority of countries, the study protocol of a clinical trial must be submitted to an ethical committee and get an approval of competent authorities.

Phase I clinical trials usually include 20–80 healthy volunteers and determine especially the maximum tolerated dose.

Phase II clinical trials focus on 30–300 people. They consist in more detailed pharmacokinetic and pharmacodynamic studies in healthy volunteers, in some subgroup with specific conditions (aged, renal or liver insufficiency) or in patients with the targeted disease. They help to determine the optimal conditions for the prescription of a drug in a specific formulation: dosage, rhythm of administration and duration and to decide whether the promising efficacy of the substance justifies to move to the next phase of development.

Phase III clinical trials evaluate the efficacy of drugs. The approval of a drug is based on the evaluation of (1) its chemical, biological, and microbiological quality, (2) its safety profile: frequency and severity of side effects, precautions and contraindications, and (3) its expected efficacy and efficiency: therapeutic effects claimed by the industrialist, indications and recommended dosages.

The drug approval is a crucial decision because it embodies the recognition of a new drug by national or supra-national competent authorities and justifies its availability for professionals and patients. The drug approval has no impact on its level of reimbursement by health insurances. The approval is issued after notice of a commission of independent experts with relevant expertise to hear data provided by the manufacturer. Phase IV studies come after the drug approval, including all pharmacovigilance studies. Despite all the efforts made to control risks, ensure transparency and the independence of the experts, recent news show that decision-making in the field of drug policy remains a difficult and sensitive issue (Blake et al. 2012). Organizing pharmacovigilance and pharmacoepidemiology remains also a topical issue in many countries (FDA 2005, Lis et al. 2012).

ICTs are widely used by the pharmaceutical industry and clinical research organizations (CRO) to support the logistics of multi-centre and often international trials, to manage the randomization of patients and the collection of data using electronic case report forms (eCRF), to control the quality of data and manage queries from the data-management, to perform the statistical analysis and the publication of the results (FDA 2007). Meta-analysis syntheses the results of trials related to a same treatment (www.cochrane.org) and are useful for decision making. ICTs have come recently to help to improve transparency and decrease publication bias on the negative trials or on the side effects.

Manufacturers are urged to register all new clinical therapeutic trials in national or international public registers (http://clinicaltrials.gov/, https://www.clinical-trialsregister.eu/). The major international scientific journals also agreed to require this registration as a prerequisite to any publication of the results of a trial (Zarin et al. 2011). Pharmacovigilance studies may require an ad-hoc collection of data or reuse data from existing information systems, registries or cohort studies. The epidemiological observation instruments reported below can also be used. Last, medico-economic studies are essential to assess the medical service and cost-effectiveness ratios of new drugs.

10.2.6   Transplantation of Organs, Tissues, Cells and Blood Transfusion

Blood centers, tissue banks, cell banks and organizations responsible for the procurement and the allocation of organs all have in common to process human body products and to deal with: the quality of procedures related to transformation, preservation and storage; and the traceability, biovigilance and safety (Pruett et al. 2012). The shortage of resources does not always permits to supply in time population needs which can also be constrained by rules of compatibility between donor and recipient. Social, ethical and legal considerations governing these activities are important matter. Due to the context of organ shortage, decisions or changes in the allocation of organs raise sensitive issues. ICTs are also widely used in this field to notify all potential donors, to register the recipients on the waiting list, to allocate retrieved organs, to follow up patients outcomes and to evaluate transplantation results (Strang et al. 2005). ICTs also offer support to traceability and transparency.

10.3   The Specifics of Decision Making in Public Health

To support decision making in public health, the ideal situation is to provide decision-makers with relevant, high level of evidence, scientific or technical knowledge summarized from medical, biological, economic, or social literature, and with reliable health statistics relevant or transposable to the population and to the issue involved (Victora et al. 2004). Health statistics are indeed crucial to anticipate changes in healthcare needs and supply, to assess the performance of health care systems and public health policies, to simulate alternative scenarios and to manage change.

Apart from clinical research where it is possible to establish a causal link between a health status and experimentally controlled factors (Greene 2009), public health statistics usually rely on observational data, allowing at best to establish an association between medical or non-medical factors and a given health status. For many determinants, health consequences are difficult to establish: this is the case, for example in environmental health and occupational medicine for chronic exposure to toxicologically low doses of a given substance or for diseases that occur after a long period of latency after exposure. It is therefore often on the basis of a beam of scattered and incomplete data, and according to experts that policy guidelines and graded recommendations are established (Black 2001; GradeWorkingGroup 2004).

In situations of uncertainty, four key principles guide the decision-making in public health. The "precautionary principle", derived from the environmental law, considers that "the absence of certainty, given scientific and technical current knowledge, should not postpone the adoption of effective and proportionate measures to prevent a risk of serious and irreversible damage to the environment at an acceptable economic cost" (Florin 1999). The "assessment principle" aims to ensure that professionals, health authorities and managers have a capacity of detection and analysis of risks and the ability to assess a priori and a posteriori their action.

The "principle of impartiality" of health decision requires the management of conflicting interests, stakes and stakeholders. This has consequences on the organization of health institutions and administrations, leading to separate services responsible for the development of an activity of those responsible for control, inspection, funding or resource allocation. This affects also the organization of medico-technical expertise. The "principle of transparency" is intended to promote an adversarial debate and the expression of multiple views to try to make the best (or least bad) decision in a context of uncertainty.

10.4 Methods and Observation Tools in Epidemiology

Providing a better understanding on the morbid condition distribution in a population, epidemiological knowledge is indeed a major support for evidence-based public health decisions (Davis et al. 2012). Epidemiology is interested in the distribution of health states and events and their medical and non-medical determinants in the population (Rose and Barker 1978a).

Descriptive epidemiology permits to quantify the importance of a health problem through its frequency, to monitor its temporal and spatial variations and to situate it by its distribution according to different characteristics of the population.

Frequency measures used in epidemiology are rates that require a precise definition of cases to be counted in the numerator and of the population from which they are derived to be counted in the denominator (Szklo and Nieto 2012). Incidence and prevalence are the two major relevant metrics. Incidence measures the rate of occurrence of a health event over a period of time in an exposed population. Prevalence is an indicator of morbidity which measures the frequency of health one given day in a defined population. When death is the studied health event, incidence measures the mortality rate. When health event is a disease occurrence, incidence is an indicator of morbidity (Rose and Barker 1978b).

Analytical epidemiology attempts to identify disease risk factors and contributes to etiological research. It uses multivariate statistical models to test the existence of a statistical link between an explained variable measuring an health state or event and a set of explanatory variables measuring risk factors and potential confounding factors one wants to control. Fitting a model to observed data uses optimization techniques that allow to estimate model parameters. The predictive value of a model fitted over a training data set can be evaluated over a set of validation data. One then enters the field of predictive epidemiology. Clinical epidemiology aims at producing from observational data and multivariate models, prognostic and diagnostic scores that could be used by clinicians for decision support. Geographical epidemiology more specifically focuses on spatial variations and their modeling.

Four main observation instruments are used in epidemiology (Pierce 2012):

- Cross sectional studies, so-called prevalence studies, the faster to achieve, are used to calculate prevalence rates, to study their variations, to quantify the importance of a health problem and to formulate hypotheses on the related risk factors.
- Case–control studies are also relatively easy and quick to perform; they can be retrospective if unbiased databases are already available; they compare the exposure to a risk factor between cases and controls and calculate an excess risk.
- Cohort studies may be prospective or retrospective, and closed or open as the inclusions are limited or renewed in time; they are studying the outcomes of a group of people over time, taking account of potential confounding factors and suspected risk factors; they are heavier, longer and more expensive to realize.
- Registries perform the continuous, long-lasting and comprehensive registration of a given health event in a population and generally include the active or a passive (by merging data with other databases) follow-up of identified cases as in cohort studies (Dos Santos Silva 1999). Although repeated point prevalence surveys may allow in some situations to get a picture of a health problem and to monitor its evolution, registries and cohorts are more relevant instruments if one needs to sustainably monitor a given health process or study a decision-consistent health care segment.

Registries and Cohorts

A registry organizes the continuous and comprehensive registration of a health event occurring in a population defined on a geographical basis: for example, the occurrence of digestive cancer within the population of a region (Jensen et al. 1991). A registry is designed to provide new insight about a health event defined by the identifiable occurrence of a significant event affecting the health of an individual. Many registries still work with paper for the notification of cases. Once the database constituted, patient follow-up can be passive, by data merging with existing demographic, medical or administrative databases.

Registries are interesting not only for epidemiological knowledge they produce on the investigated health condition. They also provide a support for decision making on the involved health care segment as soon as they also record influencing factors and indicators related to treatment modalities, quality of care, clinical performance and patients outcomes.

ICTs offer relevant tools for registries: a dynamic web application accessible via a secure portal is an interesting mean to organize data flow to the database with a decentralized data input. This implies in practice to develop a specification for an epidemiological electronic case report form accessible via the internet. Modules for the control of data quality and data-management can be added.

A cohort (Breslow and Day 1987; Dos Santos Silva 1999) is a group of patients or subjects exposed to one or more risk factors, whose outcome is monitored over a period of time appropriate to the risk of occurrence of events one wants to study. The inclusion criteria defining the cohort are various: generation born during a period of time, subjects with a specific occupation, subjects with a given exposure or with some special characteristics, or even subjects enrolled on a geographical basis and included over a given time period.

A cohort can be prospective or retrospective, respectively, depending on whether the study protocol is designed before or after the data collection. The prior existence of medical information systems facilitates retrospective cohort studies by limiting the risk of bias. Sometimes one needs to integrate information from several sources to get the required inclusion criteria, the targeted risk factors, adjustment variables and studied events. A cohort can be open or closed depending on whether inclusion of patients is renewed. A cohort study is an ideal instrument for analytical epidemiology to study the link between exposure factors and the occurrence of a specified event. Nowadays cohorts often combine observational data registration to biological sample storage in bio-banks. This approach is designed to assess very quickly the predictive value of emerging bio-markers using already available epidemiological and biological data.

10.5   The Relevance of Information and Communication Technologies for Decision Support

Document and Knowledge Management Systems described in Chap. 3 are widely used to access to technical and scientific knowledge issued from medical, pharmaceutical, biological, economic or social sciences, and used to guide decision making. In many developed countries, authorities or independent public bodies have been set up to formalize literature review, deal with medical expertise, elaborate recommendations, promote evidence-based medicine, and perform medico-economic evaluation of health care strategies. This is the case for instance with the Haute Autorité de Santé in France (www.has-sante.fr), the National Institute for Health and Clinical Excellence in the UK (www.nice.org.uk) or with the Institut für Qualität und Wirtschftlichkeit im Gesundheitswesen (www.iqwig. de) in Germany. Evidence data can be arranged within Health Information Resources management systems (HIR) and accessible to professionals, patients and their families (see e.g. www.health.gov/nhic/). Statistical software is also of great importance in the treatment of health data but this topics is not discussed here. This section specifically develops ICTs which can be used to provide decision makers with epidemiological health statistics.

10.5.1   Architecture of a Decisional Support System

A Decisional Support System (DSS) is an information system defined by its ability to transform raw data into both historical, current or predictive, and synthetic and cross-sectional views of all the activities in a business, to provide them to the relevant users and managers and to facilitate the emergence of new insights in a perspective of operational, tactical and strategic decision support. A DSS is a type of architecture used in the field of Business Intelligence that regroups organization, processes and methods that companies use to collect, maintain, and organize corporate-knowledge.

A data warehouse (DWH) is a kind of DSS where decision relies mainly on enterprise quantitative timed and spatial data series. It is a repository supported by a database used for data integration, analysis and reporting. It is therefore appropriate to support health statistics for public health decision.

10.5.1.1   Data Integration

Strategic data from multiple sources can be integrated and stored in a warehouse (Poole et al. 2002). Data load in the warehouse is a critical, complex, often costly and ad-hoc step in the building of a DWH. It depends on the organization and heterogeneity of data sources, and targeted indicators and evaluation end-points that drives warehouse data structure. Data integration most often requires the transformation of primary data; it usually relies on an extraction, transformation and loading (ETL) process of data from sources to integrate: e.g. data from operational databases, and geographical, referential or terminological resources. As a data warehouse is a tool for data integration, it can be used to constitute a multi-source information system (MSIS) (BenSaïd et al. 2005).

10.5.1.2   Data Storage: Data Warehouse and Data Marts

A data warehouse comprises a database in which strategic data integrated from different sources are transformed, reorganized and aggregated as to facilitate their analysis according to different predetermined "dimensions". A data warehouse can be divided into data marts which have specific thematic focus.

Data are referred to as "non-volatile": once entered the data warehouse, they can no longer be changed or deleted by users so that a query made at different times gives the same results. They are also referred as to "historical" as they permit to maintain the traceability of indicators and decisions made over time. The organization of data is driven by the targeted analyses and analysis axes. Data is organized in a multidimensional structure - a hypercube where each edge represents one of the dimensions according to which one wants to study ("to drill down") a given activity (Chaudhuri and Dayal 1997; Frendi and Salinesi 2003). A hypercube is composed of cells containing one or more quantitative measures (indicators) related to studied activity, such as for example the number of deaths observed a given year in a given district. The temporal dimension can be the month, quarter, year or decade, and geographic dimension be the commune, county or region. One could also drill-down data by age or socio-professional categories.

10.5.1.3   Data Analysis and Reporting

A data warehouse is also provided with query, analysis and reporting tools. Data processing such as "On Line Analytical Processing" (OLAP) is used to consolidate, drill-down, and slice data, as to enable users to analyze data interactively from multidimensional analytical perspectives (Chaudhuri and Dayal 1997). These tools enable previously trained users (1) to interactively select data, analysis dimensions and indicators; (2) to filter data according to relevant criteria and (3) to sort, group, or split them, (4) to compute simple statistics (total, average, standard deviation, . . .), and finally (5) to present results in tables or graphs within business documents. Reporting tools also give less experienced users access to predefined business documents, which they can interactively modify according to certain selection criteria or dimensions.

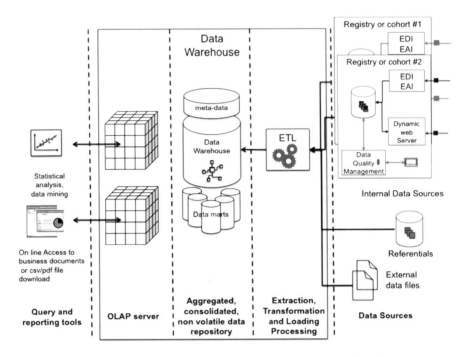

Fig. 10.2 Architecture of a epidemiological evidence-based decisional information system

10.5.1.4   Adaptations and Limitations of DWH for Public Health Decision Making

The application to public health of an architecture initially conceived for business requires some adaptations and encounters some limitations that regional or national health authorities will have to take into account. A public health DWH essentially enables non-statisticians to more easily access to simple descriptive health statistics related to health economics, epidemiology, demography or geography of health. Health statistics enable to produce a picture of population health status and care using quantified and measurable, simple or composite indicators (Fig. 10.2).

Simple indicators (e.g. case numbers, percentages) resulting mostly from counts are easily implantable in a DWH. Some composite indicators such as standardized rates or even survival rates can also be integrated into a DHW. But decisions implying to take into account the influence of medical and non-medical determinants or to control confounding factors will still fall in the scope of analytical and predictive epidemiology and will continue to rely on multivariate statistical analysis. The same holds true for complex composite indicators (i.e. cumulative incidence for competing risks, quality of life, QALY).

Complex composite indicators that can't be computed from original data sources using DWH techniques, can be computed outside of the DWH using statistical software, and integrated into the DWH as external data sources including analytical dimensions. A public health DWH thus covers only a portion of the needs for the decision support. Health DSS project managers and developers must be aware of these limitations and not underestimate the complexity of the public health domain. They also have to know about main simple epidemiological and medico-economic indicators that can be reasonably implemented in a DWH.

10.5.1.5 Epidemiological and Medico-economic Indicators

To assess health needs in a region, epidemiology focuses on incidence or prevalence rates, requiring to count patients who reside in the region, regardless of their place of treatment, and report it to the regional population size for crude rates (number of cases per million inhabitants), or adjust it on a reference population for standardized rates that allow comparisons between regions (Rose and Barker 1978c). To assess resources required to supply patient needs, care providers and public health authorities are much more concerned with the crude number of patients treated in the region, regardless of their place of residence. These quantities are indeed more concrete and easily turned into health expenditure or income: "one does not care standardized rates, but patients". Epidemiological and medico-economic indicators indeed are not contradictory but complementary. Both medico-economic and epidemiological approaches must be carried out to support health care optimization to get a better understanding of a health process and its medical and non-medical determinants. Public Health decision support system designers are faced with ambivalence when the same data are aggregated in a different way to build close indicators, related to epidemiology or health economics. Simple cross-sectional and longitudinal key epidemiological indicators that can be implanted in a DSS are shown in the box below. The main medico-economic indicators for the decision support are discussed in Chap. 9.

> **Simple Cross-Sectional Health Indicators**
>
> Most usual simple cross-sectional indicators rely on the number N_{tD+} of cases presenting a given health status, risk factor, treatment modality or disease D at time t. This number depends on status change flows (Rose and Barker 1978b). It is relevant for health conditions or risk factors longstanding enough with regard to their frequency in the population. It is not relevant for rare and quickly recovering conditions. The picture given of the studied health condition is a snapshot. It can be renewed over time to have a time series $(N_{t1},\ldots, N_{tn})$ giving an idea of the trend.

(continued)

Simple Cross-Sectional Health Indicators (continued)

Health economics will focus for example on the crude number N_{TFtD+} of individuals receiving a treatment T on one day t in a given health care facility F for a given disease D, regardless of the place of residence, as well as the regional total N_{TRtD+} of patients treated within the region for the same disease. These two indicators will enable decision-makers to assess the frequency and health care workload related to a given health problem and anticipate resources to supply according to trends in health needs. The rate N_{TFtD+}/N_{TRtD+} corresponds to the share of healthcare supply covered by the health care facility F for the given disease D within the region R. The number of people living in the region and treated in another region is used to evaluate patient «leakage» rate (patients being treated in another region/district). The number of patients treated in the region but resident in another is used to set the attractiveness rate. In health systems where patients have a free choice of health care facilities, it is not possible to define the reference population affiliated to a facility.

Epidemiology is interested in the frequency of a stated disease in a targeted population P, typically defined on a geographical basis (Szklo and Nieto 2012). It will focus for example on the number N_{RtD+} of patients resident in a given region R, presenting a given disease D on day t or treated for this disease whatever the treatment facility/region on day t. N_{RtD+} defines the number of prevalent cases of D within the regional population R on day t. Either the regional population numbers N_R or the number N_{RtD-} of disease free individuals on day t is required to compute the crude regional point prevalence P_{RtD+} of disease D: $P_{RtD+} = N_{RtD+}/N_R = N_{RtD+}/(N_{RtD+} + N_{RtD-})$.

Point prevalence is a cross-sectional morbidity indicator that can be used to assess health needs and measure the burden of a chronic disease/health status. Indeed, prevalence depends on the kinetics of health status changes and migrations in the targeted population: disease onset, recovery, remission, death, and migration. Prevalence can be adjusted, for example on age and gender, to compare the situation of a region or a country to others, tacking into account variations in sex and age structure of the populations. It is then referred to as the age- and sex-adjusted prevalence. One can also compute a specific prevalence in a sub-population: child, man/woman. The number of prevalent cases N_{RtD+} quantifies regional health needs. If a health care facility F supplies to the needs of N_{FRtD+} prevalent cases resident within the region, then rate N_{FRtD+}/N_{RtD+} corresponds to the proportion of regional population health needs covered by facility F for the disease D.

Simple Longitudinal Indicators

Most usual simple longitudinal indicators focus on the number of cases $N_{\Delta tE+}$ presenting during an observation period Δt a given health event E: birth, occurrence of a disease or disability, start of a health care procedure, healing, remission, occurrence of a complication, relapse, or death. When an event is likely to occur multiple times for the same individual, only one occurrence is registered, usually the first one unless one is interested precisely in recurrences. In epidemiology, the speed of occurrence of a new condition within a period of time in a specified population is referred to as incidence (Breslow and Day 2000). Various indicators are used to compute incidence according to available data.

When the size population is limited and the duration of risk exposure d_i, expressed using a unit of time T, is known for each individual i, one can compute the incidence rate (also known as incidence density or person-time incidence): $I_E = N_{\Delta tE+}/PT$ in pondering the number of cases $N_{\Delta tE+}$ by the sum of risk exposure durations d_i expressed in 'person-time': $PT = \sum d_i$. For example, for 50 at risk persons (i = 1–50) all observed for risk exposure durations $d_i = 2$ years, the sum of risk exposure durations is 100 person.year. If five new cases were observed, then the incidence rate is 5 per 100 person. year.

When the duration of risk exposure is not known for all individuals, as it is often the case for large populations and open cohorts, it is possible to compute an approximation admitting that time to event has a uniform distribution over the period of time and using the size of the population in the middle of the period. The crude incidence rate is then expressed per 100,000 or per million inhabitants (pmh). Such indicators mainly requiring counting tools can be implemented using DHW analytical techniques.

[NB: The instantaneous incidence rate λ_E is the incidence rate when time period Δt tends to 0 (Esteve et al. 1994). It is also referred to as hazard function in survival models. If the observed data are consistent with a constant effect over time, then $\lambda_E = I_E$. The cumulative incidence is the number of cases occurring during the period Δt divided by the total number of people still at risk at the time $t_0 + \Delta t$. These indicators requiring survival analysis tools are usually computed using statistical software.]

Care providers and health economics pay more attention to patients flows (admissions and discharges) on patient active list and length of stay in a care unit, care facility or group of health care facilities. Admissions concern subjects beginning treatment or newly supported for a particular disease, regardless of their place of residence. Admissions are not necessarily incident

(continued)

> **Simple Longitudinal Indicators** (continued)
> cases as they may have been treated before in another facility. Discharges can
> be the end of treatment session, a transfer to another health care facility, and
> include deaths. Cases can also be geolocated to get an idea of their spatial
> distribution, either from their address of residence, either from GPS
> coordinates. Admission and discharge indicators can also be implemented
> using DHW analytical techniques.

10.5.1.6   Health Data Sources

Relevant data sources depend on strategic objectives and involved authorities
legitimacy to access required health data. Unlike industry which usually owns
operational databases populating a DWH, health DSS often integrate external
data sources. Data access must conform to security and confidentiality requirements
depending on national regulation rules, especially when nominative data are
registered and merged from distinct sources (Bernstein and Sweeney 2012). Regulatory arrangements are sometimes made to ensure data access right for public
health authorities, as is the case in France for regional health agencies to access to
any regional medico-administrative data and for the Institut de Veille Sanitaire to
access to any data required for health surveillance.

In the design of a public health DSS, strategic objectives should ideally predominate in the selection of the relevant health care segment and definition of relevant
indicators of whom required data will be derived. External sources from health
facilities, demographic and economics institutes can then be selected. Existing data
sources must be considered, especially those related to hospitals, health resources
and procedure reimbursement/financing by health insurance systems (Tuppin et al.
2010). They are presented in Chap. 9.

When such data are available, a critical question is their reusability. Strategic
objectives can differ from those that motivated data collection. Indeed, data considered as relevant in an operational funding context are neither always necessary
nor sufficient to build an epidemiological, evidence-based public health DSS.
Interests of stakeholders to optimize their resources and income can also lead to
an information bias. Last, the statistical continuity required for sustainable health
DSSs can be compromised by changes in external sources coding referential or data
scheme, driven by too practical objectives, with limited consideration to further
evaluation and statistical analysis (Quan et al. 2008).

When data are not available, or do not allow to compute targeted indicators, do
not meet quality and reliability requirements, do not guarantee statistical continuity,
one then must organize data registration. ICT designers must conform to epidemiological standards used for registries or cohort studies (Dos Santos Silva 1999)
(Fig. 10.3). In view of cost incurred by the building of such epidemiological
observation tools (registries or cohort studies), a critical point is to assign them

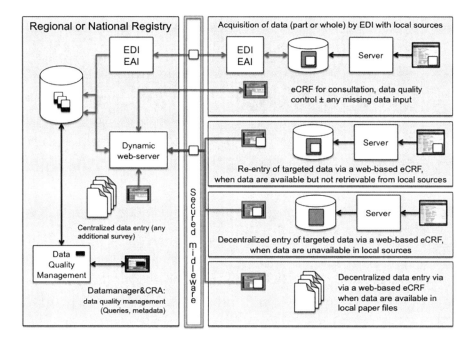

Fig. 10.3 ICTs as a support for registries, with data acquisition modalities relevant to the level of computerization of local data sources. *EDI* electronic data interchange, *EAI* enterprise application integration, *eCRF* electronic case report form, accessible via the internet and a standard browser

valorization objectives for both epidemiological research and public health decision support from their initial conception. Such a comprehensive approach permits to provide public health decision makers with both simple health descriptive statistics and multivariate studies based either on conventional statistical or data mining methods.

10.5.2 Geographic Information Systems

Health status of individuals and populations results from an interaction between humans, genetics, environment and current health care system. The interest of the Geographic Information Systems (GIS) for public health is to provide a better understanding of the spatial distribution of diseases, their determinants and treatment modalities (Robertson and Nelson 2010). The use of GIS for health has disseminated from the early 1990s. They allow health authorities and epidemiologists to improve and facilitate their work on monitoring and/or planning health care needs and supply.

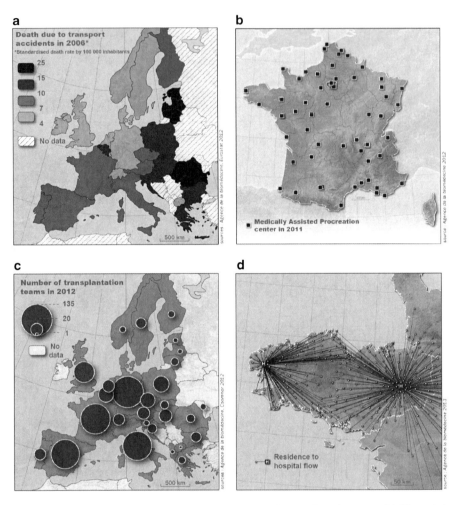

Fig. 10.4 Cartographic representations of health information. (**a**) Location map. (**b**) Choropleth map. (**c**) Proportional circles map. (**d**) Flow map

Descriptive mapping of health care needs and supply provides a useful support for decision-making in public health. Various types of cartographic representation are used: e.g. either location, choropleth, proportional circles or flow maps (Fig. 10.4). Animated maps permit for example to monitor extension of outbreaks in time and space (www.sentiweb.org). Spatial modeling is sometimes required to optimize health care. A better understanding of the geographic variation in diseases frequency or disparities in access to health care also has a significant impact on health policy. Last, geographical databases related to environmental, social or economic characteristics can for example be used to establish ecological correlations with morbid events recorded in registries or cohorts if registered patients are geolocated (Esteve et al. 1994).

The Renal Epidemiology and Information Network Information System

The Biomedicine Agency in France maintains a decisional multi-source information system to support the surveillance of end-stage renal disease (ESRD) treated by dialysis or kidney transplantation (KTx) within the Renal Epidemiology and Information Network (REIN) (Couchoud et al. 2006).

Data are collected manually via dynamic web applications (Cristal for KTx, Diadem for dialysis) accessible via a secured portal (Strang et al. 2005) or automatically via an electronic data exchange (EDI) platform receiving data from dialysis units software. This platform uses a communication protocol implementing the HL7 V3 standard (Worden and Scott 2011). Data from Cristal and Diadem databases as well as demographic and referential data are integrated in a data warehouse and populate several data-marts specialized on incidence, prevalence, and patient course among dialysis modalities and KTx. Reporting tools provide tables, figures and reports giving a transversal views on aggregated data about the registered populations.

REIN-IS is an example of a successful implementation of DSS that promotes epidemiological evidence-based decision-making for public health and professional practices. REIN supports the translation between epidemiological research, public health and professional practices (BenSaïd et al. 2005) that potentiates ESRD management innovations. It is used by regional health authorities to set up and monitor their strategic health care scheme.

Principles emerging from the REIN registry for the design of Public Health DSS based on epidemiological data are: (1) a clear specification of evaluation end-points and indicators to guide data aggregation and data warehouse design, (2) a valid communication standard that takes into account aspects related to semantics and data-quality metadata (3) a precise analysis of business/health care processes and strong collaborations with end-users (4) a compromise between the nature of the use of data and the complexity of model underlying information.

This classical architecture of DSS appeared to be a necessary but not sufficient condition to drive changes in the complex societal, technological, medical, and scientific context of an end stage organ failure treated by transplantation or dialysis – a context with major human, economic, and ethical issues. The building of simulation tools revealed to be a crucial complement: (1) to evaluate "in silico" the potential impact of various change scenarios; (2) to inform the debate with professionals, patients associations, and administrations; (3) and to manage change.

10.5.3   Simulations

Computer simulation allows "in silico" experimentations on the behavior of a system, in our case a disease or health care process, using a model assumed to provide a good reproduction of the reality (Maria 1997). Simulation techniques are used in situations where real experimentations are not possible for ethical, temporal, budgetary or practical reasons. This is often the case in public health where observational data sometimes help to identify conditions to improve; but do not help to decide the type of change to promote; and moreover do not anticipate possible impact so that change might encounter resistance among patients, professionals and involved stakeholders, or lead to undesired collateral events (Elveback and Varma 1965; Fone et al. 2003; Lewis et al. 2013). To reach a scientific value, the simulation model should of course be a close reproduction of real world, and based on relevant theoretical foundations. Simulation is also relevant in a second type of situations with no solid theoretical foundation, and where specifically one aims at developing a theory that will make account of observed data thanks to simulation techniques. One can precisely define the possible practical implications of different theoretical models and determine which provides the best approximation of reality. In health policy, simulation is also a support to decision, to the conceptualization of health process and a tool to help managing change with professionals.

10.6   Summary and Perspectives: Translational Research in Public Health

This chapter gives an overview of potential applications for business intelligence, simulation tools and geographic information systems in public health. These technologies can be combined as it is the case for the REIN registry (BenSaïd et al. 2005). The use of ICTs to optimize public health decision-making ultimately aims at accelerating the dissemination of innovations for the benefit of patients and the maintenance of an optimal health status for the populations. Such an approach is consistent with the "translational research" paradigm in public health.

Translational research was defined (Zerhouni 2005) as an effort to accelerate the application of the most recent researches for the benefit of patients, ensuring a continuum between basic research, clinical research and patients care, and to ensure the dissemination of diagnostic and therapeutic innovations. A broader and bidirectional meaning emerged gradually: observation and bioclinical facts are also the source of questions for basic research. Physiological or pathological mechanisms assumed to reflect some biological facts and/or their importance in vivo will be supported or challenged by their confrontation with the clinical reality.

This feedback from the applied clinical research to basic research is a key aspect in type 1 translational research ("from bench to bedside"). This paradigm applies equally to traditional curative, preventive or predictive medicine (Ziegler et al. 2012). Accelerating the application of medical innovations for the benefit of populations implies to integrate the development of health policy and strategies into the field of type 2 translational research ("from bedside to population") (Rychetnik et al. 2012). Population aging and its consequences on resources to supply for the maintenance of health along gained life years makes the increase of disease free survival a key issue for health policy. The sustainable management of the physiological capital of individuals appears as the ultimate most comprehensive objective for health policy. Extending the scope of translational research to the maintenance of an optimal physiological and psychological status in population is also placing health promotion in the field of translational research.

10.7   For More Information

We recommend that readers consult websites mentioned in this chapter.

Readers from computer science community should especially spend time in reading WHO public health related free documents and IARC epidemiological references.

Readers from medical community should especially spend time in consulting the many websites about decisional support system and data warehouse techniques.

Exercises

Q1 What is the WHO definition of a "well-functioning health information system"?

Q2 From the table below giving number of patients treated for a chronic disease by region, what is the prevalence for each region? (mha = million inhabitants). This disease affects predominantly older female. What should you do to compare regional prevalence rates?

Region	Treated within region on Dec 31	Non resident treated within region on Dec 31	Resident treated outside region on Dec 31	Regional population on Dec 31	Prevalence
R1	130	50	80	3.2 mha	?
R2	320	60	20	7.0 mha	?
R3	190	10	20	4.0 mha	?

Q3 You are designing a data warehouse for a national authority that is in charge of chronic disease survey. The number of treated cases is given by regional registries that populate the DWH. The prevalence table given in

(continued)

Exercises (continued)

Q2 is to appear in one of the business documents. What external data do you need to compute crude and standardized prevalence?

R1 "A well-functioning health information system is one that ensures the production, analysis, dissemination and use of reliable and timely information on health determinants, health system performance and health status." (WHO 2007a)

R2 (pmh = per million inhabitant). Because the disease predominates in female and older patients, one needs to adjust Prevalence on the sex and age structure of the population and to compute standardized prevalence rates before regional comparison.

Region	Prevalence
R1	$(130 - 50 + 80)/3.2 = 50$ pmh
R2	$(320 - 60 + 20)/7 = 40$ pmh
R3	$(190 - 10 + 20)/4 = 50$ pmh

R3 One needs to integrate demographic data into the DWH.

References

Altman DG, Bland JM (1994) Diagnostic tests 1: sensitivity and specificity. BMJ 308:1552

AMIA (2013) www.amia.org/applications-informatics/public-health-informatics. Consulted on 24 June 2013

Andermann A (2008) Revisiting Wilson and Jungner in the genomic age: a review of screening criteria over the past 40 years. Bull World Health Organ 86(4):317–319

Ben Saïd M, le Mignot L, Mugnier C et al (2005) A multi-source information system via the internet for end-stage renal disease: scalability and data quality. Stud Health Technol Inform 116:994–999

Bernstein AB, Sweeney MH (2012) Public health surveillance data: legal, policy, ethical, regulatory, and practical issues. In: CDCs vision for public health surveillance in the 21st century, morbidity and mortality weekly report MMWR 61, Suppl; July 27, pp 30–4

Black N (2001) Evidence based policy: proceed with care. BMJ 323(7307):275–279

Blake KV, Devries CS, Arlett P et al (2012) Increasing scientific standards, independence and transparency in post-authorisation studies: the role of the European Network of Centres for Pharmacoepidemiology and Pharmacovigilance. Pharmacoepidemiol Drug Saf 21(7):690–696

Bourdillon F, Brücker G, Tabuteau D (2007) Définitions de la Santé Publique. in Traité de Santé Publique, 2d Edition; 1–4; Médecine-Sciences Flammarion Ed

Breslow NE, Day NE (1987) The role of cohort studies in cancer epidemiology. In: Breslow NE, Day NE (eds) Statistical methods in cancer research. IARC Scientific Publications, Lyon

Breslow NE, Day NE (2000) Fundamental measures of disease occurrence and association. In: Breslow NE, Day NE (eds) Statistical methods in cancer research, 8th edn. IARC Scientific Publications, Lyon

Chaudhuri S, Dayal U (1997) An overview of data ware-housing and OLAP technology. In: SIGMOD Record Web edition, vol 26(1), pp 65–74. (http://www.sigmod.org/publications/sigmod-record/9703/index.html/?searchterm=Chaudhuri)

Corogeanu D, Willmes R, Wolke M et al (2012) Therapeutic concentrations of antibiotics inhibit Shiga toxin release from enterohemorrhagic *E. coli* O104:H4 from the 11 German outbreak. BMC Microbiol 12(1):1–1

Couchoud C, Stengel B, Landais P et al (2006) The renal epidemiology and information network (REIN): a new registry for end-stage renal disease in France. Nephrol Dial Transplant 21 (2):411–418

Danchin N, Neumann A, Tuppin P et al (2011) Impact of free universal medical coverage on medical care and outcomes in low-income patients hospitalized for acute myocardial infarction: an analysis from the French National Health Insurance system. Circ Cardiovasc Qual Outcomes 4(6):619–625

Davis FG, Peterson CE, Bandiera F et al (2012) How do we more effectively move epidemiology into policy action? Ann Epidemiol 22(6):413–416

Dos Santos Silva I (1999) Cancer epidemiology: principles and methods, Overview of study designs. IARC Scientific Publications, Lyon, pp 83–101

Food and Drug Administration (2007) Computerized systems used in clinical investigations. 2007 May, pp 1–13. Available from www.fda.gov/downloads/Drugs/GuidanceCompliance-RegulatoryInformation/Guidances/UCM070266.pdf. Consulted on 12 Dec 2012

Food and Drug Administration (2005) Good pharmacovigilance practices and pharmacoepidemiologic assessment. 2005 Mar, pp 1–23. Available from www.fda.gov/downloads/Regulatory Information/Guidances/UCM126834.pdf. Consulted on 12 Dec 2012

Elveback L, Varma A (1965) Simulation of mathematical models for public health problems. Public Health Rep 80(12):1067–1076

Esteve J, Benhamou E, Raymond L (1994) Statistical methods in cancer research. IARC Scientific Publications, Lyon

Florin D (1999) Scientific uncertainty and the role of expert advice: the case of health checks for coronary heart disease prevention by general practitioners in the UK. Soc Sci Med 49:1269–1283

Fone D, Hollinghurst S, Temple M et al (2003) Systematic review of the use and value of computer simulation modeling in population health and health care delivery. J Public Health Med 25 (4):325–335

Frendi M, Salinesi C (2003) Requirement engineering for data warehousing. In: Proceedings of the 9th international workshop on requirements engineering: foundations of software quality (REFSQ'03), Klagenfurt/Velden, Austria, 16–17 June 2003

Frenk J, Bobadilla JL, Stern C et al (1991) Elements for a theory of the health transition. Health Transit Rev 1(1):21–38

Gordon RS (1983) An operational classification of disease prevention. Public Health Rep 88 (2):107–109

Grade Working Group (2004) Grading quality of evidence and strength of recommendations. BMJ 328:1490–1494

Graham H (2010) Where is the future in public health? Milbank Q 88(2):149–168

Greene T (2009) Randomized and observational studies in nephrology: how strong is the evidence? Am J Kidney Dis 53:377–388

Hussey P, Anderson GF (2003) A comparison of single- and multi-payer health insurance systems and options for reform. Health Policy 66(3):215–228

Jensen OM, Parkin DM, MacLennan R et al (1991) Cancer registration: principles and methods. IARC Scientific Publications, Lyon

Lewis B, Eubank S, Abrams AM et al (2013) In silico surveillance: evaluating outbreak detection with simulation models. BMC Med Inform Decis Mak 13(1):12

Lis Y, Roberts MH, Kamble S et al (2012) Comparisons of food and drug administration and European medicines agency risk management implementation for recent pharmaceutical approvals: report of the international society for pharmacoeconomics and outcomes research risk benefit management working group. Value Health 15(8):1108–1118

Maria A (1997) Introduction to modeling and simulation. In: Proceedings of the 97 winter simulation conference, Atlanta, 7–10 Dec, pp 1–7

Minkler M (1989) Health education, health promotion and the open society: an historical perspective. Health Educ Behav 16:17–30

NIH (2013) http://www.nih.gov/health/clinicaltrials/basics.htm. Consulted on 24 Jun 2013

Pierce N (2012) Classification of epidemiological study designs. Int J Epidemiol 41(2):393–397

Poole J, Chang D, Tolbert D et al (2002) Common Warehouse Metamodel, an introduction to the standard for data warehouse integration. Wiley, New York

Pruett TL, Blumberg EA, Cohen DJ et al (2012) A consolidated biovigilance system for blood, tissue and organs: one size does not fit all. Am J Transplant 12(5):1099–1101

Quan H, Li B, Duncan Saunders L et al (2008) Assessing validity of ICD-9-CM and ICD-10 administrative data in recording clinical conditions in a unique dually coded database. Health Serv Res 43(4):1424–1441

Robertson C, Nelson TA (2010) Review of software for space-time disease surveillance. Int J Health Geogr 9:16

Rose G, Barker D (1978a) What is epidemiology? BMJ 2:803–804

Rose G, Barker D (1978b) Rates. BMJ 2:941–942

Rose G, Barker D (1978c) Comparing rates. BMJ 2:1282–1283

Rychetnik L, Bauman A, Laws R et al (2012) Translating research for evidence-based public health: key concepts and future directions. J Epidemiol Community Health 66(12):1187–1192

Strang WN, Tuppin P, Atinault A et al (2005) The French organ transplant data system. Stud Health Technol Inform 116:77–82

Szklo M, Nieto FJ (2012) Measuring disease occurrence. In: Epidemiology: beyond the basics, 3rd edn. Jones & Barlett Learning Publishers, Burlington

Tuppin P, de Roquefeuil L, Weill A et al (2010) French national health insurance information system and the permanent beneficiaries sample. Rev Epidemiol Sante Publique 58(4):286–290

Vanhoorne MN, Vanachter OV, De Ridder MP (2006) Occupational health care for the 21st century: from health at work to workers' health. Int J Occup Environ Health 12(3):278–285

Victora CG, Habicht J-P, Bryce J (2004) Evidence-based public health: moving beyond randomized trials. Am J Public Health 4(3):400–405

Walters L (2012) Genetics and bioethics: how our thinking has changed since 1969. Theor Med Bioeth 33(1):83–95

Winslow C-EA (1926) Public health at the crossroads. Am J Public Health 16(11):1075–1085

Worden R, Scott P (2011) Simplifying HL7 Version 3 messages. Stud Health Technol Inform 169:709–713

World Health Organization (1998) Therapeutic patient education. http://www.euro.who.int/document/e63674.pdf. Consulted on 12 Dec 2012

World Health Organization (1999) Overview of the environment and health in Europe in the 90s. www.euro.who.int/__data/assets/pdf_file/0003/109875/E66792.pdf. Consulted on 12 Dec 2012

World Health Organization (2007a) Everybody's business: strengthening health systems to improve health outcomes. www.who.int/healthsystems/strategy/everybodys_business.pdf. Consulted on 12 Dec 2012

World Health Organization (2007b) The world health report 2007: a safer future. www.who.int/whr/2007/whr07_en.pdf. Consulted on 12 Dec 2012

Zarin DA, Tse T, Williams RJ et al (2011) The ClinicalTrials.gov results database – update and key issues. N Engl J Med 364(9):852–860

Zerhouni E (2005) Translational and clinical science – time for a new vision. N Engl J Med 353 (15):1621–1623

Ziegler A, Koch A, Krockenberger K et al (2012) Personalized medicine using DNA biomarkers: a review. Hum Genet 131(10):1627–1638

Chapter 11
Security, Legal and Ethical Aspects of Computerised Health Data in Europe

C. Quantin, F.-A. Allaert, C. Daniel, E. Lamas, and V. Rialle

Abstract The European directive 95/46/CE has defined the legal framework of all personal data collection and treatment and the right of patients about the processing of their personal medical information. The development of telemedicine and domotics are real services provided to the patients but the raising new questions about the share of liability in case of medical litigation and the protection of patients' privacy when a camera is permanently at home. Solutions protecting privacy may be found by using security tools based on cryptography and therefore then main technical principles must be known.

Keywords Liability • Data protection • Security • Personal data • Telemedicine • Patient's right • Cryptography

C. Quantin (✉)
Bourgogne University, 1 Boulevard Jeanne d'Arc, BP 77908, Dijon 21079, France
e-mail: catherine.quantin@chu-dijon.fr

F.-A. Allaert
Health Claim Medical Chair ESC, 29 rue Sambin - BP 50608, 21006 DIJON Cedex, France

C. Daniel
CCS Domaine Patient AP-HP, 05 rue Santerre, 75012 Paris, France

E. Lamas
National Institute for Medical Research (INSERM), 101 rue de Tolbiac, 75654 Paris, Cedex 13, France

V. Rialle
CHU de Grenoble / Pôle de Santé Publique – Responsable de l'UF ATMISS & Laboratoire AGIM (AGe, Imagerie, Modélisation), FRE 3405 CNRS-UJF-EPHE-UPMF/équipe AFIRM, Boulevard de la Chantourne, 38700 La Tronche,

A. Venot et al. (eds.), *Medical Informatics, e-Health*, Health Informatics,
DOI 10.1007/978-2-8178-0478-1_11, © Springer-Verlag France 2014

After reading this chapter, you should, in the framework of the European legislation:

- Be able to cite the main principles that need to be applied for the automated processing of personal information.
- Be able to explain in what context patients can gain access to their personal medical data, and what procedures they need to follow.
- Know the principles that guide the distribution of responsibilities and obligations between the different actors involved in telemedecine
- Be able to cite the different tools and methods that can be used to make health data secure
- Know how to explain in what way technologies for healthcare at home and autonomy open up new opportunities in the care of chronically-ill patients taking the French legislation or applications as an example when possible.

11.1 Introduction: Data Protection and Security of Personal Medical Information: From Law to Security Tools

As doctors are now able to exchange electronic information easily and often work in multidisciplinary teams, they feel a growing need to communicate with others to follow their patients. In most cases, these exchanges of information concern nominative or at least identifiable data, which jeopardizes data security. To ensure that these exchanges can occur while patients' rights are preserved, it is necessary to be familiar with the principal concepts of data security and to define the principles that govern it. Integrity, availability and confidentiality are the three fundamental aspects of data security. Though there is no doubt about the importance of the first two aspects, most attention is paid to the third aspect since the full rigour of the law can be applied if it is not respected. The risk of infringing personal freedoms and invading a person's private life brings into play a legislative arsenal dominated by concerns about protecting patients' rights with regard to medical information. While for doctors, patients' rights seem to translate into obligations and responsibilities.

New information technologies and multimedia communication will probably provide doctors with a tool to achieve better health care. Health care generates and uses a massive amount of information and all of the professionals in the field are on 1 day or another confronted with the difficulty of collecting, managing and communicating information. All are now obliged to be trained in using information technology or at least to find out what IT could bring to their field of activity and responsibilities.

Diffusing personal medical information on telematic networks, that is to say information concerning an identified or identifiable physical person as defined by the European Directive 95/46/CE on the protection of physical persons with regard to the processing and the free circulation of personal data, benefits both doctors and patients. Doctors are able to escape from their usual isolation to make decisions, while patients benefit from better quality health care if the available tools are used appropriately. In most cases, this new way to practice medicine requires contractual provisions that guarantee the respect of patients' rights and define the responsibilities of doctors.

Even though personal medical information is highly confidential, it is in most cases destined to be provided to authorised third parties provided the patient agrees. These authorised third parties can vary: doctors and other health care professionals involved in providing care, doctors who are responsible for making sure health insurance claims are justified, and doctors working as health inspectors. Protection mechanisms, which may be complex (Arney et al. 2011; Chen et al. 2012; Chryssanthou et al. 2011; Das and Kundu 2012; Hsu and Lu 2012; Quantin et al. 2009), must be established to reconcile the notions of security and communication, which appear to be rather contradictory (Kun et al. 2007; Rey and Douglass 2012).

It is impossible to deal with all the legal aspects of data privacy and security in few pages. Therefore we decided to focus on some specific aspects which often worry people who are not familiar with the subject: European directives on data protection, the legal aspects of the re-use of clinical data in the context of international multi-centric clinical research, medical responsibility in telemedicine, the ethical aspects of domotics and an introduction to cryptographic tools.

11.2 Legislation Concerning Privacy and Protection of Personal Data: The Example of Europe

As the French law n°78-17 of the 6th January 1978 relative to informatics, files and freedom, was one of the oldest laws in this domain, we often refer to this text in this section.

Since the first proposal presented by the European Union Commission on the 15th April 1992 to final approval resulting from the joint decision of the European Parliament and the Council on the 24th October 1995 more than 3 years of effort were necessary to achieve publication of Directive 95/46/CE relative to 'the protection of persons with regard to the processing of personal data and the free circulation of these data'.

The aim was to define common obligations for protection for all countries in the European community. This protection needed to be compatible with national specificities, contribute to the free exchange of information in internal markets and allow cross-border flow of information with countries outside the EU.

Directive 95/46 on the protection of individuals with regard to the processing of personal data and on the free movement of such data was enacted in October 1995, which was still in the pre-Internet period. The Directive is characterized by particularly centralized and local data protection. These days, during both patient care coordination and research activities, information and communication technologies (ICT) developments make it possible to process a great deal of data over large distances in a very short time. Automated processing of data has become the daily routine and it is very important for health data to be processed with respect for the privacy and rights of the patient and healthcare professional. Social network sites for example, process and transfer large quantities of data very quickly all over the world. Even in healthcare the Internet is increasingly used for processing health data. As a consequence, it is also important to respect the ePrivacy Directive 2002/58 which was amended (Hustinx 2009) leading to the new e-Privacy Directive 2009/136/EC.

11.2.1   The Treatment of Personal Data in the European Union

11.2.1.1   The Aim of the European Directive of 24th October 1995

Data processing was primarily regulated under the provisions for the Internal Market. Therefore, the general Data Protection Directive 95/46 combines two goals: protecting the fundamental right to data protection and ensuring the free flow of personal data within the internal market.

> Though in article 1 the Directive deals with *"the protection of fundamental freedom and rights of persons, notably with regard to their private lives"*, it must also be in keeping with the pursuit of European integration based on the free circulation of goods and persons. The second paragraph of the same article states *'Member states cannot restrict or forbid the free circulation of personal data between member states for reasons relative to the protection provided for in paragraph 1'*.

Most of the difficulties encountered in drafting the Directive, which aimed to harmonise protection provided by national legislation, came about because of the balance between these two principles.

Rather than opt for a maximalist solution that would have imposed as the standard the highest level of protection provided for in the different national legislations, the European legislator seems to have preferred a compromise, which was more difficult to establish, but complied more closely with the community view. It sought the highest level of protection that was acceptable without jeopardizing the protection of individual freedom and was yet compatible with the possible short-term evolution in the legislation in countries that provided the weakest protection to their residents. The resulting text, like all compromises, may have appeared to the most demanding countries as a step backwards with regard to the guarantees provided to citizens. However, it reflects the high priority the European Union gives to enabling less developed countries or countries with very different cultures to integrate the community process without too much antagonism. A very similar notion in the second article of paragraph 1 of the Directive was already present in Convention n° 108 of the Council of Europe, which became substantive law in France on the 15th November 1985.

> In paragraph 2 of article 12, the convention stated *'a Party cannot, with the sole aim of protecting privacy, forbid or demand special authorization for the cross-border flow of personal data destined for the territory of another Party'*.

The circulation of personal data between member states is of course necessary for patients who decide to receive healthcare in other EU member states, according with the EU Directive on cross-border healthcare.

11.2.1.2 The Field of Application

The Directive defined 'personal data' as: *"any information concerning an identified or identifiable physical person"*. The notion of identifiable corresponds to the indirectly nominal nature of certain information provided for in the French law n° 78-17 of the 6th January 1978 relative to informatics, files and freedom, which was one of the oldest laws in this domain.

The same terms 'directly and indirectly' are used in both texts. In contrast, the Directive limits the possibilities of an excessively restrictive jurisprudential interpretation of the term 'directly or indirectly' by indicating in the text the very wide meaning that had to be used *'notably by referring to an identification number, or various specific elements, particular to the person's physical, physiological, psychic, economic, cultural or social identity'*

11.2.1.3 Administrative Formalities

> The Directive states *'The person responsible for the processing, or if the case may be, his or her representative, must notify the appropriate authorities'*. In addition, it provides for a 'simplified notification', notably for *'categories of processing that are not likely to prejudice the rights and freedom of the person concerned'*

The content of the notification (outcomes, categories of persons and the data concerned, safety measures for the processing ...) appear to be similar to those in French law and in daily practice is subject to national regulations.

Each member state must set up a supervisory authority, an independent body that will monitor the date protection level in that member state, give advice to the government about administrative measures and regulations, and start legal proceedings when data protection regulation has been violated (art. 28). Individuals may lodge complaints about violations to the supervisory authority or in a court of law.

The controller (i.e. the person responsible for the processing) must notify the supervisory authority before he starts to process data. The notification contains at least the following information (art. 19):

- The name and address of the controller and of his representative, if any,
- The purpose of purposes of the processing,
- A description of the category or categories of data subject and of the data or categories for data relating to them,
- The recipients or categories of recipient to whom the data might be disclosed,
- Proposed transfers of data to third countries,
- A general description of the measures taken to ensure security of processing.

Nonetheless the main innovation of this European text lies in the fact that it authorizes member States to envisage derogations to the obligation to notify, in particular when *'when the person responsible for the processing designates, in compliance with the national law to which he is subject, a person responsible for the protection of personal data notably:*

To ensure in an independent manner the internal application of national provisions (...);

To keep a record of the processing steps carried out (...);

And in so doing to guarantee that the processing is not likely to prejudice the rights and freedom of the persons concerned (...).

This possibility to have a derogation does carry a risk of weakening the protection granted to persons and introduce fears about the efficacy of a guarantee of public freedom may be justified when it is in the hands of a private individual whose independence may be compromised.

11.2.1.4   Information and Consent of the Persons Concerned

The person concerned (i.e. the patient for health data) has the right to be informed when his personal data is being processed. The controller must provide his name and address, the purpose of the processing, the recipients of the data and all other information required to ensure the processing is fair (art 10 and 11).

Data may be processed only under the following circumstances (art 7).

– When the person concerned has given his consent
– When processing is necessary for the performance of or the entering into a contract
– When processing is necessary for compliance with a legal obligation
– When processing is necessary in order to protect the vital interests of the person concerned
– When processing is necessary for the performance of a task carried out in the public interest or in the exercise of official authority vested in the controller or in a third party to whom the data are disclosed
– When processing is necessary for the purposes of the legitimate interests pursued by the controller or by the third party or parties to whom the data are disclosed, except where such interests are overridden by the interests for fundamental rights and freedoms of the person concerned. The person concerned has the right to access all data processed about him. The person concerned even has the right to demand the rectification, deletion or blocking of data that is incomplete, inaccurate of isn't being processed in compliance with the data protection rules (art 12).

It is stated that *'processing can only be carried out (...) if the person has unequivocally given his or her consent'* but the strength of this condition is considerably reduced by the adverb 'unequivocally', The word 'unequivocally' leaves considerable room for interpretation even though the definition of consent in article 2-h is precise and describes it as the manifestation of *'free, specific and informed'* willingness. For this measure of protection to be effective, in cases of litigation, it should be up to the manager of the file rather than the person concerned to prove the unequivocal consent of that person. This would limit the temptation to consider abusive interpretations.

A range of derogations to this obligation of patient's consent have been envisaged. Some do not raise any questions. Thus, the necessary processing *'on the execution of a contract in which the person concerned is one of the parties'* appears to be logical with regard to contract law. In the same way, when the processing is imposed by a *'legal requirement'* there is a derogation to avoid any conflict between the texts. Other derogations are difficult to apply, when, for example, it is necessary to define the essential nature of the processing *'in the vital interest of the person concerned'*, or *'in carrying out a mission in the public interest'*, and especially *'for the accomplishment of the legitimate interest of the person responsible for the processing(...) provided that they do not override the interests, or rights or fundamental freedoms of the person concerned'*

As in the French law of 6th January 1978, data that *'concern racial or ethnic origins, political opinions, religious or philosophical convictions, union membership'* benefit from reinforced protection in article 8 of the Directive as do *'data relative to health and sexuality'*. The sensitive nature thus attributed to the last two categories is one of the major contributions of the Directive to the protection of individual freedoms. The processing of such information is in principle forbidden, but certain derogations have been envisaged. Thus, the processing of such data is no longer forbidden if the person has given *'his or her explicit consent'*, but once again the interpretation of the term 'explicit' is debatable.

For data related to health, the principal derogation is that provided for in paragraph 3 of article 8 which indicates that the prohibition does not apply:

> *when the data needs to be processed for the purposes of medical prevention, medical diagnosis, the provision of health care (...) and provided that the data is processed by a health care professional subject to national law or to the regulations established by law by the competent national authority for professional confidentiality, or by another person also subject to an equivalent obligation for confidentiality.*

The obligation for professional confidentiality that applies to persons who process data therefore appears to justify the absence of the explicit consent of the persons concerned. In contrast, information must not be gathered without informing the persons concerned, who, on the contrary, must be duly informed so that they can exercise their right to oppose the collection of such data as provided for in article 14 of the Directive.

11.2.1.5 The Rights of the Persons Concerned

Sections IV, V, VI and VII of the Directive define for the person concerned the right to be informed about the processing of his or her personal information as well as the right to oppose the processing of the data and a right of access.

The modalities of the right to be informed defined in the Directive vary depending on whether the data are collected directly (article 10) or not (article 11) from the person concerned. In both situations, the Directive introduces a new notion; to waive the procedure to inform the person, if the person has already been informed. But how can one prove that the person was duly informed, and on the contrary what about the risk of seeing file controllers consider abusively that the person had been informed.

> When data are not collected directly from the person concerned, the person must be informed at the latest *'when the data are communicated for the first time'*, but is it not already too late for the person to use his or her right of opposition? Finally, paragraph 2 of article 11 states that the obligation to inform does not apply to processing for statistical purposes (...) *when it proves to be impossible to inform the person concerned, or if it requires a disproportionate amount of effort'*. The right of opposition is not based only on a simple "legitimate reason" but according to paragraph 'a' of article 14, one must have *'substantial and legitimate reasons relative to one's particular situation'* and *'justified'*.

This provision may give the feeling that the protection granted to citizens has been weakened, but it also states that for files used for commercial purposes, opposition need not be justified.

As for the right of access, this must be provided *'without constraint at regular intervals and without excessive cost or delay'* and, in the body of the text, there are no particular restrictions concerning medical data.

> However, in "whereas" 42, the Directive indicates that *'member States may, in the interest of the person concerned (...) limit the right of access and information; that they may, for example, specify that the access to medical data can only be obtained through the intermediary of a health-care professional.*

Of course, it is regrettable that this indication is only mentioned in a "whereas" and not in the normative part of the text, which effectively takes away its strength. The European directive indicates also that the data shouldn't be kept in a form which permits identification of data subjects for longer than is necessary for the purposes for which the data were collected or for which they are further processed. Member states shall stipulate appropriate safeguards for personal data stored for longer periods for historical, statistical or scientific use (art 6).

11.2.2   The Transfer of Personal Data to Countries Outside the European Union

11.2.2.1   General Principle: The Requirement of an Adequate Level of Protection

> The regulations applicable to transfers to a third country of *'personal data that has been processed or will be processed after the transfer'* are defined in chapter IV of the Directive. Article 25 states the fundamental principle by indicating that such transfers can only take place if *'. . ./, the third country in question can guarantee an adequate level of protection'*

Immediately, there seems to be a fundamental difference with the text of Convention n° 108 of the Council of Europe concerning the protection of persons with regard to the automated processing of personal data, which up to now, governed the cross-border transfer of nominative data. The Convention text appears to be more in favour of free exchange than does the Directive text.

> The Convention maintains a principle of free cross-border transfer: *'one party cannot, for the sole purpose of the protection of privacy, forbid or require special authorization for the cross-border transfer of personal data destined for the territory of another party'* while the Directive is clearly restrictive and only authorizes free circulation if there is an adequate level of protection.

Nonetheless, analysis of the meaning of *'an adequate level of protection'* shows that the European Directive could eventually prove to be more tolerant than the Convention, and even be permissive.

In the European Convention, the fundamental principle is not to obstruct the free circulation of personal data if the regulations in the country of destination provide guarantees that are equivalent to those in the country of origin.

The difference between the requirements of the Directive and those of the Convention is considerable. In the Convention, the requirement is *'equivalent regulations'*. This means examination of the legislation in the foreign country, and, first of all, to verify the existence of texts whose purpose is to directly or indirectly guarantee the privacy of persons with regard to the automated processing of their personal data. In the Directive, the term *'level of protection'* no longer refers to a statuary requirement, but to an overall appreciation, about which we know very little, neither what it covers nor on what it is based. The term *'adequate'* suggests an evaluation of the conditions governing the free circulation of nominative data on a case by case basis, depending on their nature and the processing they will be subjected to. It is always difficult to make such assessments in the absence of sufficient documentation on the reality of practices in information technology and the control of these practices in certain countries outside the EU. The change from the assessment of *'equivalent legislation'* to *'an adequate level of protection'* may diminish the guarantees provided to persons concerning the automated processing of their personal data. To prevent deviation of the purposes of the processing, under the cover of an official procedure, whether or not the level of protection is adequate could be judged by assessing it against technical standards established by the European Committee for Standardization of the European Union. In addition, this approach, as we will see later, opens the door to the abandon of the penal nature of offences against individual freedom.

11.2.2.2 Derogations: The Individual Interests of Persons and Contractualisation

When countries outside the EU do not provide an adequate level of protection, member States see in the Directive that the transfer of personal data can nonetheless take place under derogation for reasons that aim to protect the individual interest of persons and for the execution of contractual obligations.

Such derogations are motivated not only by economic considerations, but also for medical purposes. For example, if a French citizen is hospitalized in a country outside the EU that does not provide an adequate level of protection for personal data, the medical information in the patient's medical record still needs to be transferred.

It is possible to transfer this information according to three derogations included in paragraph 1 of article 26 – that is to say provided that:

the person concerned has unequivocally given his or her consent for the future transfer

the transfer is necessary for the execution of a contract between the person concerned and the person responsible for the processing, or for the execution of pre-contractual measures taken at the patient's request.

the transfer is necessary to protect the vital interests of the person concerned.

In the same way, the absence of an adequate level of protection cannot block the creation of international registries of diseases, or the organization of multinational therapeutic trials, if, as stated in paragraph 2 of article 26 of the Directive:

The person responsible for the processing provided sufficient guarantees with regard to the protection of both the privacy, freedom and fundamental rights of persons and the exercise of the corresponding rights; these guarantees can notably result from appropriate contractual clauses.

The drawback of contractualisation lies in the fact that it excludes the penal dimension of the repression of offences with regard to individual freedoms.

11.2.3 Re-using Clinical Data in International, Multicentre Research

Epidemiologic and clinical studies are increasingly planned and conducted at the European level. The same is true for new research activities in public health (e.g. comparative effectiveness). Moreover, in the framework of translational research, new ICT are developed in order to map diverse data sources characterizing a given phenotype to omics data sets (genomics, proteomics, transcriptomics, metabolomics) to achieve functional interpretation of these omics data sets.

There is a need to strike a balance between encouraging advances in scientific knowledge and the best use of innovative services and taking care of the associated risks with regard to privacy and security.

A major hurdle remains in the lack of harmony among regulations in different countries regarding for example, the use of health data for scientific purposes:

This is the case of the use of health data of Clinical Information Systems or Clinical Data Warehouses for scientific purposes.

This can be overcome by coordinating regulations at the European level. In all types of research activities, patients across Europe must be guaranteed equal rights in terms of the privacy of their personal health data.

The globalization of health care actors requires more harmonized rules for health data processing, particularly as the exchange of data between European e-health actors will not be limited to the context of patient care coordination – the data may also be processed for evaluation, research or statistical purposes. Currently, harmonized rules on this further use are lacking. Several Member States have formulated strict rules for the processing of medical data for research purposes, while other Member States of the European Union have more flexible rules. Article 8 of the Directive leaves too much room for different legislation in the Member States of the European Union, which is not good for the establishment of an internal market in which international quality review projects, epidemiological studies, clinical trials and post-marketing surveillance projects are emerging. It is regrettable that Article 8 does not contain more specific rules for the processing of medical data for research purposes, as more specific rules at the European level are needed.

11.2.3.1 Different Situations in Which Health Data Are Reused in Research

With the development of different Clinical Information Systems (CIS) in different healthcare facilities in private practices, and hospitals (Electronic Healthcare Records (EHR), Laboratory or Radiology Information Systems (RIS or LIS)

Hospital Information Systems (HIS)), or within health networks or at the national level (Personal Electronic Healthcare Record (pEHR)), clinical and biological data collected for healthcare purposes (diagnosis, lab test results, genetic analyses, medical imaging, prescriptions, etc. ...) have become more accessible and potentially usable for research and public health purposes. There are more and more initiatives to define the modalities for the creation of Clinical Data Warehouses and the integration of Clinical Information Systems with Clinical Research and Public Health systems at the level of the healthcare establishment, nationally and/or internationally.

These modalities for the integration of Information Systems for healthcare and research raise, among other things, new ethical, legal and deontological issues related to the development of shared personal medical data, and in particular the reuse of computerized personal data for research purposes. Currently, the reuse of computerized personal data for research purposes poses a number of problems because existing legislation is not specific to this type of research and therefore presents serious shortcomings.

Any proposal to set up a system to regulate ethico-legal issues regarding the reuse of personal medical data must imperatively include aspects such as confidentiality, security and the rights of both patients and healthcare personnel.

Figure 11.1 shows different situations in which health data are reused in various research contexts (evaluation of activity and quality of health care provided ('comparative effectiveness'), biological research (phenotype-'omics' databases, biobanks), epidemiological research, public health research, and clinical research). Data processing (access, integration) must be planned at the local, regional, national and international level.

11.2.3.2 Towards Setting Up an Ethical and Legal Framework for the Reuse of Health Data for Research Purposes

Respect for the Privacy and Rights of the Patients Regarding the Reuse of Health Data for Research

Processing involving files that include personal data, and in particular health data, must be conducted in the context of a legal framework that guarantees the rights of patients (patients' information, data security and confidentiality).

Though the legal framework for processing health data in the context of healthcare or interventional research is clearly defined by current legislation, it is not the case when the final purpose of the processing is modified (reuse of health data for research purposes). Because of this legal void, the actors (healthcare professionals, researchers, managers of healthcare information systems) have to identify the ethical and legal aspects to respect in this new context.

In France, the French Commission for Data Protection (CNIL) and various Committees for the Protection of Persons (CPP) have put forward several proposals to establish an agreement protocol concerning the sharing of data within hospitals.

Clinical Information Systems (IS) (Electronic Health Records, Personal Medical Records, etc)/Clinical Data Warehouses			
Evaluation of activity and quality of healthcare	Biological research	Epidemiology and public health	Clinical research
LOCAL (Doctor's surgery, hospital) « *healthcare team* » / *outside the* « *healthcare team* »			
e.g. Indicators of hospital care quality *(e.g. proportion of hospitalization reports sent within 8 days (aggregated data)*	e.g. Integration of data from phenotype-«omics» and Biobanks of the establishment	e.g. Feasibility studies aggregated data/ Collection of cohort data, single-centre registries	e.g. Feasibility studies *(e.g. nb of diabetics with HBA1C >6%)* (aggregated data/ Collection of data from single-centre studies
REGIONAL/NATIONAL			
e.g. Indicators of national quality (Indicators to Improve care quality and safety (IPAQSS) – HAS)	e.g. Regional or national databases for phenotype-«omics» and Biobanks	e.g. National registries and observatories (e.g. InVS)	e.g. Feasibility studies / Collection of data from national multi-centre studies
INTERNATIONAL			
e.g Indicators of antibiotherapy surveillance of the European Antimicrobial Resistance Surveillance Network (EARS-Net)	e.g. International databases for phenotype-«omics» and Biobanks (e.g. European Bioinformatics (EBI), (EBML))	e.g. International registries and observatories (e.g. European Centre for Disease Control)	e.g. Feasibility studies / Collection of data from international multi-centre studies (e.g. project Electronic Healthcare Record for Clinical Research (EHR4CR))

Fig. 11.1 Typology of the reuse of health data in Clinical Information Systems (CIS) and clinical data warehouses (CDWs) according to the type of the research and the geographic scope of data exchange

Certain hospitals implemented, following approval from the CNIL and the CPP, a health-data sharing surveillance committee within the establishment. The role of this committee is to provide information on and apply the current legislation by approving and authorizing such uses.

Access to clinical data contained in patients' records for use for research or public health purposes must be defined in such a way that a balance between individual concerns (linked to the efficacy of the individual management of the patient) and the collective concerns of research or Public health is maintained.

In addition, in this new context, the crucial question is to obtain the consent of the patient to ensure the protection of personal data while reconciling the imperatives of quick, easy access to the data on the one hand and guarantees of confidentiality and security on the other. In biomedical research, the universal values relative to the protection of persons are the same in most countries. These values have led to the requirement for written informed consent as a true doctrine. It is the 'Gold Standard' of bioethics.

Therefore, so that patients can give their written informed consent, they must be clearly informed about how the clinical data to be used in a context of research will be collected and processed (form and final purpose of the processing, the persons responsible for the processing). However, when the data are reused in one or several research projects, questions arise about how consent can be obtained. Should patients be asked to give global consent? Should various levels of consent be proposed? How should the request for consent be worded (opt in/opt out)? How should the absence of consent be notified and taken into account? How can the patient withdraw consent? How should the guarantee of data confidentiality be worded, and implemented? What about the management of traces?

In order to reconcile patients' rights and the development of research with regard to the reuse of computerised data, it is relevant to consider the possibility of organising the 'donation of health data' (collection of patients' information and consent) and by so doing protect data confidentiality while authorizing the sharing and linkage of data (between healthcare teams and establishments at the national or international level).

The notion of 'donating de-identified clinical data' for research or public health purposes is emerging among both patients and researchers. Before using patients' records for the common good, it is necessary to define the term 'common good' and to establish the conditions to foster trust and cooperation between actors. It is interesting to note that the study 'Consultation paper on protection of personal health information', reported that more than 71 % of the people interviewed though that it was unnecessary for healthcare professionals to ask for specific authorization to use data for every new clinical situation or research project.

Respect for the Privacy and Rights of Healthcare Professionals Within
Reusing Health Data for Research

Like patients, healthcare professionals have the right to be informed about and give
their consent for the sharing of data in this new context of using health data for
research and public health purposes. In concrete terms, it means that the producers
and users of data as well as representatives of civil society need to establish new
deontological rules for the sharing of health data for research purposes. The way in
which producers of data are involved when the data are used in different healthcare
contexts need to be specified.

Among the challenges associated with these new practices is professional
secrecy. This is an essential element in the relationship between healthcare
professionals and patients. Legislation on this subject in the different member states
of the European Union varies widely. These differences with regard to professional
secrecy have led to extremely complex legal issues concerning the cross-border
transfer of data (cf. Sect. 11.2.2.1). The legislation in force concerning data
protection in this new context must therefore be analysed carefully. Like for
patients, it would be worthwhile examining the possibility of organising the 'donation of health data'.

11.3   Medical Responsibility in Telemedicine

Telemedicine rises medico-legal problems when a patient suffers injury. The doctor
participating in telemedicine activities must be aware of the different aspects of his
professional responsibility in the area, and the conditions in which he could be
asked to demonstrate diligence

11.3.1   The Assignment of Responsibilities

The responsibilities of doctors must be clearly identified to avoid their
dilution, which would be prejudicial to the interests of the victim of the
injury. This identification must take account several factors: the principle of
care, the competence of the doctors, their unequal access to the relevant
information, their knowledge of the operation and the limitations of the
telemedicine system, and also of possible technology malfunction.

11.3.2   The Principle of Care

The determination of the respective responsibilities of doctors who contribute to the diagnostic and therapeutic decisions is a classical aspect of the legal and ethical analysis during proceedings instituted by a patient in which several doctors are involved.

The object of this analysis is the search for the medical behaviour which led to the error, since it is this behaviour that determines the possible responsibility of the doctors, and not the diagnostic or therapeutic error itself. It is accepted in the jurisprudence that the duty of the practitioner is only a duty of means. If the means, technical or intellectual, normally used by a competent and diligent professional have not been used, this represents criminal negligence.

A doctor cannot be sanctioned for not having been able to make a difficult diagnosis, for example in studying an x-ray film or a pathology examination slide. On the other hand, if the lesion is common and obvious, the facts show that the professional has not given the care based on current scientific knowledge. In other words, doctors are expected to know what should be known by any competent, skilled practitioner. They also have the duty to require the help of a colleague when they think they are not competent enough.

11.3.3   The Level of Competence of the Doctors

Three situations are possible in telemedicine. The doctors could be a general practitioner and a specialist, two specialists from different disciplines or two specialists from the same discipline. When a general practitioner asks for an opinion from a specialist, it seems legitimate that the latter takes responsibility for his reply. He is effectively being appealed to because of his expertise, and the requester would normally follow the advice he gives. If, on the contrary, he takes the risk of not following it, the general practitioner could see his conduct censured. The situation is identical if a specialist asks for an opinion from a specialist in a different discipline.

When the two doctors practice in the same specialty, it would be tempting to analyze the situation in the same way by considering that the doctor asking for the opinion is acknowledging the superiority of his colleague in a peculiar field of their common specialty. This analysis would implicitly lead to the creation of a category of "super specialists" in a variety of "sub specialities" which did not officially exist. Faced with a difficult medical case that he cannot resolve, he is obliged either to decline to give an opinion, or to ask an advice from a third party while completely and personally assuming his responsibility as specialist (Laske 1996).

This duality which governs the behaviour of the specialist is reflected in the approach taken by the health insurance schemes. Although they recognize the existence of two different medical acts when a general practitioner asks for a specialist opinion or when a cardiologist refers his patient to a nephrologist, this not the case if a cardiologist refers his patients to other cardiologists. The specialist must justify his status, which is the reason for his being consulted by the patient. He personally assumes his responsibility towards his patient in case of patient's litigations and secondary will take action against a referent doctor who has eventually badly advised him. This recurrent action will be conducted according to the terms of the contract they should have established between them for medical advisory.

> The case of telemedicine differs from the normal condition of a request for a specialist opinion in that the specialist cannot examine the patient and does not have access to all the information. The doctors who request and give opinions are thus not in the same situation with respect to the information on which the diagnosis is based.

11.3.4   The Unequal Access of Doctors to Information

During a telematic consultation, the conduct of the doctors must also be evaluated as a function of the respective roles they have in reaching the diagnosis. The practitioner requesting opinion and the referent practitioner do not have the same access to patient's information. The requester of the opinion has access to all the available information, while the referent in general receives only the part of that information that was selected by the first doctor. This selection must be made by a competent person, who is able not only to choose information that is relevant for the diagnosis, but also to interact effectively with the referent. This is the most common situation in telemedicine, where the two doctors are of the same specialty and share the same knowledge. The use of the method by doctors from different specialties is justified in particular by emergency situations and difficulty in access to the specialist.

> The fact of not having all the available information does not exonerate the specialist from his responsibility with respect to the advice he gives. In cases of doubt or of difficulty in diagnosis, it is up to him to ask for additional information, and to decline to give an opinion if he thinks that the data is insufficient, or if he feels not competent to do so.

11.3.5   The Command of the Telemedicine System

The use of telemedicine system also requires each doctor to have full knowledge of its use and limitations.

> If the quality of an image is poor, or if the relevant information of the lesion has not been captured, both the referent and the requester could be held responsible if this lack of quality leads to an error. It is relevant to note that radiologists, for example, have obligations concerning the quality of the technique used to take an x-ray.
>
> Doctors may also be held responsible if the transfer of data leads to information that is distorted, damaged or communicated to unauthorized third parties, whether this breach is due to them, or to a third party.

The main difficulty in analysing these responsibilities could be to distinguish between problems that result from incomplete control of the system and those due to its malfunction.

11.3.6   Equipment Malfunctions

According to the jurisprudence, it seems that the technical malfunction of a telemedicine system calls into question the responsibility of the doctor. The patient has the right to expect that the instruments used by the doctor are not defective. However, in this context, it is possible for the doctor to take action against the vendor. In telemedicine, the suitability must be expressed in terms of quality of production and transmission of information. The compression and decompression of images, which aims to increase transmission speeds on the network and to reduce the size of memory required, must therefore not do so at the expense of their readability. It is therefore necessary to evaluate teletransmission systems to know if their use is justified and to define the conditions and limitations of such a use.

> This scientific evaluation of the quality of the product, and of its suitability for the service, which it is supposed to fulfil, will be a decisive factor in cases of litigation between the manufacturer and the doctor who uses it. The manufacturer could be sanctioned for not evaluating the product, but the responsibility of a doctor who had bought a system without inquiring about its guarantees could also be questioned.

In addition, the supplier is obliged to provide an instruction booklet written in comprehensible terms.

11.3.7 The Burden of Evidence

The determination of responsibilities requires a precise analysis of the facts and their context, i.e. the exact content of the telematic exchange at the origin of an injury. This implies that this information should be archived and that the validity of its archiving as an element of proof before a court must be studied.

11.3.8 Records in Telemedicine

In the case of medico-legal litigation, doctors must present proof that they have behaved diligently and in accordance with the usual guidelines by producing the images transmitted and the replies given.

The quantity of information generated by image storage in practice requires the use of optical memories, since only these have the enormous storage capacity to store several tens of images per day. For the legal reasons discussed in the next paragraph, non-rewritable CD-ROM is the preferred option. This stored information nevertheless does not provide evidence of the telematic transaction, particularly in the case of disagreement between doctors. They could in fact deny sending or receiving the images or the replies in question. There must therefore be proof that the images and the replies have been sent and received by the doctors in question (Coatrieux et al. 2011a, b; Wei et al. 2012).

The health professional's card, which will shortly be available, allows the electronic signature of transmitted documents, and thus ensures the identification and authentication of the sender, so that he cannot deny being the author. This is referred to as non-repudiation. The electronic signature also enables the content of an electronic exchange to be sealed, and guarantees its integrity (Engelbrecht et al. 1995).

The coupling of the transmission to a time standard testifies to the time of the request and the delay before the reply is given. Finally, an authenticated acknowledgement must be sent on the reception of images and replies.

The complexity of the security mechanisms will mean that, with the general-ization of telemedicine, the majority of electronic transactions will be made not directly from doctor to doctor, but via a central server. It is this server that will receive requests for opinions, direct them to the available referent doctors, implement security procedures and play the role of third-party guarantor.

Although, from the technical point of view, these solutions offer better security than the exchanges by letter used up until now, their legal recognition has not yet been clearly established. Analysis of existing legislation, however, shows that there are few obstacles.

11.3.9 The Legal Value of Electronic Records

The law should soon take formal notice of the necessary extension of hard-copy media to non-material media by defining records as a set of documents, irrespective of their date, form and material support, but the fact that non-material documents correspond to the notion of records is not sufficient to establish their value as evidence. The rules applicable to evidence moreover vary depending on the juris-diction, but nowadays all contain elements favouring the legal value of records on non-rewritable optical disks and the electronic signature.

11.3.10 Remuneration of Doctors

"all effort deserves reward" and telediagnostic equipment is expensive. It would thus be unreasonable to expect a doctor in the future to give diagnostic opinions to his colleagues repeatedly without financial reward, either for himself or his parent establishment.

This association of expertise and the resulting financial exchanges must be included in a contractual framework, which clearly eliminates any risk or suspicion of divergence from medical ethics.

This contract must stipulate that experts send, with a pre-defined frequency, a summary of examinations performed for each requesting doctor. In return, the requesting doctors must pay the appropriate amount for the acts performed, less any justifiable expenses for the administration of samples and administrative documents.

However, the number of requests for opinions should not exceed a given volume. Excessive use would be a misapplication of the purposes of telemedicine, and could create "medical" structures which would be no more than sites for sample collection or the taking of radiological images, for example. A maximum volume of two thirds of the total volume of acts carried out by the requesting doctor could be fixed.

11.3.11 Conclusion

Apart from the reluctance of practitioners with regard to unfamiliar computer technology, the fear of legal grey areas that surround telemedicine is a cause of hesitation by potential users.

The clarification of the legal framework in which telemedicine is practiced, and in particular the identification of responsibilities and the specification of doctors' remuneration within a contractual framework, seems now to be an essential condition for its more widespread use.

11.4 Healthcare Technology for Home Use and Autonomy

11.4.1 The Arrival of Healthcare Technologies in the Home and Autonomy with Regard to Healthcare Politic

Ethical issues concerning healthcare data with regard to Technologies and Services for Healthcare at Home and Autonomy (TSHHA) (cf. Chap. 17 of this book) must respond to the double challenge of longevity and the economic crisis. This multi-faceted challenge can be considered in the context of two major phenomena that are on a collision course: the decrease in medical and social resources and the increase in needs for the management of chronic disease, including the imminent dependence of a substantial proportion of elderly persons. Preponderant cofactors of health in chronic disease, home-care and autonomy have been given considerable importance. They now play a major role and have explicit links with technology. Chronic heart failure, arterial hypertension, cancer, the surveillance of diabetes, high-risk pregnancies, the care of frail elderly people at home, and the return to home of such patients following hospitalization: the list of 'telehealth' applications is long, and now lies at the heart of policy considerations throughout the world and especially in Europe. In this respect, the study CARICT ("ICT-based solutions for caregivers: assessing their impact on the sustainability of long-term care in an ageing Europe", 12.2010–12.2011) is one of the most interesting and informative studies on telehealth. Jointly funded with DG Connect (Directorate General "Communication Networks, Content and Technology" of the European Community) and run by the European Centre for Social Welfare Policy and Research in Vienna, this study systematically documented the extent of ICT-enabled services to support carers, and explored in depth how this type of initiative works. The goal of the study was to understand how ICT can play role in the future mix of services and supports available to the elderly and their carers, and how they facilitate the creation of a sufficient number of skilled and motivated domiciliary caregivers in European society. The CARICT Project Summary Report (Carretero et al. 2012) stresses the profound medico-economic and organisational upheaval and the demand for ICT to help the transition to this future mix of services and supports in order to meet the needs and expectations of society regarding home care and autonomy.

Given the diversity of chronic diseases and the medical needs of patients, a considerable amount of data is necessary for the medical follow-up of a patient. Now that sensors and computerised recording devices are readily available, these data can be legally fed into networks under strict conditions of confidentiality so that they can be interpreted remotely as authorized by the law in an increasing number of countries worldwide. The vocation of telemedicine and health networks is now to implement and regulate these new practices, under the auspices of ethical governance.

11.4.2 The Question of Ethics

11.4.2.1 Why Ethics?

These practices respond to various international declarations, including the Universal Declaration of Human Rights, which stipulates in article 27-1 that: 'Everyone has the right freely (...) to share in scientific advancement and its benefits'. The recommendation R(98)9 of The Committee of Ministers of the Council of Europe to the Member States on dependence (adopted by the Committee of Minister on 18thSeptember 1998) prolongs and clarifies this declaration by stipulating that: 'All dependent persons or persons who may become dependent, regardless of their age, race and beliefs and the nature, origin and degree of severity of their condition, should be entitled to the assistance and help required to enable each of them to lead a life as far as possible commensurate with their ability and potential. They should therefore have access to services of good quality and to the most suitable technologies'.

Nonetheless, the development of TSHHA and in particular medical or medico-social remote monitoring of patients, made necessary by intense economic pressure and required by institutional and legal processes, could not move forward without careful reflection involving all levels of society including patients, families, healthcare and social services professionals, business professionals and those responsible for health policy. The general framework for this wide-ranging social reflection was ethics. As ethics is the "expression of the measure" in the words of Pr Jean Bernard, it automatically became the founding principle; what Paul Ricoeur summarised in his generic concept: "Let us define 'ethical intention' as aiming at the 'good life' with and for others, in good institutions" (Ricoeur 1992) and what Edgar Morin expressed in these terms: "the ultimate goal of ethics has two complementary facets. The first is resistance to cruelty and barbarity. The second is the accomplishment of human life." (Morin 2004).

11.4.2.2 The Opposing Faces of Technological Innovation

From this point onwards, ethical considerations will become both the motivation for the reasoned and controlled uses of technology and a rampart against the dehumanisation of healthcare and the many ways in which technology can lead to abuse if it is deviated from its initial objectives, that is to say for health care and social support. It is therefore in the ethical objectives of scientific and technological development that these two opposing facets become apparent in home care and autonomy:

1. Positive facet: providing medical supervision, assisting, helping, answering calls for help, restoring links, improving quality of life, optimising expenditure, reducing costs, encouraging care and support, avoiding or delaying placement in an institution.
2. Negative facet: imposing decisions without respecting the wishes or obtaining the consent of the person concerned, intervening in the person's private life, violating the dignity of the person, dehumanising the relationship with the patient or the elderly person in loss of autonomy, giving rise to all sorts of abuse and ill-treatment, replacing professionals and voluntary caregivers by automatic devices when not necessary, etc.

The debate on these ethical risks has been under way for several years and frequently leads to the same remark: these risks are less a result of the technologies themselves than of their disorganised and uncontrolled use in a context in which market forces predominate.

Voices are being raised in anticipation of the misuse of these technologies and to avoid the backlash of such misuse, which would consist in depriving those who most need them of the benefits they provide. The ethical nature of these innovations has thus become the *sine qua non* condition to obtain public confidence in them, and thereafter ensure their success in a democratic society governed by the principle of public debate, which itself will be enlightened by scientific data.

11.4.2.3 For a Citizen-Based Control of Use

The demand for ethical control in a world characterised by its complexity and its contradictions has developed considerably in the fields of health and technological innovation.

Europe has financed entire research-action projects concerning the ethics of technologies for home care and autonomy, such as: PHM-Ethics (Personalized Health Monitoring Ethics) and HIDE (Homeland Security, Biometric Identification & Personal Detection Ethics). Ethical initiatives around the world and ethical codes in telehealth are also key issues: e-HealthCode of Ethics, eHealth Ethics Initiative 2000, etc.

11.4.2.4   Towards an Ethical Evaluation of Technologies and Services

Talking about fears is one thing, taking ethical decisions that have implications on the economy and health is another. Any judgment must therefore be based on established facts, indicators of the service rendered and satisfactory levels of evidence. The evaluation thus has a major place in public health under the auspices of an ethical evaluation. The application of these ethics in the field of health technologies is necessarily multiaxial and multidisciplinary, with the following six distinct dimensions (TEMSED): (1) The purely *t*echnical dimension (robustness, reliability, maintenance ...), (2) the *e*rgonomics (relationship between the users and the technical equipment), (3) The *m*edical service rendered (new therapeutic approaches), (4) the *s*ocial utility (improvement in/creation of personal services for individuals ...), (5) the *e*conomic dimension (costs and real savings) and (6) *d*eontological aspects (respect of ethics and the law, responsibilities and recourse) (Rialle 2009).

11.4.2.5   Conclusion: Conditions for Good Practices

A certain number of conditions for good practices for TSHHA can be suggested in the light of elements succinctly presented in this section: (A) *Evaluation*: precise indicators of the medical and social service rendered must be available, recommendations for good practices must accompany any technology used; (B) *Prescription*: professionals must carry out a methodical evaluation of the disabilities and disadvantages of patients with a chronic disease, or with a handicap or frail elderly persons (gerontechnology consultation, use of the International Classification of Functioning, Disability and Health-ICF ...) including the state of the persons place of residence, and this must form the basis of any prescription of or recommendation for these technologies and the associated services; (C) *Use*: these technologies must be used in the spirit of and respect for fundamental principles of healthcare ethics (Humanity and dignity, Solidarity, Equity and justice ...). Once such technologies have been prescribed, a follow-up must be organised by a medical or medico-social team. The use of these technologies must involve all of the actors concerned with chronic disease or gerontology (doctors, nurses, psychologists, social workers, physiotherapists, occupational therapists, dieticians, speech therapists, chiropodists ...), in close collaboration with the family aids; (D) *Economic equity*: equality of access to these technologies and services must be guaranteed, notably by ensuring appropriate financial support thanks to a mutual welfare system; (E) *Training*: persons involved in caring for patients or frail elderly or disabled people must benefit from training in the possibilities and limits of these technologies and services; (F) *Debate*: the place of technology in the care of frail or socially disadvantaged persons must involve deciders and healthcare professionals, and beyond that, every citizen, and constitute an active theme for bioethics; a forum for open, accessible debate must accompany the use of this technology.

11.5   Technologies for Data Security

11.5.1   Ciphering or Encryption, the Corner Stone of Confidentiality

To encrypt a message, is to apply a transformation function that will make it illegible to others. This function is applied using an encryption algorithm (Cormen et al.). In order to personalize the encryption, a key is used (Fig. 11.2). If we take the example of an exchange of information between a hospital and a private healthcare practice, the hospital doctor will be sure that the general practitioner to whom the message is sent will be able to read it, since he or she is the legitimate recipient and will therefore be the only person to know the decryption key.

We suppose that the algorithm is public and that the confidentiality is only ensured by the user's key, which must therefore be difficult to find, even for an experienced cryptanalyst. A good encryption algorithm will be an NP-complete algorithm; that is to say that the inverse calculation (corresponding to the decryption of the message) will only possible by testing all of the possible values of the key.

An encryption algorithm is said to be symmetric or secret-key when a single key is used for both encryption and decryption. This is the case, for example, of the Data Encryption Standard (DES) algorithm adopted as the official standard of the USA government in 1977. The problem posed by the use of this type of algorithm is the fact that the sender and the recipient share the encryption key. In contrast, asymmetric (or public key) algorithms, which were developed as early as 1976, require two keys: the first is a public key that anyone can use to send an encrypted message to a given recipient; the second is a private key that is known only by the given recipient, and this key alone is able to decrypt the message. This procedure precludes the problem of sharing a key as only the legitimate recipient, the holder of the private key, is able to decrypt the message. The most widely known public-key algorithm is the RSA algorithm (Rivest et al. 1978). The security of this algorithm is based on the hypothesis that factoring a large number in products of prime numbers is long and difficult.

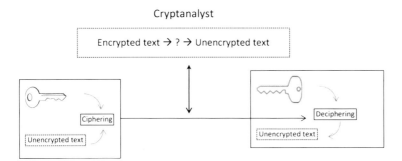

Fig. 11.2 Encryption, decryption and cryptanalysis

11.5.2   Electronic Signatures and Verification of Integrity

The second level concerns the use of methods involving an electronic signature to allow the recipient doctor to authenticate the identity of the doctor who sent the message. In the example above, this means that the general practitioner will be able to make sure that the message was really sent by that particular hospital doctor. The legal value of electronic signatures was recognized by French law n° 2000-230 of the 13thMarch 2000 bearing on adaptation of the law of proof to information technologies and the electronic signature (Allaert et al. 2004).

This mechanism brings together two procedures: the signature of a set of data and the verification of this signature. The signature of a message is based on a characteristic key held by the sender. The requirement is that only the signer can produce the signature and that the verification does not make it possible to reproduce the signature. Generally, public key algorithms like the RSA are used. The use of the electronic signature also makes it possible to guarantee the integrity of the message, that is to say, to ensure that the message was not modified during the transmission.

In Fig. 11.3, we see that the sender creates a fixed-size fingerprint in the message, which itself is of a variable size, using a hash function (Cormen et al.; Quantin et al. 1998a, b). The hash function is a new comer in the world of modern cryptology. They were developed in particular to allow the elaboration of techniques for secure electronic signatures. Hash functions are said to be one-way if it is impossible to calculate the inverse in a "reasonable" time using current technology. The hash function transforms a clear text of a given length into a fixed-length hash value, often called the fingerprint. Among the many hash functions proposed by cryptologists, the function considered the most secure is the Secure Hash Algorithm (SHA) recognized as the USA standard by the National Institute

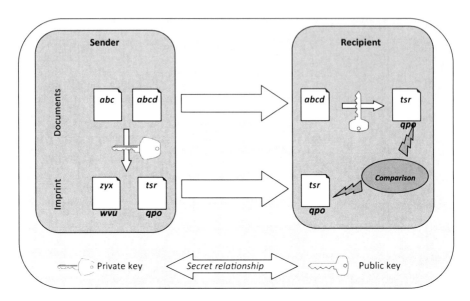

Fig. 11.3 Electronic signature

for Standard and Technology (NIST). This hash function is incorporated into the DSA (Digital Signature Algorithm), which was proposed by the NIST in 1991.

In the first step, the message to be hashed is completed by a string to make its size a multiple de 512 bits. Each block of 512 bits is then cut into 16 sub-blocks of 32 bits, each of which is then transformed into 80 words of 32 bits to which 80 operations are applied. The result of the SHA is a fingerprint, that is to say a fixed-size message of 160 bits. The fingerprint is therefore specific to the message. In particular, a slight modification of the message will lead to a radically different fingerprint. The sender sends both the clear text and the encrypted finger print. To be sure of the origin and the integrity of the message, the recipient will first of all recalculate the fingerprint of the message using the same hash algorithm as that used by e the sender, will then compare the fingerprint obtained with the fingerprint that had already been decrypted. The recipient can thus be certain that the sender was indeed the signer of the message received, since only the sender knew the private key used to encrypt the fingerprint, and that the corresponding public key was the only one to enable decryption.

11.5.3   The Use of Hashing Techniques to Guarantee the Anonymity of Personal Information

The third use of cryptographic techniques concerns the gathering of medical information within a structure outside the establishment where the patient was treated.

> The problem of linking nominative medical information to conduct multi-centre epidemiological studies is arising more and more frequently, for example, in the context of collaborative studies involving private-sector practitioners and hospital doctors. According to the recommendations of the French Commission for Data Protection (CNIL) (Vulliet-Tavernier 2000), it is now preferable to use cryptographic techniques that guarantee the irreversible transformation of data, such as one-way hash methods as proposed by the DIM of Dijon CHU in 1996 (ANONYMITY software) (Quantin et al. 1998a, b). Indeed, unlike encryption methods, which need to be reversible so that the legitimate recipient can decrypt the message, one-way hashing methods are irreversible.

The result of hashing is a perfectly anonymous code (impossible to return to the identity of the patient) that is always the same for a given individual, meaning that data for the same patient can be gathered (Quantin et al. 2005b; Turchin and Hirschhorn 2012). The SHA is, to our knowledge, the most secure public hashing algorithm in the face of attempts at decryption. Unlike legislation concerning encryption functions, which can be very strict depending on the size of the key, a simple declaration is all that is required for the use of hash functions. As these functions are irreversible, they cannot be used by secret organizations that wish to exchange information. Nonetheless, even though hashing is irreversible, it does not guarantee perfect information security. As the algorithm is public, hashing can be applied to a vast number of identities. It could be possible to compare the codes obtained with the code of a given individual for the hashed file and find the identity, a so-called dictionary attack. To counter such attacks, double hashing has been suggested. If, for example, one wishes to gather files from various sources, every sender would use a first key called K1. This 'key' K1, used by every data-collection centre to hash the identities would make it possible to protect the information with regard to persons who are not involved in the study and therefore do not know the key. However, as all of the centres taking part in the study must use the same key, it is necessary to ensure the security of centralized information, even with regard to collection centres that know K1. The information received by the processing centre that links the data is therefore hashed a second time using the same hashing algorithm, but with a second set of keys K2. Following this double hashing of identity data, performed first at the collection centre and then at the processing centre, the anonymity of the files is definitively preserved (Quantin et al. 2005a; Quantin et al. 2007a; Quantin et al. 2007b). Instead of double hashing identity, another solution is combining hashing and enciphering analysis as implemented for medical and administrative data collection in Switzerland (Borst et al. 2001).

11.5.4   Combining Hashing and Enciphering Algorithms for Epidemiological Analysis of Gathered Data

To conduct epidemiological studies, it is often necessary to compile individual records which come from different sources. In Europe, the linkage of nominal files within the framework of medical research is subject (Couris et al. 2006) to the European directive of October 24th, 1995, concerning nominal data processing, which requires information to be rendered anonymous before its transfer for linkage purposes (Blakely et al. 2000). In order to respect this legislation, there is a need for mechanisms that enable patient data to be anonymized. The proposed solution (Quantin et al. 2008) makes use of an non-reversible cryptographic method that will be applied to each nominal file before linkage: hashing can be performed in order to render data anonymous, in particular health data, (Churches and Christen 2004; Quantin et al. 1998b). As hashing functions are available to the public at large, dictionary attacks remain the most important security issue, especially when the information is collected from many sources and gathered at a regional or national level (Bellare et al. 1996). As the algorithm is public, hashing can be applied to many identities. It is thus possible to compare the codes obtained with those of a given individual from the hashed file and find his or her identity. Although irreversible, hashing does not guarantee the complete security of information. The hash can be made dependent on a secret key, applying the hash function to the patient identity concatenated with the hash key. For example, in France, all hospitals use the same hash key in order to anonymize discharge abstracts so that administrative data for any individual patient can be linked at the national level. However, as the same hash key is used by different hospitals, it would be possible for one of these hospitals, by means of a dictionary attack, to gain access to patients' identities and records stored at the national level. Hence, any hospital that collects data would be able to identify patients of other hospitals, which may, in certain cases, be competitors.

   To avoid this inconvenient, double hashing (one at the level of the sources and another one at the centralised level) was used. This is why the French National Commission for Data Protection (CNIL) recommended the use of a second anonymous level, again through a hash function, to better protect access to data stored in the national centre for data processing, i.e. the same hash function with a key known by the national centre only. Moreover, this commission recommended that different keys (especially for the second hashing), be used for each study. Unfortunately, this recommendation is in contradiction with the common requirement in public health and biomedical research regarding the ability to link records that refer to the same entity in separate data collections (Armstrong and Kricker 1999). This linkage requires separate data collections to share a common identifier.

To overcome these difficulties, the combination of hashing and enciphering techniques can be proposed. This methodology (Quantin et al. 2008) reinforces the balance between the two pillars of information security, which are the protection of confidentiality and the availability of information. An important question is: what is the usefulness of storing masses of very well protected data if they cannot be used for public health purposes? Even though implementing the solutions proposed here would be rather time-consuming, it is extremely important to make the stored data available for use in the field of public health for the benefit of patients and, possibly, under the guardianship of a new family of national trusted third parties such as the Data-Matching Authority (Australian DMA) (King et al. 2012).

11.5.5   *Medical Record Search Engines, Using Pseudonymised Patient Identity: An Alternative to Centralised Medical Records*

The concept of empowerment can be defined as a "social process of recognizing, promoting and enhancing people's abilities to meet their own needs, solve their own problems, and mobilize necessary resources to take control of their own lives" (Jones and Meleis 1993).

In the health care context, patient empowerment means promoting autonomous self-regulation so that the individual's potential for health and wellness is maximized. Patient empowerment begins with information and education and includes seeking out information about one's own illness or condition, and actively participating in treatment decisions (Lau 2002). It points out the passage from an hold model for care based on patients' "compliance" with a health care professional's "directives" to a new paradigm based on patients' "adherence" to health care professional's "recommendations", through, among other things, active participation of the patient in the management of his personal medical records. That means also to move from the traditional patient's medical record managed by health professionals and used under their supervision and authority to a Patient Controlled Health Record (PCHR) (Mandl et al. 2001).

As described by L. Rostad (Rostad 2008) a PCHR contains data from multiple care sites, and the patient is in complete control of the information. According to a previous study conducted by SE Ross (Ross and Lin 2003) and to the results of clinical trials, the main benefit of the giving patients direct access to their medical records is the improved communication between the doctor and the patient. Other benefits concerning only modest improvements in adherence, patient education, and patient empowerment were found in certain randomized controlled clinical trials, but not in all. However this lack of efficacy could result from the fact that patients had only *access* to their medical records, which maintains them in a passive situation; access alone is insufficient as a real active patient *control* over their personal health records is needed. By empowering him of his personal health information, we may expect that patient will become a real key manager of his own health, beside the Medical Practitioner (MP). But at the same time, responsibility must be shared according to the knowledge of each actor. This mixed management implies sharing responsibilities between the patient and the Medical Practitioner (MP) by making patients responsible for the validation of their administrative information, and MPs responsible for the validation of their patients' medical information.

One of the proposed solutions (Quantin et al. 2011b; Quantin et al. 2011a; Quantin et al. 2011c) is to gather and update patients' administrative and medical data in order to reconstitute patients' medical histories accurately. This method is based on two processes. The aim of the first process is to provide patients administrative data, in order to know where and when they received care (name of the health structure or health practitioner, type of care: outpatient or inpatient). The aim of the second process is to provide patients' medical information and to validate it under the responsibility of the MP with the help of patients if needed. During these two processes, the patients' privacy will be ensured through cryptographic hash functions like the Secure Hash Algorithm, which allows the pseudonymization of patients' identities. A Medical Record Search Engine (Quantin et al. 2011b; Quantin et al. 2011a; Quantin et al. 2011c) will be able to retrieve and to provide upon a request formulated by the MP all the available information concerning a patient who has received care in different health structures without divulging the patient's true identity. Associated with strong traceability of all access, modifications or deletions, this method can lead to improved efficiency of personal medical record management while reinforcing the empowerment of patients over their medical records.

11.6   For More Information

Regarding data protection and security of personal medical information: from law to security tools

Allaert FA, Blobel B, Louwerse K, Barber B. Security standards for healthcare informations systems. Studies in Health Technology and informatics, IOS Press. 2002.

Regarding the legislation concerning privacy and protection of personal data

Directive 95/46/EC on the protection of individuals with regard to the processing of personal data and on the free movement of such data.

http://ec.europa.eu/justice/policies/privacy/docs/95-46-ce/dir1995-46_part1_en.pdf

Convention n°108 (28 January 1981) for the Protection of Individuals with regard to Automatic Processing of Personal Data.

http://conventions.coe.int/Treaty/en/Treaties/Html/108.htm

French law "Loi informatique et libertés", Act n°78-17 of 6 January 1978, on information technology, data files and civil liberties.

http://www.cnil.fr/fileadmin/documents/en/Act78-17VA.pdf

"Post i-2010 priorities for new strategy for European information society (2010–2015)," from http://ec.europa.eu/information_society/eeurope/i2010/docs/post_i2010/090804_ipm_content.pdf

See also European Commission Recommendation of 2 July 2008 on cross-border interoperability of electronic health records'.

http://ec.europa.eu/information_society/newsroom/cf/itemlongdetail.cfm?item_id=4224

See also European Commission Recommendation of 30 April 2004 on 'e-Health – making healthcare better for European citizens: an action plan for a European e-Health Area.

http://eur-lex.europa.eu/LexUriServ/LexUriServ.do?uri=COM:2004:0356:FIN:EN:PDF

Regarding clinical research and re-using clinical data in international, multicentre research

Markman JR, Markman M. Running an ethical trial 60 years after the Nuremberg Code. Lancet Oncol. 2007 déc;8(12):1139–46.

Neff MJ. Institutional review board consideration of chart reviews, case reports, and observational studies. Respir Care. 2008 oct;53(10):1350–3.

Petrini C. Ethical issues in translational research. Perspect. Biol. Med. 2010;53 (4):517–33.

Regarding medical responsibility in telemedicine

Implementing secure healthcare telematics applications in Europe. The ISHTAR Consortium (Eds). IOS Press, 2001.

The new navigators: from professionals to patients. Proceedings of MIE2003. Studies in Health Technology and Informatics 95. Editors: Baud R, Fieschi M, Le Beux P, Ruch P. IOS Press, 2003.

Regarding healthcare technology for home use and autonomy

FIEEC and ASIP: Etude sur la Télésanté et Télémédecine en Europe (2009):

http://esante.gouv.fr/sites/default/files/Etude_europeenne_Telesante_FIEEC_ASIPSante_0.pdf

European Commission.eHealth Benchmarking III – Final Report and related annexes (2011):

http://ec.europa.eu/information_society/newsroom/cf/item-detail-dae.cfm?item_id=6952

Regarding technologies for data security

Thomas H. Cormen, Charles E. Leiserson, Ronald L. Rivest, et Clifford Stein (2010). Introduction to Algorithms – third edition.

Website of the French Network and Information Security Agency (Agence nationale de la sécurité des systèmes d'information): www.ssi.gouv.fr

Website of the National institute of standards and technologies: www.nist.gov

Exercises

On the legislation concerning privacy and protection of personal data

Q1 What are personal data according to the on la European directive of 1995?

Q2 What types of information are covered by professional secrecy in the field of health?

Q3 What is the automated processing of personal data?

R1 Any personal information relative to a physical person who is identified or who can be identified, directly or indirectly by referring to an identification number or to one or several elements particular to that person.

R2 Personal data as defined in the law in accordance with the European directive of 1995. It is important to bear in mind that such information is not always medical.

R3 Processing of personal data concerns any operation or series of operations on such data whatever the procedure, notably the collection, recording, organisation, storage, adaptation or modification, extraction, consultation, utilisation, transmission, diffusion or any other way of making the data available, the linking or interconnection, as well as the blocking, deletion or destruction.

On healthcare technology for home use and autonomy

Q4 Cite 6 conditions for good practices with regard to Technologies and Services for Health at Home and Autonomy.

Q5 Cite the six dimensions specific to multiaxial and multidisciplinary evaluation of Technologies and Services for Health at Home and Autonomy.

Q6 Explain in what way the evaluation process is one of the ethical conditions for health data with regard to Technologies and Services for Health at Home and Autonomy.

Q7 Describe succinctly the two contrasting facets of technological innovation.

Q8 Cite two major international declarations that legitimize the use of Technologies and Services for Health at Home and Autonomy

R4

1. Evaluation: precise indicators of the service rendered whether medical or social must be available; recommendations for good practices must accompany any diffused technology;
2. Prescription: Professionals must make a methodical evaluation (consultation in gerontechnology, use of the ICF . . .) of the disabilities and disadvantages

Exercises (continued)

of persons with chronic disease, or with a handicap or frail elderly people, including the state of their place of residence. This evaluation must be the basis of a prescription or a recommendation for these technologies and the associated services;

3. Use: these technologies must be used in the spirit and the respect of the fundamental principles of healthcare ethics (Humanity and dignity, Solidarity, Equity and justice...). Following the prescription of such technologies, a medical or medico-social team must follow the patient. The use of such technology must concern all of those involved in chronic disease or gerontology (doctors, nurses, psychologists, social workers, physiotherapists, ergotherapists, dieticians, speech therapists, podologists ...), working closely with family aids;

4. Economic equity: equality of access to these technologies and services must be guaranteed, notably by a solidarity-based system to cover costs;

5. Training: and person involved in the care of people who are sick or frail because of age or handicap must have the opportunity to receive training in the possibilities and limits of these technologies and services;

6. Debate: the role of technology in the care of persons who are frail or socially disadvantaged must involve decision-makers and healthcare professionals, and even every citizen, and become an active theme of bioethics; forums for open debate must go along with their use.

R5

1. The purely technical dimension (robustness, reliability, maintenance),
2. The ergonomic characteristics: the relationship between the apparatus and their users,
3. The medical service rendered (new avenues for therapeutic practices),
4. The social service rendered (improvement/creation of personal services...),
5. The purely economic dimension (costs and real gains)
6. The deontological aspects (respect of ethics and legislation, responsibilities and recourse)

R6 Ethical decisions that involve the economy and health require any judgment to be based on proven fact, indicators of services rendered and sufficient levels of evidence. Evaluation is therefore of major importance in public health. This is referred to as evaluation ethics.

(continued)

Exercises (continued)
R7

1. Medically supervising, assisting, helping, rescuing, restoring links, improving quality of life, optimising expenditure, reducing contributions, encouraging and providing support for those who remain at home, avoiding or delaying placement in an institution.
2. Imposing things without considering the wishes or the consent of those concerned, interfering in private lives, violating personal dignity, dehumanising the relationship with the sick or elderly person who is losing autonomy, giving rise to all sorts of abuse and ill-treatment, replacing as much as possible professionals and voluntary helpers by machines, etc.

R8

1. The Universal Declaration of Human Rights, which specifies in article 27-1: 'Everyone has the right freely to participate in the cultural life of the community, to enjoy the arts and to share in scientific advancement and its benefits.'
2. Recommendation R(98)9 of The Committee Of Ministers To Member States relative to dependence (adopted by the Committee of Ministers 18th September 1998) lengthens and clarifies this declaration by stipulating that: 'All dependent persons or persons who may become dependent, regardless of their age, race and beliefs and the nature, origin and degree of severity of their condition, should be entitled to the assistance and help required to enable each of them to lead a life as far as possible commensurate with their ability and potential. They should therefore have access to services of good quality and to the most suitable technologies'.

On technologies for data security

Q9 What is the difference between symmetric and asymmetric encryption?

Q10 Briefly explain the concept and the implementation of a digital signature.

R9 In symmetric encryption, the same key is used for both encryption and de-encryption. This key must therefore be kept secret to guarantee the confidentiality of the exchange. In asymmetric encryption, encryption is achieved using a so-called 'public' key. This key, even though it is known to all, corresponds to a single recipient who alone can read the message, which ensures confidentiality. However, this system can be used for the digital signature. In this case, encryption is achieved using the sender's so-called 'secret' key. The fact that it can be read using the sender's 'public' key confirms that the message was really sent by this particular sender. In this case, confidentiality is not protected.

> **Exercises** (continued)
>
> **R10** The digital signature makes it possible to make sure that the person who claims to be the author of a message is in fact the author. The author hashes his message with his private key (known to him alone). This hashing generates a fingerprint: a series of fixed-size bit string, which is incompre hensible to people, but corresponds to a sort of mathematical concentration of the message. Any modification (even of a single character) of the message leads to a radically different fingerprint. The recipient receives both the plain text and the fingerprint. By recalculating a fingerprint from the plain text and comparing this with the fingerprint attached to the message, the recipient can ensure that the sender was indeed the signatory of the message.

References

Allaert FA, Le Teuff G, Quantin C et al (2004) The legal acknowledgement of the electronic signature: a key for a secure direct access of patients to their computerised medical record. Int J Med Inform 73(3):239–242

Anonymous 'Convention n°108 (28 January 1981) for the Protection of Individuals with regard to Automatic Processing of Personal Data'

Anonymous 'French law "Loi informatique et libertés", Act n°78-17 of 6 January 1978, on information technology, data files and civil liberties'

Anonymous 'Post i-2010 priorities for new strategy for European information society (2010–2015).' http://ec.europa.eu/information_society/eeurope/i2010/docs/post_i2010/090804_ipm_content.pdf

Anonymous 'Since EHR systems may contain a large amount of data over a long period of time, the new European legal framework should also foresee, among other things, the need for a comprehensive logging and documentation of all processing steps that have taken place within the system, combined with regular internal checks and follow-up on correct authorization, and regular internal and external data protection auditing. See also European Commission Commission Recommendation of 2 July 2008 on cross-border interoperability of electronic health records', C (2008) 3282 final, 2 July 2008, Point 14(k). It will also be an important challenge for legislators to guarantee that all groups in society (including single parents, homeless persons, the elderly and disabled, isolated communities, etc.) have equal access to electronic health records. See also European Commission, 'e-Health – making healthcare better for European citizens: an action plan for a European e-Health Area', COM (2004) 356 final, 30 April 2004, 15'

Anonymous 'Directive 95/46/EC on the protection of individuals with regard to the processing of personal data and on the free movement of such data'

Armstrong BK, Kricker A (1999) Record linkage – a vision renewed. Aust N Z J Public Health 23(5):451–452

Arney D, Venkatasubramanian KK, Sokolsky O et al (2011) Biomedical devices and systems security. Conf Proc IEEE Eng Med Biol Soc 2011:2376–2379

Bellare M, Canetti R, Krawczyck H (1996) 'Message authentication using hash functions', the HMAC construction. RSA Laboratories'CryptoBytes 2:1–5. http://www.cs.ucsd.edu/users/mihir/papers/hmac.html/

Blakely T, Woodward A, Salmond C (2000) Anonymous linkage of New Zealand mortality and census data. Aust N Z J Public Health 24(1):92–95

Borst F, Allaert FA, Quantin C (2001) The Swiss solution for anonymously chaining patient files. Stud Health Technol Inform 84(Pt 2):1239–1241

Carretero S, et al (2012) Can technology-based services support long-term care challenges in home care?: analysis of evidence from social innovation good practices across the EU, in European Commission – Joint Research Centre. (http://www.epractice.eu/files/Can%20Technology-based%20Services%20support%20Long-term%20Care%20Challenges%20in%20Home%20Care_%20Analysis%20of%20Evidence%20from%20Social%20Innovation%20Good%20Practices%20across%20the%20EU_%20CARICT%20Project%20Summary%20Report_0.pdf)

Chen YY, Lu JC, Jan JK (2012) A secure EHR system based on hybrid clouds. J Med Syst 36(5):3375–3384

Chryssanthou A, Varlamis I, Latsiou C (2011) A risk management model for securing virtual healthcare communities. Int J Electron Healthc 6(2–4):95–116

Churches T, Christen P (2004) Some methods for blindfolded record linkage. BMC Med Inform Decis Mak 4:9

Coatrieux G, Quantin C, Allaert FA et al (2011a) Lossless watermarking of categorical attributes for verifying medical data base integrity. Conf Proc IEEE Eng Med Biol Soc 2011:8195–8198

Coatrieux G, Auverlot B, Roux C (2011b) Watermarking – a new way to bring evidence in case of telemedicine litigation. Stud Health Technol Inform 169:611–615

Cormen TH, Leiserson CE, Rivest RL, Stein C. Introduction to algorithms, 3rd edn. Edition Dunod. Collection: Sciences Sup

Couris CM, Gutknecht C, Ecochard R et al (2006) Estimates of the number of cancer patients hospitalized in a geographic area using claims data without a unique personal identifier. Methods Inf Med 45(5):515–522

Das S, Kundu MK (2012) Effective management of medical information through a novel blind watermarking technique. J Med Syst 36(5):3339–3351

Engelbrecht R, Hildebrand C, Jung E (1995) The smart card: an ideal tool for a computer-based patient record. Medinfo 8(Pt 1):344–348

Hsu CL, Lu CF (2012) A security and privacy preserving e-prescription system based on smart cards. J Med Syst 36(6):3637–3647

Hustinx P (2009) Data protection in the light of the Lisbon Treaty and the consequences for present regulations. In: 11th conference on data protection and data security – DuD 2009, Berlin

Jones PS, Meleis AI (1993) Health is empowerment. ANS Adv Nurs Sci 15(3):1–14

King T, Brankovic L, Gillard P (2012) Perspectives of Australian adults about protecting the privacy of their health information in statistical databases. Int J Med Inform 81(4):279–289

Kun L et al (2007) Improving outcomes with interoperable EHRs and secure global health information infrastructure. Conf Proc IEEE Eng Med Biol Soc 2007:6159–6160

Laske C (1996) Legal liability issues in health care telematics. Med Inform Europe 34:942–945

Lau DH (2002) Patient empowerment – a patient-centred approach to improve care. Hong Kong Med J 8(5):372–374

Mandl KD, Szolovits P, Kohane IS (2001) Public standards and patients' control: how to keep electronic medical records accessible but private. BMJ 322(7281):283–287

Morin E (2004) La méthode, tome 6: Ethique, ed. Le Seuil, Paris

Quantin C, Benhamiche AM, Faivre J et al (1998a) How to ensure data security of an epidemiological follow-up: quality assessment of an anonymous record linkage procedure. Int J Med Inform 49(1):117–122

Quantin C, Bouzelat H, Allaert FA et al (1998b) Automatic record hash coding and linkage for epidemiological follow-up data confidentiality. Methods Inf Med 37(3):271–277

Quantin C, Allaert FA, Gouyon B et al (2005a) Proposal for the creation of a European healthcare identifier. Stud Health Technol Inform 116:949–954

Quantin C, Binquet C, Allaert FA et al (2005b) Decision analysis for the assessment of a record linkage procedure: application to a perinatal network. Methods Inf Med 44(1):72–79

Quantin C, Allaert FA, Fassa M et al (2007a) How to manage secure direct access of European patients to their computerized medical record and personal medical record. Stud Health Technol Inform 127:246–255

Quantin C et al (2007b) Interoperability issues regarding patient identification in Europe. Conf Proc IEEE Eng Med Biol Soc 2007:6161

Quantin C, Trouessin G, Allaert FA et al (2008) Combining hashing and enciphering algorithms for epidemiological analysis of gathered data. Methods Inf Med 47(5):454–458

Quantin C, Coatrieux G, Allaert FA et al (2009) New advanced technologies to provide decentralised and secure access to medical records: case studies in oncology. Cancer Inform 7:217–229

Quantin C, Benzenine E, Allaert FA et al (2011a) Medical record search engines, using pseudonymised patient identity: an alternative to centralised medical records. Int J Med Inform 80(2):e6–e11

Quantin C, Benzenine E, Auverlot B et al (2011b) Empowerment of patients over their personal health record implies sharing responsibility with the physician. Stud Health Technol Inform 165:68–73

Quantin C, Jaquet-Chiffelle DO, Coatrieux G et al (2011c) Medical record: systematic centralization versus secure on demand aggregation. BMC Med Inform Decis Mak 11:18

Rey J, Douglass K (2012) Keys to securing data as a practitioner. J Med Pract Manage 27(4):203–205

Rialle V (2009) Quelques enjeux de l'éthique évaluative en gérontologie. Revue Francophone de Gériatrie et de Gérontologie 16(156):262–266

Ricoeur P (1992) Oneself as another (trans. Blamey K), ed. University of Chicago Press, Chicago

Rivest RL, Shamir A, Adleman L (1978) A method for obtaining digital signatures and public key cryptosystems. Commun ACM 2:10

Ross SE, Lin CT (2003) The effects of promoting patient access to medical records: a review. J Am Med Inform Assoc 10(2):129–138

Rostad L (2008) An initial model and a discussion of access control in patient controlled health records. Proceedings of the 2008 Third International Conference on Availability, Reliability and Security. IEEE Computer Society, Washington, DC, USA, pp 935–942

Turchin MC, Hirschhorn JN (2012) Gencrypt: one-way cryptographic hashes to detect overlapping individuals across samples. Bioinformatics 28(6):886–888

Vulliet-Tavernier S (2000) Discussion about anonymity in health data processing. Médecine et droit 40:1–4

Wei J, Hu X, Liu W (2012) An improved authentication scheme for telecare medicine information systems. J Med Syst 36(6):3597–3604

Chapter 12
Hospital Information Systems

P. Degoulet

Abstract Hospital information systems (HIS) are computer systems designed to ease the management of all the hospital's medical and administrative information and to improve the quality of health care. After a brief historical review, this chapter describes the overall strategy to select, deploy and then evaluate the deployed solution. Clear IT governance and IT alignment strategies are prerequisite to HIS deployment success. Emphasis is put on business process analysis/reengineering to select the business components which integration will constitute the overall HIS. In the clinical domain, main business components include the Patient Identification (PID), Admission/Discharge/Transfer (ADT), the electronic health record (EHR), the computerized provider order entry (CPOE) and the resource and appointment scheduling (AS) systems. Deployment phases are considered from basic ancillary services to integration into broader system and HIS data reuse for research. HIS return on investment (ROI) is described from a clinical, organizational and financial point of view.

Keywords Hospital Information Systems (HIS) • Clinical Information Systems (CIS) • Strategic alignment • IT governance • Health business components • Electronic Patient Record (EPR) • Electronic Health Record (EHR) • Computerized Provider Order Entry (CPOE) • CCOW • HL7 • IHE • Risk analysis • HIS evaluation • HIS return on investment (ROI)

P. Degoulet (✉)
CRC INSERM U872 eq 22, René Descartes University, 15 rue de l'Ecole de Médecine, Paris 75006, France
e-mail: patrice.degoulet@egp.aphp.fr

A. Venot et al. (eds.), *Medical Informatics, e-Health*, Health Informatics,
DOI 10.1007/978-2-8178-0478-1_12, © Springer-Verlag France 2014

After reading this chapter you should be able to:

- Explain the principal objectives of a hospital information system (HIS) and summarize the main processes that can benefit of computerized information processing;
- Summarize the workflow within a hospital of information relating to biological and imaging investigations, and describe the different steps that can benefit from computerization and/or automation;
- Summarize the workflow within a hospital of information relating to drug prescription and dispensing and explain which steps can benefit from computerization and/or automation;
- Cite the main sub-systems and sets of functionalities of a HIS;
- Cite the benefits and drawbacks of HIS architectures based on business processes (horizontal applications), hospital structures (vertical applications), and mixed;
- Quote the main outcomes that can be expected from a fully computerized hospital information system as regards the quality of care and the hospital productivity;
- Discuss the main issues and the barriers observed when deploying an integrated HIS

12.1   Introduction: Brief History

A hospital information system (HIS) may be defined as "a computer system designed to ease the management of all the hospital's medical and administrative information and to improve the quality of health care (Degoulet and Fieschi 1999)". A HIS is an integrated system by vocation (i.e. within the hospital and with the outside environment), that could also be called integrated hospital information and communication system (HICS).

The first HIS were developed in the mid-1960s in the United States and a few European countries, such as the Netherlands, Sweden or Switzerland (van de Velde 2003). Their development has followed the general evolution of computer technology (Fig. 12.1): main frames, microcomputers replacing passive terminals, minicomputers tied together into distributed systems, internet-based applications, and more recently cloud computing.

The architecture of the earliest mainframe-based systems allowed the connection of health professionals through passive terminals to databases of shared patient records. The emergence of the mini-computer industry was at the origin of multiple applications dedicated to structures smaller than an entire hospital, in particular in ancillary departments such as biology, radiology, pharmacy, and in some specialized medical units (e.g., intensive care, cardiology or oncology). Development of microcomputers was a chance for entire generations of end-users to get a glimpse to applied information technology (IT). More user-friendly interfaces

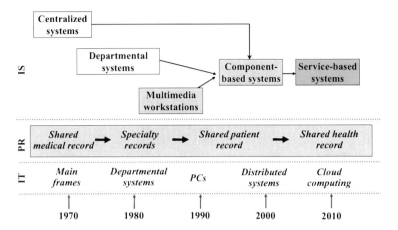

Fig. 12.1 Evolution of HIS info- and infrastructures. *IS* information system, *PR* patient record, *IT* information technology

facilitated the access to HIS functions but also to personal applications such as document or spread sheets processing that justifies the term Personal Computer (PC). Multimedia capabilities of PCs appeared as an opportunity in the health domain where images and signal processing are of crucial importance for patient care. PCs became multimedia workstations hiding the complexities of the underlying infrastructure (Degoulet et al. 1994). In the years 2000 information systems are progressively built up as sets of software-components like Lego pieces exchanging standardized messages through communications channels (software bus) (Sauquet et al. 1994). With the development of the Internet and high speed wide area networks, application servers can quit the hospital environment to be hosted in secured clouds. Patient records that were centralized in main-frame applications, distributed in departmental systems could be unified again and eventually be consolidated at multisite, regional or even wider levels.

Although several hundred HIS are available on the market, few hospitals have reached a deployment and maturity level sufficient to allow the transparent sharing and communication of information between health professionals as well as access to contextual knowledge to foster the use of good medical practice rules. Heterogeneity of tasks and for hospitals professional involved, diversity of provider organisations structures, and the complexity of the management of change (Lorenzi and Riley 2010) mainly explain the difficulties encountered.

Nevertheless the computerisation process is now considered as a necessity both by deciders and health professionals in a planning strategy that should take into account the following preconditions for success:

– A clear definition of the IT governance to be aligned on the overall institution governance;
– An IT "urbanisation" road map following a precise analysis of health business processes;

- An adequate estimation of the financial and human resources necessary to reach the defined HIS target;
- A step by step deployment strategy with realistic calendars taking into account the long term objectives, the resources engaged, and the management of change;
- A good understanding of the sociology of healthcare institutions;
- An analysis of the risk associated with any large IT project and a precise plan to guaranty the continuity of services and the adequate recovery in case of system failure;
- An internal and external communication strategy.

12.2 The IT Governance Strategy

12.2.1 e-Governance Concepts

Financial scandals of the last 20 years such as the 2001 Enron in the US or the 2008 *Société Générale* in France have clearly shown the necessity for a precise governance of private and public companies including internal and external control procedures to guaranty the respect of legislation, social norms and rules of good practice in the field. The Sarbanes-Oxley Act of 2002 (SOX) enacted by the 107th United State Congress gives a framework to guaranty accounting transparency, the independence of auditors during review processes, and mechanisms to raise alerts (US Government 2002). It has been adapted to many different country legislations.

IT governance or eGovernance represent the adaptation to IT of general governance principles. Eight dimensions are described for IT governance in (Georgel 2009): strategic alignment, IT management, IT resource management, IT risk management, IT performance and services management, IT audit and control, IT value, and IT maturity. COSO, CobiT and ITIL are significant examples of reference frames for IT governance (COSO 2013; Isaca 2012; ITIL 2013).

12.2.2 IT Strategic Alignment

IT alignment consists in making the IS objectives fully coherent with the main objectives of the enterprise, i.e., providing optimal care at minimised cost (Fig. 12.2).

Quality of care is defined by the Institute of Medicine (IOM) as "the degree to which health services for individuals and populations increase the likelihood of desired health outcomes and are consistent with current professional knowledge". Development of decision support systems is a major contributing technology that can be adapted at each step of health intervention (prevention, diagnosis,

Main objectives	Contributing objectives
Improve the quality and continuity of care	Standardize medical processes Help decision-making Reduce professional errors Improve outcomes
Control costs	Optimize processes Reduce administrative overhead Reduce the length of inpatient stays Facilitate strategic decisions

Fig. 12.2 Major objectives of a hospital information system

therapeutics, and prognosis) with the target of improving outcomes and reducing inadequate behaviour and medical errors (IOM 1994).

Strategic alignment is the first step of a value generating process at the enterprise level, which for a hospital might mean better attractiveness and greater productivity (Fig. 12.3).

12.2.3 IT Management

IT management is organised in many hospital around two streams of decision workflow, the administrative information management from one side (e.g., billing, accounting, payrolls) under the direction of a chief information officer (CIO) and the health information management under the direction of a chief medical information officer (CMIO) (Fig. 12.4, left column).

In depth analysis of a hospital information system may show that this duality is antagonist to the key objective of a HIS. A data element, e.g. patient age can be considered an administrative or a medical item depending on the context of use. In a university hospital all pieces of information can support clinical or translational research.

For a hospital, IT management can be controlled by an IT governance committee attached to the hospital governance or strategic committee that will guaranty the alignment of the IT strategy with the overall hospital strategy.

The IT commission is the place where the IT manager(s) should discuss with end-user representatives of the application modalities of the IT strategic plan. Ad hoc groups can be created for any given specific project.

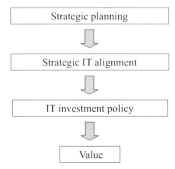

Fig. 12.3 Strategic alignment of the information system is intended to generate value (Adapted from Georgel 2009)

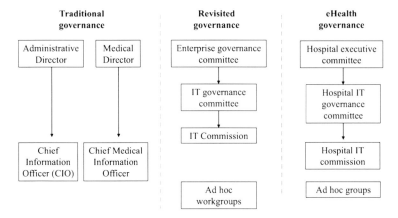

Fig. 12.4 IT management structures

12.3 HIS Urbanisation Plan

12.3.1 The Four Quadrants Model of Urbanisation

Under the term urbanisation, several steps can be considered from the information system analysis up to the selection of software components and HIS deployment (Fig. 12.5) (Le Roux 2009):

- Understanding of the enterprise missions (i.e. the hospital) and the actors involved (external and internal);
- Analysis of main activities and of the flow of tasks (business processes);
- Analysis of data and information exchanges;
- Selection of software components and their integration to build up the hospital information system;
- Selection of a deployment strategy and management of change (Tomas 2007).

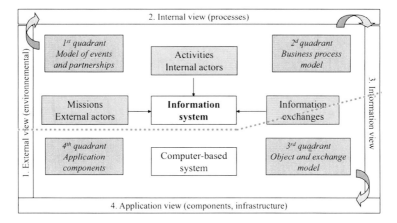

Fig. 12.5 The four quadrants urbanisation model (Adapted from Le Roux 2009). Activities under the *dotted line* are highly IT-oriented

12.3.2   External View: The Information System Environment

In an environmental model, the enterprise is considered as a black box that interacts with the HIS (general practitioners, insurances companies, etc.). Outpatients visits, admission/discharge, medical advices are the main interaction with the hospital environment. Figure 12.6 illustrates the multiplicity of actors that will interact with the HIS. External actors include government bodies, the industry (e.g. pharmaceutical, software vendors), health insurances, and media. Patients interact as clients but also as lobbies through consumer groups or social networks. Hospital missions cover patient care but also teaching and research.

12.3.3   Internal View: Structures and Business Processes

To achieve the internal view, the hospital is then considered as a "glass box" with actors and structures. Hospital actors include the healthcare professionals (physicians, nurses, paramedics, biologists, pharmacists, engineers) and various administrative groups (finances, logistics, etc.) (Fig. 12.6).

In France hospital structures are organised into two categories (administrative and medical) with their associated human and material resources (beds, equipments) and a strong hierarchy (Fig. 12.7) (Degoulet and Fieschi 1999). In North America the material and space resources are likely shared between the various human resources (departments). In term of information system every structure becomes a resource available to any other structures or to the outside, that performs acts, produces information and consumes resources.

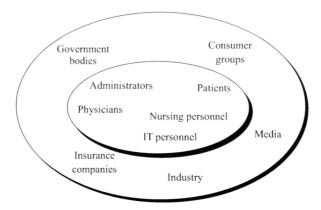

Fig. 12.6 Main actors of a HIS (Adapted from Degoulet and Fieschi 1999)

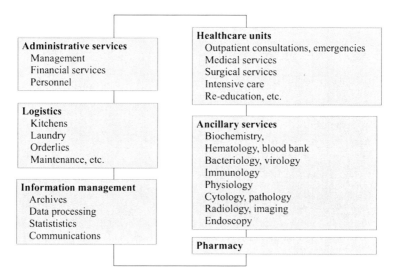

Fig. 12.7 Hospital structures (Adapted from Degoulet and Fieschi 1999)

The functional analysis starts from the hospital activity to describe information circuits that will determine IS functions and then select the functions that will benefit from an automated and/or computerised approach.

A business process is a set of structured activities or tasks that support the missions of an enterprise. Operational processes are directly connected with the core business of the enterprise (i.e., the patient care for a hospital). Supporting processes are facilitators for the operational processes (e.g., supplies, accounting, patients' billing or professional recruitment). Management processes are associated with the management of the enterprise (e.g. strategic planning, financial forecasting or research support).

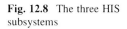

Fig. 12.8 The three HIS subsystems

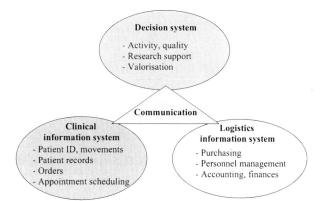

Business processes can be decomposed into several sub-processes. Achievement of a process implies that the right organisation is installed. An appropriate analysis of enterprise objectives and business processes leads to some forms of reengineering where existing situations are subjects to criticisms and alternate ways proposed. Certain structures might be considered as overrepresented and conversely other ones underrepresented (e.g. the structures dealing with the management or reengineering processes).

Grouping of the three major categories of processes allow considering three main HIS subsystems: the clinical information system, the logistics information system, and the management/decision system (Fig. 12.8). This chapter mainly covers the clinical information system that directly corresponds to the core business of the hospital.

IS urbanisation is particularly concerned with transversal processes that concern a multiplicity of actors and structures within the hospital. A significant example is given by the cycle of biological investigations. Figure 12.9 shows the different steps within the three hospital structures concerned: – the clinical units for medical orders, blood sampling and result display; – laboratories for biological analyses and results production; – administrative departments for billing and activity measurements. Most tasks can benefit of automation (e.g. analysis automaton) and/or computerised processing (e.g., physician order entry).

The imaging production cycle shares many similarities with the biological process as shown in Fig. 12.10. Most steps can benefit from computerised processing. Production of image and their pre-processing can benefit from automation and informatics (e.g., scanners, MRI, echography, video endoscopy). At the interpretation steps computer programs are susceptible to facilitate image visualisation, 3D reconstructions from sets of 2D slides, automatic detection of specific zones (e.g., breath calcifications) or multiple types of measurements (distances, volumes, densities, etc.). Vocal dictation and recognition is widely used by radiologists for the automated production of imaging reports.

The drug cycle correspond to one of the most complex process processes in a hospital, in particular in France where physician orders need to be validated

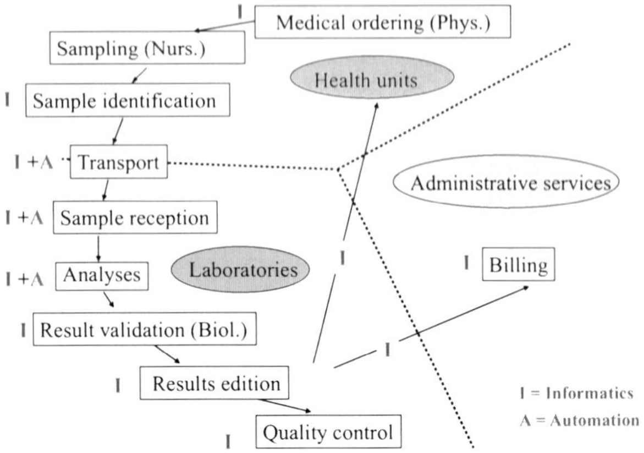

Fig. 12.9 Biological orders management process. *Phys.* physician dependent task, *Nurs.* nurse dependent task, *Biol.* biologist dependent task

Fig. 12.10 The imaging order management process. *Phys.* physician dependent task, *Rad.* radiologist dependent task

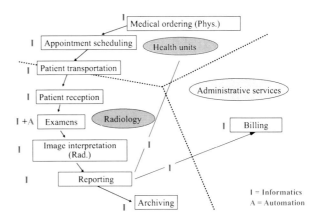

by pharmacists. This means an interaction flow between physicians and pharmacists in case of disagreement with the final decision being at the responsibility of the physician (Fig. 12.11). Drugs are administered to patients by the nurses. All steps can benefit from a computerised approach. Robots can be used for drug preparation at the pharmacy level (e.g. perfusions) and or transportation from the pharmacy to the clinical units. Barcoding or radiofrequency identification of unit doses (RFID) given to the patient can close the medication loop, avoid drug distribution to the wrong patient and help to validate in the IS the step of drug administration.

Business process modelling allows the grouping of the different prescription order processes into a general order process including other investigation and nursing orders. Its computerisation will generate the *Computerised Provider Order Entry* (CPOE) business component. This grouping is all the more justified that orders of different nature are commonly combined (e.g., biology and imaging, biology and drugs) to constitute orders sets or more complex protocols that include temporal and/or patient value dependent constraints.

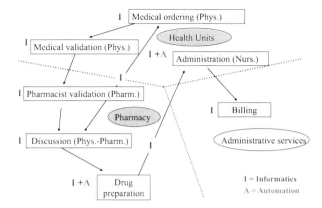

Fig. 12.11 The drug order management process. *Phys.* physician dependent task, *Pharm.* pharmacist dependent task, *Nurs.* nurse dependent task

12.3.4   The Informational View

At each execution step within a process, information is generated, used or exchanged. The informational view gives a conceptual model of the business objects that are individualised during the business modelling phase and the interactions between the business objects.

A business object is an abstract concept that supports a meaning for the actors and structures of the enterprise. A business object has both static properties (e.g. a patient identification, a date of prescription, a drug dosage) and dynamic properties that describe the behaviour of the object and its interactions with the other objects. The UML (Unified Modelling Language) is the most commonly used language to describe the static and dynamic properties of objects.

12.4   The Computerisation Strategy

12.4.1   Component Selection

Transformation of business objects into computerised components or applications represents the last step of IS urbanisation/reengineering approach. The adequate selection of components and/or application conditions the quality and life expectancy of the future HIS. The objective is to integrate in a coherent structure sets of components/applications that will work together and exchange standardised messages. Inappropriate selection and integration will lead to technological dead ends, organisational issues and finally useless expenses.

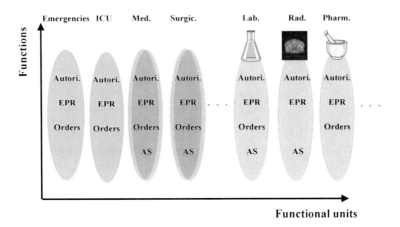

Fig. 12.12 Structure-oriented approach and vertical silos. *Autori.* autorization

12.4.2 *Structure-Oriented Approaches and Vertical Silos*

This approach, too frequently used in hospitals consists in selecting the computer applications on the basis of the hospital structures and/or professional lobbies (Fig. 12.12). For the clinical information systems, this means the selection of applications for the different categories of clinical units (e.g., urgencies, intensive care units, cardiology or oncology) or investigation units (e.g. biology, radiology, endoscopy, pharmacy, pathology). For the clinical units this strategy has given rise to specialised or even disease-oriented medical records that can be shared between the same specialties of different institutions and facilitate epidemiological studies and translational research. It has however the major drawback of splitting the patient records between the different specialised records, a situation that can rapidly constitute a barrier for the coordination and continuity of care for a given patient. It leads to the deployment of redundant applications such as different ordering or appointment functions (e.g. the one in a clinical system and the one in a radiology system) and possibly incoherent information (i.e. between the different pieces of patient records).

The number Nb of interfaces needed to integrate n vertical applications increases with the square of n, approximately:

$$Nb = n^*(n - 1)/2$$

if one considers that every application needs to be connected with all the remaining applications $(n - 1)$ and that connecting the application x with the application y allows at the same time to connect y with x. Complexity increases significantly when two hospitals are merged to constitute a multi-site environment since the applications n_A of site A are added to the applications n_B of B. Integration of

Fig. 12.13 Process-oriented
approach and horizontal
components

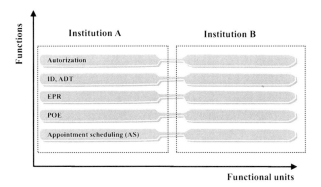

applications exchanging standardised messages through a software bus allows a
significant reduction of interfaces to the total number *n* of applications to integrate.

12.4.3   Process-Oriented Approaches: Horizontal Components

The horizontal approach consists in individualising transversal processes that can
benefit from computerisation and deploy the component that will support these
processes (Fig. 12.13). If the number n' of processes to computerise remains small
($n' \ll n$), then the number of interfaces to develop remains under control, in
particular if a software bus is used. When several hospital are merged, the number
n' of components does not increase as shown on Fig. 12.13 since the clinical
processes concerned do not change. Technical and human risks (security, confidentiality) increase however with the number of sites concerned and need to be
considered with caution.

If the electronic patient record is considered as a process, giving rise to an
Electronic Patient Record (EPR) component, then the HIS will be patient centred
and the unique and shared EPR will a become coordination and continuity of care
tool. If the appointment/scheduling (AS) process is considered a component, then it
becomes possible to optimise the patient trajectory within one or several hospitals.

12.4.4   Mixed Approaches

Mixed approaches consist in using a horizontal approach for clinical units and a
vertical approach for investigation units (Fig. 12.14). The horizontal approach for
healthcare units allows an institution to benefit from a shared EPR, a common
computerised order entry (CPOE) and appointment and scheduling system. The
vertical approach for technical units corresponds to the pre-processing of data

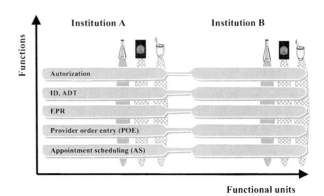

Fig. 12.14 Mixed horizontal/vertical approaches

production (e.g., connection of automatons in biology, production of image from the imaging modalities, reconstitution of complex oncology perfusions for the pharmacy). Transversal functions that are shared by an institution (e.g., appointment scheduling, report dictation) need however to be deactivated in the vertical application and taken in charge by the transversal components. In case of major computer crash, the application in the technical units can be used temporarily as stand-alone applications.

Complexity is intermediate between a fully integrated horizontal approach and a distributed vertical approach with a total number of application/components n" between *n* and *n'*. It is frequently recommended by international organisms such as HIMSS for which the ancillary department applications (biology, radiology, and pharmacy) constitute the first phase of a global computerisation process.

12.5 The Business Components of a Clinical Information System

12.5.1 Healthcare Business Components

Figure 12.15 illustrates the set of components that are required to build up a fully integrated clinical information system (CIS) (Van de Velde 2003). Four major components are specific of the hospital environment. Others are more generic and required in many other activity sectors.

Patient identity (ID) and movement management are essential to guaranty the uniqueness of the patient record within one or several institutions and that each professional will access a unique record whether the patient is examined as inpatient, as outpatient, or given advice at distance of the healthcare facility. Computerised tools allow the search for close identities (name, given name, birth date, etc.) to avoid the constitution of duplicate identities. Temporary identities need to be created (e.g., urgency situation) for not yet identified patients but also

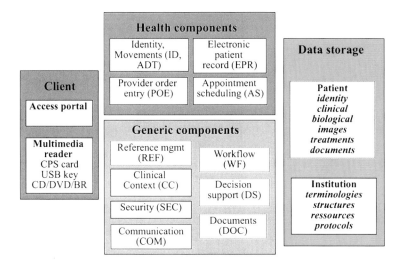

Fig. 12.15 Major components of a clinical information system

programs to merge pieces of records belonging to a unique patient. The ADT (Admission/Discharge/Transfer) component is necessary to instantly localise patients inside the institution.

The unique and shared multimedia electronic patient record (EPR) component is the key component of any CIS. It allows the recording of patient-related data and information whatever their source (physician, nurse, medical device, etc.) or their nature (e.g., clinical, biological, imaging, nursing, "omics" data, procedures). Data protection and data/functions access procedures (i.e., access rights) need to be defined according to the profession of the user (e.g., physician vs. nurse vs. secretary, senior vs. junior), the point of access (inside/outside the hospital), the role of the professional as regards the care of the patient, and the presence/ absence of the patient in the institution at the time of access. The traceability of accesses should be guaranteed with a granularity sufficient in case of controversies between the patient and the institution.

The act management component (*Computerised Provider Order Entry* or CPOE) allows, according the user profile (e.g. physician, nurse), to prescribe individual acts (e.g. serum potassium, vital sign follow up), groups of acts constituting an order set (e.g. ionogram), or more complex associations that will be specific of a clinical situation (e.g. urinary infection, coronary dilation) or a disease stage (e.g. chemotherapy for a typed cancer). Branching rules (e.g. starting and ending condition for a drug) and/or loops can be described in the most complex protocols (e.g., a chemotherapy protocol). Orders are stored in the EPR and allow the production of the different care plans for a patient or a group of patient (e.g. set of beds in a clinical unit).

The appointment and resource scheduling component (AS) allow the efficient management of a single or a set of appointments for a patient (e.g. radiotherapy or dialysis sessions) or a group of patients. Managing the different appointments (inpatient/outpatient visits, specialist advices, investigations, operating rooms) by the same component facilitates the optimal use of resources and the integration of the patient preferences and geographical constraints (e.g. time to go from point A to point B within a hospital) when planning the appointments (e.g. a 1 day surgical intervention).

12.5.2   Generic Components

In this category are included various components that are found in non-healthcare sectors of activities but might be reused or adapted to the hospital environment (Fig. 12.15). They are nevertheless essential to guaranty the integrated and smooth functioning of a HIS.

The reference manager is a generic component that needs to be fed with health-related specific nomenclatures or terminologies. For example diagnosis codes rely on the 9th or 10th revision of the international classification of diseases (ICD9/ICD10). LOINC is the current standard for biological orders terms. SNOMED-CT is the most comprehensive nomenclature of medical terms (IHTSDO 2013). DICOM is the imaging standard in medicine including structures for their acquisition conditions, the storage of images and the associated reports) (DICOM 2013).

The CCOW (Clinical Context Object Workgroup) component based on the HL7 corresponding standard keeps among other functions track of the patient and user identification on a given terminal (the context of work), to help the possibility of making decision errors by displaying the records of several patients at the same time (CCOW 2013).

The decision support (DS) component can be triggered by most major components and particularly the EPR and CPOE components. Domain-independent business rules syntaxes such as SBVR (Semantics of Business Vocabulary and Business Rules) can be used in addition to more domain dependent business rule syntaxes such as Arden or GLIF (OMG 2008; Boussadi et al. 2011).

Integration can be achieved by access to common database structures (e.g. the patient record structures) or by messaging. Communication tools are necessary in case of message communication. HL7 is being considered as the international messaging standard (e.g. admission/discharge/transfer procedures, exchange between the EPR and the laboratories or the pharmacy) (HL7 2013). IHE (Integrating the Healthcare Enterprise) promotes the coordinated use of established standards through integration profiles (IHE 2013). When different terminologies coexist in a CIS, the reference manager may help in solving semantic interoperability issues by mapping terms between the different terminologies. Integration of ancillary applications in the case of a mixed approach may need the development of various plug-in on top of the applications and semantic interoperability tools such as semantic mediators (Degoulet et al. 1998).

12.6 Project Management

12.6.1 Deployment Phases

Deployment of a clinical information system is a lengthy and resource consuming process (3–10 years) that will considerably change the organisation of an institution. It should be considered as a succession of steps or phases, phase $n + 1$ being started when phase n is completed. The HIMSS organisation in the US considers seven stages of electronic medical record adoption (EMR), the term EMR being considered to the broad sense used in this chapter as clinical information system (HIMSS Analytics 2013).

- Stage 0: All three ancillary clinical systems (i.e., Laboratory, Radiology, Pharmacy) not installed
- Stage 1: All three ancillary clinical systems installed
- Stage 2: Major ancillary clinical systems feed a clinical data repository (CDR) where physician can access investigation results. The CDR contains a controlled medical vocabulary. Information from the imaging system may be linked to the CDR at this stage.
- Stage 3: Nursing/clinical documentation (e.g. vital signs, flow sheets) is integrated with the CDR. Implementation of the first level of clinical decision support (CDS) (e.g. drug/drug, drug/lab conflicts).
- Stage 4: Computerised Practitioner/Provider Order Entry (CPOE) for use by authorised professionals and second level of CDS related to evidence based medicine protocols.
- Stage 5: Close loop medication administration with barcoded or radio frequency (RFID) identification of administered drugs.
- Stage 6: Full physician documentation with structured templates. Third level of CDS with variance and compliance alerts. Fully integrated PACS
- Stage 7: Paperless hospital. Clinical data warehousing to analyse pattern of care and improve the quality of care. Data continuity between internal and ambulatory care is guaranteed.

For the first 2013 quarter, the HIMSS group estimated that 78.0 % of US hospitals had reached stage 3 and over, and 41.7 % stage 4 and over. Only 4.2 % of hospitals had only the three ancillary clinical systems installed (stage 1). 1.9 % of hospitals were at stage 7 and 9.1 % at stage 6 (HIMSS Analytics 2013).

At each stage, the deployment should be as global as possible (Fig. 12.16, right column) to avoid the constitution of computerised islands inside an institution, the professionals and the patients being confronted with different levels of computerisation when going from one unit to the other.

Once entered in the EPR, data can be reused for other purposes that immediate care, in particular when "omics" are coupled with clinical data. Examples include quality of care measurements, epidemiologic studies (phenome-wide/genome-wide association studies), and clinical and translational research (e.g. patient selection in

Fig. 12.16 Deployment
strategies

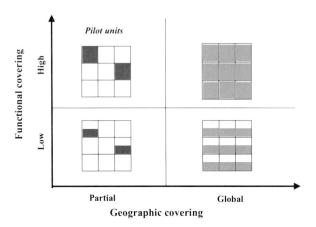

the context of a more personalised practice medicine). Direct querying from
operational database is considered to generate a risk on the smooth functioning of
the CIS (Fig. 12.17). Querying on mirrored database is less prone to side effects.
Building of clinical data warehouse seems more promising and has the advantage of
possible sharing of anonymised information on a national or international context
such as the experiences with i2b2, tranSMART or similar platforms (Kohane et al.
2012; tranSMART 2013).

12.6.2   Required Resources

Estimation of investment and running costs for HIS vary from one country to the
other due to the large discrepancies between licensing, adaptation and localisation
costs. Difference legislation may add extra costs that are of an order of magnitude
greater that the only cost of translation. IT running budgets in Europe are frequently
between 1 % and 3 %, inferior by a two to three factor to the costs observed in
North-America (e.g. 3–10 %)

Human resources are of particular importance when deploying a HIS. Health
professionals (physicians, nurses, secretaries) need to invest to the adaptation of the
clinical information system to the local context (e.g. reference terminologies,
orders, protocol and guidelines formalisation, structured data entry in specific
domains). Referents or "super-users" in each clinical unit that will receive a more
intensive training are great facilitators for the success of a CIS deployment.

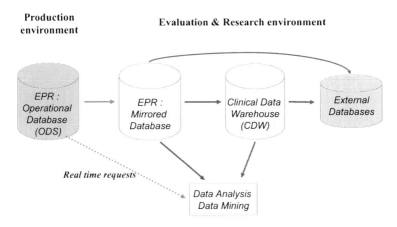

Fig. 12.17   Reuse of clinical data and data warehousing

12.6.3   Risk Analysis and the Unexpected Effects of CIS Deployment

Deployment of a CIS changes considerably the organisation of an enterprise and the relationships between professionals that will become more and more dependent on the smooth functioning of the IT applications. Unexpected consequences of IT implementation have now been studied such as the ones observed after the introduction of a CPOE (Ash et al. 2004, 2009; Koppel et al. 2005).

The most significant observed changes fall into the following categories:

- CIS allows the health professional to work asynchronously by comparison with the traditional visits for inpatients where senior and junior physicians as well as nurses and clinical pharmacists can interactively participate to the daily decisions. A physician prescribing from his/her office has not the guaranty that his/her orders will be fully understood by the nurses and/or the patient.
- Entering an order in a CPOE is frequently more time-consuming than dictating to an assistant but less prone to transcription errors
- New types of errors are observed with current computer interfaces (e.g. typing error such as an extra zero for heparin units that can remain in the treatment window, wrong selection in a scrolling menu).
- Never ending demands requested from the software providers generate frustration (delays of development), are source of programming bugs and possibly functional regression (i.e. a function that was routinely used become non-functional)
- Change in the power structure of the organisation where the junior physician and the nursing staff become more computer-competent than the senior physician and managers are source of wrong behaviour.

– As with any new technology, excessive emotional reactions in front of computer-associated workflow issues or bugs can generate aggressiveness and rejection of the entire computer system (Sittig et al. 2005).

Awareness of the unexpected effects of a computerised system should be considered in the larger framework of the different categories of IT associated risk. IT risks are either human (intentional or not) or technological (e.g. disponibility, integrity, confidentiality breaches). A risk management plan needs to be settled before any HIS deployment as well a crisis plan to be applied in case of severe computer system failure (ISACA 2012). IT risk increases with the size of the institution and the number of component/applications involved.

12.6.4 Change Management

Change management concerns all actions that will facilitate the transition for an institution from an initial IS state A to a target state B. Following elements are considered to be facilitators for that transition:

– The constitution of a visible eGovernance structure
– User participation to the different steps of IS implementation, i.e. from the selection to the deployment and evaluation stages
– Individualisation of opinion leaders that will foster the project
– An objective presentation of the expected benefits, difficulties, and barriers to overcome
– Constitution of mixed teams involving health professionals and IT specialists
– An adequate training of each future end-user, adapted to the specificities of the profession involved and the constitution of referents or "super-users" that will remain close to their less trained colleagues
– A permanent evaluation of the user satisfaction
– A good reactivity of IT personnel to answer end-user demands and repair/replace broken materials
– Maintenance contracts to guaranty 24/24 h 7/7 days are in place.

12.7 Evaluation of Hospital Information Systems

12.7.1 Levels of Use

Statistics such as the ones provided by the HIMSS organisation show a progressive penetration of the EHR and HIS based solutions within hospitals. However the fact that a HIS is put at the disposition of professionals does not mean that the system is used, and if used that it is correctly and efficiently used (Fig. 12.18). For example, it

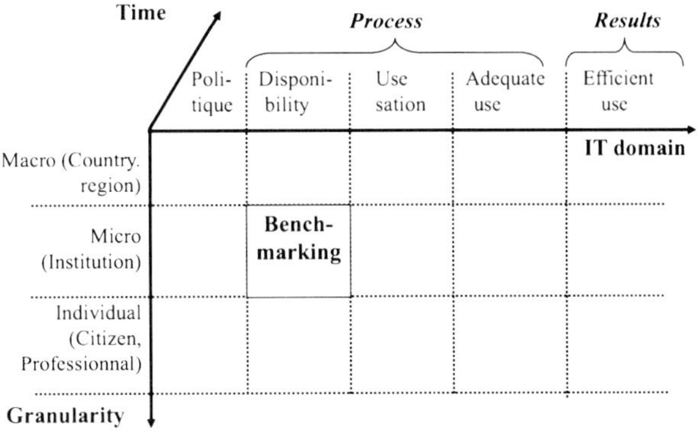

Fig. 12.18 The different
levels of use of an IT solution

seems important for the orders in a CPOE to be entered by the responsible professionals (e.g., physicians for drug orders) and not delegated to a data entry personnel (e.g.., nurses, secretaries) with the risk of transcription or misinterpretation errors. Production of (adequate) use indicators is essential during all the life of a deployment project to motivate end-users, track resistant groups and do comparisons at a national or international level (benchmarking).

12.7.2 Return on Investment (ROI)

An adequate and efficient use of a clinical information system can be translated in term of clinical, organisational and financial benefits (Fig. 12.19). Certain benefits are more tangible than others and can be expressed in monetary value. Other are more difficult to quantify, but nevertheless important as the increased visibility and attractiveness of a highly automated and computerised institution.

In term of clinical benefits, significant results have been observed in term of reduction of medical errors, in particular those related to drug orders or investigation orders. Generalisation of prescription through protocols facilitates professional adherence to rules of good medical practice.

The CIS can be used to couple patient data with validated medical knowledge to trigger decision rules and/or generate teaching guides for the patients and/or the professionals.

Communication functions reduce transcription errors, accelerate the investigation loops (i.e., from the ordering step to the integration of results into the EHR) and by consequence reduce the length of stays.

The different categories of benefits can, in a certain extent, be translated in term of monetary units to be integrated in return on investment studies (ROI). Most recent studies tend to show that:

– Financial benefits are not immediate but postponed (3–5 years) after the initial deployment and appropriate phase often considered as counter-productive;

	Clinical	Organizational	Financial
Tangibles	↘ Iatrogenic behavior	↘ Transcription costs	↗ Revenues
	↘ Mortality/Morbidity	↘ Furnitures - Impression	↗ Productivity
	↘ Information loss	↘ Length of stays	↗ Reimbursement rate
	↗ Drug delivery		
Non tangibles	↗ Patient education	↗ Care coordination	↘ Juridical issues
	↗ Clinical pathways	↗ Information availability	↗ Partnerships
	↗ Clinical alignment	↗ User satisfaction	↗ Institution visibility
		↗ Organizational alignment	↗ Financial alignment

Fig. 12.19 IT return on investment (ROI) approaches (Adapted from Meyer and Degoulet 2008)

- There could be an investment threshold under which investments provide no return or are counterproductive (e.g., persistence of the manual and the computerise system);
- Benefits in term of activity billed to third parties, are positively correlated with the IT capital (personnel and investments taking into account depreciation and amortisation) (Meyer 2010).

12.8   Summary and Conclusions

Hospital information systems that have been developed during the last 40 years have now demonstrated an efficiency and maturity sufficient to consider their rapid generalisation (ONC 2013). In that context the main issue is to determine the most appropriate strategy to select and deploy an existing solution taking into account the technical, financial, and organisational dimensions of such a project. In a difficult economic situation, 5–10 years can represent an appropriate time scale to achieve the goal of an operational HIS. Computerisation of technical units, namely biology, imaging, and pharmacy is often considered as a useful preliminary step keeping in mind that some transversal functions should be better managed by transversal components such as provider order entry or scheduling systems. A shared unique/multimedia EPR is the key components of a CIS, facilitating the continuity of care and the coordination of health professionals. Data warehousing coupled with the CIS can efficiently facilitate quality of care management, epidemiologic studies and translational research.

12.9   For More Information

This chapter contains many internet links relating to hospital and clinical information systems. We particularly recommend the official sites of HIMSS, IOM, ONC as well as the main standard–related sites such as DICOM, HL7, HTSDO, Isaca, ITIL or IHE for their coordinate use through integration profiles.

Exercise

Q1 A 700 beds hospital has the following HIS applications: – a medico-administrative system (ID, ADT) developed in house, two laboratory subsystems (biology on one hand, and bacteriology on the other hand) feeding a result server also developed in house, a commercial pharmacy subsystem used for drug prescription and validation, a radiology subsystem including its appointment and report production modules, an urgency department system only integrated with the ID/ADT subsystem. Which applications are you ready to keep, to quit or to plan to quit?

R1 The situation described in this exercise is frequent in hospitals with the beginning of computerisation of the administrative circuit and the main ancillary departments. The in house development of IT applications is now gradually abandoned and replaced by the purchase and configuration of commercial software components/packages. The application which can disappear in short/mid-terms are the following: the ID/ADT applications replaced by the corresponding modules of a future integrated HIS, the result server and the emergency unit application replaced by the EMR component of the HIS, the radiology appointment module of the radiology application replaced by the general appointment/scheduling component (AS), the prescription part of the pharmacy system replaced by the general POE component. The substitution of the two laboratory systems by an industrial product in compliance with the international exchange standards (HL7) must be made as soon as possible.

Once a solution adopted for the clinical information system, the IT deployment is usually planned in two to four phases. During a given phase, the configuration of the component for the following phase can be made in parallel.

The four following phases represent an example of a validated strategy. Of course, the real schedule has to take into account the specificities of the selected solution, the investment level and be then adapted according to the user feedbacks from the first deployment phases.

(continued)

Exercise (continued)

- Phase 1 (12–18 months): constitution of the eGovernance structure, of the project group devoted to the urbanisation/IS reengineering project, redaction of the call for proposals; choice of an HIS/CIS solution; choice of an IT infrastructure to support the HIS/CIS (i.e. internal to the hospital and/or external by an accredited data hosting provider)
- Phases 2 (6–12 months): configuration of the basic HIS/CIS functions (ID/ADT, EPR, laboratory exams management and integration of the laboratory results into the EPR, DRG coding, in/outpatient stays report production). End-user training to the basic functions.
- Phase 3 (6–12 months): Deployment of the CIS basic functions; configuration of a second set of functions (e.g. simple orders and order sets, appointments for visits, investigations, and operating rooms reservation, simple decision rules), end-user training to the second set of functions; bugs corrections for basic CIS functions.
- Phase 4 (6–12 months): deployment of the second set of CIS functions; configuration of advanced functions (e.g. complex protocols and decision rules, structured questionnaires). Correction of bugs of phases 2 and 3.

Some vendors often propose their own laboratory, pharmacy and radiology modules either as integrated components sharing the same database structures or as integrated applications communicating through standardised message. If this is the case, the configuration of the laboratory, pharmacy and imaging systems will be concomitant with the EPR configuration of phase 2. Deployment of the hospital wide PACS functions can be planned in phases 3 or 4 of the project.

References

HIMSS Analytics (2013) http://www.himssanalytics.org/home/index.aspx. (Description of different levels of adoption of EMR solutions by healthcare providers. It includes annual survey of their deployment in various countries). Accessed 21 Apr 2013

Ash JS, Berg M, Coiera E (2004) Some unintended consequences of information technology in health care: the nature of patient care information system related errors. J Am Med Inform Assoc 11(2):104–112

Ash JS, Sittig DF, Dyskstra R et al (2009) The unintended consequences of computerized provider order entry: Findings from a mixed methods exploration. Int J Med Inform 78(suppl 1): S69–S76

Boussadi A, Bousquet C, Sabatier B et al (2011) A business rules design framework for a pharmaceutical validation and alert system. Methods Inf Med 50:36–50

CCOW (2013) http://www.hl7.com.au/CCOW.htm. (HL7 CCOW webpage). Accessed 31 Mar 2013

COSO (2013) http://www.coso.org/. (COSO official website). Accessed 31 Mar 2013

Council of the Institute of Medicine (1994) America's health in transition: protecting and improving quality. National Academy Press, Washington, DC

Degoulet P, Fieschi M (1999) Introduction to clinical informatics. Springer, New York. Corrected 2d printing. 1999 (With which this chapter shares a part of the introduction, figures 12.6 and 12.7 various ideas shared with Marius Fieschi during 30 years of fruitful collaboration)

Degoulet P, Safran C, Bowers GH (1994) Design and processing issues for the health care professional workstation. Int J Biomed Comp 34(1–4):241–247

Degoulet P, Sauquet D, Jaulent MC et al (1998) Rationale and design considerations for a semantic mediator in health information systems. Methods Inf Med 37:518–526

DICOM (2013). http://medical.nema.org/. (DICOM NEMA main website). Accessed 31 Mar 2013

Georgel F (2009) IT gouvernance, 3ème édition. Dunod, Paris

US Government (2002) http://en.wikipedia.org/wiki/Sarbanes-Oxley. (Pub.L. 107–204, 116 Stat. 745, enacted July 30, 2002). Accessed 31 Mar 2013

HIMSS (2013) http://www.himss.org/ASP/index.asp. (Site of the HIMMS association (Healthcare Information and Management Systems Society), a major organisation devoted to the understanding and promotion of professional health informatics. HIMMS organises annual international meetings on the different continents.). Accessed 31 Mar 2013

HL7 (2013) http://www.hl7.org/. (Main HL7 web page). Accessed 31 Mar 2013

IHE (2013) http://www.ihe.net/. (Main IHE web page). Accessed 31 Mar 2013

IHTSDO (2013) http://www.ihtsdo.org/fileadmin/user_upload/doc/. (SNOMED-CT documentation site). Accessed 31 Mar 2013

Isaca (2012) http://www.isaca.org/COBIT/Pages/default.aspx. (Cobit 5 framework). Accessed 31 Mar 2013

ITIL (2013) http://www.itil-officialsite.com/. (ITIL official website). Accessed 31 Mar 2013

Kohane IS, Churchill SE et al (2012) A translational engine at the national scale: informatics for integrating biology and the bedside. J Am Med Inform Assoc 19(2):181–185

Koppel R, Metlay JP, Cohen A et al (2005) Role of computerized physician order entries in facilitating medication errors. JAMA 293(10):1197–1203

Le Roux B (2009) La transformation stratégique du système d'information. Lavoisier, Paris

Lorenzi NM, Riley RT (2010) Organizational aspects of health informatics. Managing technological change. Springer, New York. (A useful textbook on the organizational aspects of HIS)

Meyer R, Degoulet P (2008). Assessing the capital efficiency of healthcare information technologies investments: an econometric perspective. Yearb Med Inform 2008:114–127

Meyer R, Degoulet P (2010) Choosing the right amount of healthcare information technologies investments. Int J Med Inform 79:235–31

OMG (2008) http://www.omg.org/spec/SBVR/1.0/. (Documents sssociated with Semantics of Business Vocabulary and Business Rules (SBVR), version 1.0). Accessed 31 Mar 2013

ONC (2013) http://healthit.hhs.gov/portal/server.pt/community/healthit_hhs_gov__home/1204. (Web site national coordinator for health IT in the US. Provides a structure approach for the evaluation of the meaningful use of EHRs). Accessed 31 Mar 2013

Sauquet D, Jean FC, Lemaitre D et al (1994) The HELIOS unification bus: a toolbox to develop client/server applications. Comput Methods Programs Biomed 45(Suppl):S13–S22

Sittig DF, Krall M, Kaalaas-Sittig J, Ash JS (2005) Emotional aspects of computer-based provider order entry: a qualitative study. J Am Med Inform Assoc 12(5):561–567

Tomas JL (2007) ERP et PGI: Sélection, déploiement et utilisation opérationnelle. 5ème édition. Dunod, Paris. (A methodological guide for the selection, deployment and use of enterprise resource planning (ERP) solutions)

tranSMART (2013) http://www.transmartproject.org/. (tranSMART project website). Accessed 31 Mar 2013

Van de Velde R, Degoulet P (2003) Clinical information systems. A component-based approach. Springer, New York. (Provides a detailed analysis of main HIS functionalities, of a building approach based on a set of components)

Chapter 13
Sharing Data and Medical Records

P. Staccini, C. Daniel, T. Dart, and O. Bouhaddou

Abstract Health Information Exchange (HIE) can be defined as the sharing of healthcare data and records. It is an answer to the challenges of the evolution of medical practice and patient empowerment. It addresses legal, ethical and technical aspects. More than a question of semantic interoperability between simple or complex computerized systems managed by health care professionals or organisations, the matter is the empowerment of the patient promoted as the key actor of the management of his/her health data and the joint responsibility for healthcare delivery of multiple providers who may not have the same health care information technology. This chapter gives an overview of the increasing social, ethical, economical and industrial challenges, as well as technical aspects of health information exchange and nationwide experiments in Europe and United States.

Keywords Health information exchange • Health record bank • Health portal • National health information network • Personal electronic health record • Personal health record • Personal medical record • Patient empowerment • Shared electronic health record

P. Staccini (✉)
Hôpital de Cimiez, D.I.M, Nice Sophia Antipolis University, Nice 06003, France
e-mail: pascal.staccini@unice.fr

C. Daniel
CCS Domaine Patient AP-HP, 05 rue Santerre, 75012 Paris, France

T. Dart
INSERM UMRS 872 Equipe 20, Paris, France
e-mail: thierry.dart@gmail.com

O. Bouhaddou
HP, San Diego, California (USA)
e-mail: omar.bouhaddou@va.gov

A. Venot et al. (eds.), *Medical Informatics, e-Health*, Health Informatics,
DOI 10.1007/978-2-8178-0478-1_13, © Springer-Verlag France 2014

> **After reading this chapter, you should be able to:**
>
> - Explain the issues related to the sharing of healthcare data and records
> - Understand the data sharing process from a patient or provider point of view
> - Describe the processes and technical components to ensure data privacy
> - Describe the components of interoperability used for data sharing
> - Quote the different approaches implemented in different countries

13.1 Introduction

Almost 40 years have passed since Drs. Shenkin and Warner wrote a paper in the New England Journal of Medicine "Sounding board. Giving the patient his medical record: a proposal to improve the system" (Shenkin and Warner 1973). The authors proposed that "complete and unexpurgated copy of all medical records, both inpatient and outpatient", would enhance patient autonomy, improve patient-physician relationships, and serve as an educational tool. They argued that would provide them with a source of information which providers would feel more obligated to make understandable. The patient could then take the record to another provider or an objective third party to have their situation assessed without great expense. By doing this, they believed that objectifying the process of care could lead to better results and make physicians' jobs more satisfactory and concrete and provide a vehicle to improve communication and coordination (Shenkin and Warner 1975).

Whether a practitioner works alone or as a member of a care team in a health care facility, the computerization of patient record and data exchange are nowadays essential tools that no one challenges or rejects to go back to paper records. These are surely healthcare information technologies that are difficult to implement, slow to generalize, but irreversible.

Medical practice has evolved. The singular "patient-physician" relationship remains the foundation of understanding of ethical principles that guide the behavior of any provider. But society (through several laws and various national agencies) requests (and checks) that health care professionals will no longer be confined to respect an "obligation of means" but are practicing to reach "outcomes" (indicators measurement). Medical practice must be demonstrably effective and safe. Patient safety depends on how the continuity and the multidisciplinarity of care are organized. Patient safety is thus directly related to the degree of continuity and completeness of the information flow that supports the actions and the decisions between the different providers.

In a health care facility, a clinical information system reflects the wish to ensure the ubiquitous availability of data. With the setting of information systems dedicated to home medical care, the thinking remains in terms of "delimited systems" in which the information flow is potentially under control. Yet, the need for collaboration between care professionals has passed the walls of hospitals. The continuity of care is open through the city and involves numerous care providers, care professionals and citizens (Vest and Gamm 2010). The challenges of health information exchange and sharing medical records are based on the understanding and the mastery of information relationships between these worlds.

Beside Electronic Health Record (EHR) for providers, personal health records (PHR) for patients have been developed. The PHR concept has been added in 2010 in the MeSH thesaurus. EHR is a longitudinal record, containing patient medical history and critical data to support it. A PHR is not a comprehensive record and does not replace the EHR for providers. The EHR supports the patient care process driven by the healthcare professionals. Medical data contained in EHR is useful for care coordination. While the EHR is created and maintained by healthcare professionals, the patient is responsible for the access and the content management of his/her own PHR.

13.2   Challenges of Sharing Health Care Data

13.2.1   Patient Point of View

In many industrialized countries, the health care sector faces major challenges: aging of the population, high prevalence of chronic diseases, economic tensions of health care system, growing imbalances of the medical demography. Therefore, it is important, in such an environment, to imagine solutions that ensure care coordination and safe, quality patient outcomes. In this respect, the development of health information systems is essential. The issues addressed by data sharing ensure the highest level of patient safety, the identification and analysis of the potential health risk factors, for both the sedentary patient and the travelling one with chronic pathology (travel for tourism or work). Within such a system, the patient is the beneficiary, but also the actor of his/her own support (Ball et al. 2007).

13.2.2   Provider Point of View

For health care professionals, the main challenges of sharing health data are to:

- Facilitate the coordination between the different health care professionals, in allowing each of them to have better information for multidisciplinary medical decision making
- Facilitate and promote personalized medicine, based on the level of evidence while having comprehensive patient data.

These challenges can be described in the following categories:

Patient safety through:

- The use of decision support systems embedded in the clinical information system (hospital or private office): drug interactions, over- and under- drug dosages, indications and precautions of use;
- Tracking of the administration of health care products and the use of medical devices;
- The review of more complete biological variables, clinical signs and symptoms;
- More effective communication among professionals. For example, the expert directly accesses patient data or the pharmacist can directly call the physician and validate the prescription concerning the product or the dosage.
- Speed and durability of access to personal information available non-stop over the Internet.

Available services such as:

- Possibility to obtain the consultation from specialists in isolated areas with no specialists, in order to improve the equity of access to the supply of care and medical services;
- Better monitoring of health events;
- Better documentation of episodes of care;
- Reduction of "administrative" activities such as events reporting, call notes, mailings, etc.

13.2.3 Legal and Ethical Aspects

Personal health data are sensitive data and their privacy must be preserved and protected. If the processing of health data is necessary (see Chap. 4), the design and development of health information systems and health data exchanges lead to situations where the current legal framework appears inadequate to warranty their privacy. The legal framework must be adapted to the development of new uses and practices. The issue is to balance new rights of patients (right of direct access to their health data, right to mask data, right to decide who may consult or not his/her medical record, right to know who has accessed their data) with the needs of health care professionals to perform their work in good conditions.

The legal framework of data exchange has to make users, patients and health care professionals confident. For the patient, it is important that the PHR remains under his/her exclusive control (creation and access rights). The patient must be the only one to have an "automatic access" to his/her PHR and be able to determine who will be allowed to access it (matrix of access rights). Every log must be tracked and all the logs can be consulted at any time. The patient must be able to choose the documents he/she wants to store in his/her PHR. This means that he/she must be able to hide documents to certain categories of health professionals and even, if he/she wishes, to ask for the withdrawal of certain documents. In some countries, for example France, the PHR system is promoted by a national agency but health care data are hosted by private and accredited third parties. The data hosting provider does not have access to these health data and its activity is monitored by a national and independent committee and controlled by the government.

13.2.4   Scientific Issues

With the development of clinical information systems and personal medical records, one question is to reuse health data collected for a specific purpose, for instance coordination of care, for another purpose, for instance, clinical research or public health evaluation. The PHR can be considered as a source of data (e.g., self-entered data). The goal is to allow the actors of clinical research or public health to detect patients with specific clinical conditions. There is also a challenge to integrate the PHR as a tool for health monitoring, clinical trials or cohorts' management. There is a need to develop advanced features to identify patients, either directly or with the help of health care professionals, who are likely to be included in clinical trials or cohorts, or, if they already take part in a clinical trial or a cohort, who have presented an event targeted by the protocol. PHR data are also potentially useful for improving patient enrollment in a clinical trial while supporting collection of clinical data.

Regarding ethical rules, reusing data collected in clinical practice for clinical research or public health purposes, changes the primary goal of computerization. The patient must give informed consent: he/she must be informed of the act of "donor of health data" for public health or clinical research and must authorize this purpose of use explicitly. In clinical research and public health, policies are needed to ensure and maintain data privacy. By creating a climate of trust, it is likely to improve the quality of the information inferred from clinical data and the analysis of samples. In such a context of coupling PHR and evaluation, guarantee of data privacy must be compatible with the constraint of individual follow-up.

13.2.5   Financial Challenges

At a societal level, one can expect a decrease of unjustified spending and greater efficiency by diminishing:

- The redundancies of examinations and medical tests by different providers;
- Unjustified visits and exams with reference to a standardization of practices established in accordance with the recommendations of national guidelines or according to the principles of evidence-based medicine;
- Disparities of health care access between different regions.

In a health care facility, accessing to the comprehensive patient record with all the data and documents from all sources that know the patient, available in a standard way, improves medical decisions and choices while reducing costs.

13.2.6 Industrial Challenges

Numerous experiments of PHR have taken place in many countries for several years. PHR projects faced difficulties such as: non-communicating information systems, underdeveloped clinical information systems, compartmentalization of private and public sectors, of hospital and ambulatory settings, heterogeneous and late standardization of commercial offerings, multiple decision makers and difficulty of project leadership, etc.

In France, a national agency has been created to meet the goal of "creating and implementing good conditions for the deployment of shared health care information systems based on a national framework: repositories of interoperability standards, national health identifier, enrollment of professionals, accompanying of users." Private companies were asked to participate in working groups that defined the framework for semantic interoperability. Since 2012, vendors of computerized solutions for health care professionals can make their products certified and indexed by a national agreement center.

13.3 Technical Considerations of Information Exchange

13.3.1 Models and Design

Many definitions of PHR exist. The American Health Information Management Association (AHIMA) defines a personal health record (PHR) as a "universally available, lifelong resource of health information needed by individuals to make health decisions. Individuals own and manage the information in the PHR, which comes from healthcare providers and the individual. The PHR is maintained in a secure and private environment, with the individual determining rights of access. The PHR is separate from and does not replace the legal record of any provider." The first PHR initiatives consisted in recording and storing personal health information in paper format (health information hand-written by individuals or healthcare professionals, copies of lab test results, clinical notes or discharge summaries, extracts printed from most electronic medical records). Probably the most successful paper-based PHRs are dedicated to maternity and childhood.

Paper-based PHRs are low cost, extremely flexible and user-friendly compared to rigid electronic systems but they are subject to physical loss and damage. Health information professionals agree that PHRs should be lifetime records of personal health information, that they will be useful as more information is stored electronically and that web-based PHRs are most effective than carrying the PHR information in USB storage (Adler-Milstein et al. 2011). USB-based PHRs currently on the market appear to have deficiencies. Tethered or web-based PHRs may be a better option for consumers at present (Maloney and Wright 2010). In this chapter, we focus on online PHRs defined as an "Internet-based set of tools that allows people to access and coordinate their lifelong health information and make appropriate parts of it available to those who need it" (Archer et al. 2011). There are multiple online PHR models that have been described in the literature (Kaelber et al. 2008; Tang and Lee 2009). PHRs range from "stand-alone" PHRs that depend only on self-entered data from patients and do not integrate with any other systems to "tethered" PHRs that provide a patient oriented view integrated with other electronic health information. Once data is in a PHR it is usually owned and controlled by the patient.

Most EHRs, however, are the property of the provider, although the content can be co-created by both provider and patient (Chumbler et al. 2011; Turvey et al. 2012). Based on the primary source of data, we distinguish four PHR models, including provider-tethered, payer-tethered, third-party, and interoperable PHRs. Provider-tethered and payer-tethered (both linked only to healthcare data within their own organization information systems), and third-party PHRs all exist today, with interoperable PHRs representing a future type of PHR based on robust standards for electronic healthcare data exchange (Kaelber et al. 2008). Some PHRs have been developed by non-profit organizations, while others have been developed by commercial ventures. Regardless of model design, the main technical challenges are related to functionalities and architecture options, privacy and security and semantic interoperability.

13.3.2 Functionalities

Generally speaking, key features of the PHR are to provide individuals access to summary of their medical history and secure e-communication with healthcare practitioners (Tang et al. 2006). The PHR is especially useful for patients navigating complex social and health care transitions and therefore participating to encounters with different healthcare providers such as for example patients with chronic disease (Earnest et al. 2004; Varroud-Vial 2011), tourists or international business travelers. The nature of the patient's illness affects preference for functionalities. PHR functionalities can be classified as infrastructure functionalities – information collection, information sharing and exchange – and application functionalities for information self-management (Kaelber et al. 2008).

13.3.2.1 Information Collection

PHRs are used by patients to keep up-to-date records. In addition to data from healthcare providers (e.g., allergies and adverse drug reactions, vaccinations, chronic diseases, illnesses and hospitalizations, imaging reports, laboratory test results, lists of current medications, discharge summaries, etc.) and monitoring devices (e.g., weight, blood glucose), a PHR could store other personal data on social status, family history, living and work environment, healthy lifestyles (e.g., diet, exercise, smoking, weight loss, and working habits) (Tang et al. 2006). Primary care physicians play a key role in patient health. PHRs are likely to be linked to physician electronic medical record systems, so PHR adoption is dependent on growth in electronic medical record adoption and usage of exchange facilities between medical records and PHRs.

13.3.2.2 Information Sharing and Exchange

PHRs facilitate information sharing so that information can be displayed for review to authorized stakeholders. PHRs also provide solutions for sending and receiving electronic messages to and from healthcare professionals (physicians, pharmacists, nurses). If physicians generally prefer telephone or face-to-face communication, patients prefer email communication for some interactions (e.g., requesting prescription renewals, obtaining general information), and in-person communication for others (e.g., treatment instructions) (Hassol et al. 2004). Some PHRs offer patients the opportunity to submit their data to their clinicians' EHRs (Cimino et al. 2002), to healthcare organizations (e.g., pharmacovigilance, surveillance of infectious diseases) or to clinical research organizations. However, while PHRs can help patients keep track of their personal health information and communicate with their providers, the value of PHRs to healthcare providers and organizations is still unclear. Information sharing and exchange functionalities directly among providers reduce the need for inter-provider communications to access PHR information (Tobacman et al. 2004). Patients may also benefit from sharing information on their conditions with other patients having similar problems using online patient communities such as PatientLikeMe or Caring Voices, e-forums, private messaging, and comments (Frost and Massagli 2008). Finally, PHRs with connectivity to a library of computerized services, containing health information and self-management tools, offer a new opportunity for developing innovative interventions to reduce risks, prevent diseases of public health concern (e.g., influenza) and improve the uptake of preventive health services (Bonander and Gates 2010).

13.3.2.3   Information Self-Management

Many PHR systems are physician-oriented, and do not include patient-oriented functionalities. These must be provided to support self-management and disease prevention if improvements in health outcomes are to be expected (Archer et al. 2011). Patient health self-management can be supported by PHRs that allow patients to record, edit, and retrieve their healthcare data, including for example weight, observations of daily living, stress scales, vital signs (e.g., blood pressure measurements) or lab test results (e.g., blood glucose) (Kaelber et al. 2008; Hess et al. 2007). Additional functionalities provided by PHRs are online appointment scheduling, e-reminders for appointments, automated advice programs, medication renewals, pre-encounter questionnaires, e-visits, viewing and accessing disease-specific information health and practice information, disease specific tools (such as Smoking Cessation Management), remote disease monitoring (Wang et al. 2004; Halamka et al. 2008). Frequent monitoring can lead to early detection of critical situations and timely intervention (Demiris et al. 2008). Self-care monitoring tools are becoming more mobile and reliable, particularly in 'smart home' applications (Martin et al. 2008).

13.3.3   Architecture

No particular architecture has yet emerged as being the most effective to meeting PHR functional requirements. Different types of computing platform – software packages and websites – have been developed in order to process or exchange healthcare data. Since, primary care or hospital electronic medical record systems (EMRs) are often used as a source of data for PHRs, linkages between PHRs and EMRs thus appear to be critical to the successful use of PHRs (Demiris 2012). Allowing patients to enter or view their own health data in their healthcare provider's EMR can convey much more to the patient than stand-alone PHRs, enabling patients to gather their entire fragmented medical history in one place (Tang et al. 2006; Tang and Lee 2009). PHRs implement different applications including transactional, analytical and content delivery capabilities of the system, such as appointment scheduling, medication renewal, patient decision support system and disease education materials. PHRs integrate computerized decision-support systems in order to recommend care protocols tailored to risk levels (Bates and Bitton 2010; Luo 2010). Another key technical issue is to integrate telemonitoring reports – such as telemonitoring Cardiovascular Implantable Electronic Devices (CIEDs) for example – to the care planner that implements guideline-driven care plans in the PHR (Yang et al. 2011). In addition, mobile communication devices such as cell phones, smartphones and other mobile tablet PCs are relatively inexpensive, portable technologies that can collect environmental and patient-entered information and transmit it via the Internet to a PHR. Combined with actionable decision support, the mobile communication devices-PHR combination, or "mPHR," can analyze aggregate data to activate mobile, patient-specific output such as medication or appointment reminders or healthy habit tips. Individuals who access such information and decision prompts from a portable communication device in an outpatient setting benefit of real-time support and can make informed health decisions.

13.3.4 Privacy and Security

Patients have a legal right in most states to request their healthcare data. Under recent USA legislation, providers using a certified EHR who want to qualify for monetary incentives will be required to provide an electronic copy to their patients. In the UK, according to the government information strategy for the NHS, every primary care practice in England will have to offer patients online access to their care records by 2015 (currently only 1 % do so). PHRs invert the long-standing paradigm of health care institutions as the authoritative data-holders and data-processors in the system. With PHRs, the individual is the center of his or her own health data universe, a position that brings new benefits but also entails new responsibilities for patients and other parties in the health information infrastructure. Ethical and regulatory issues rose from this shift. Trustfulness (i.e., health and wellness information is processed ethically, and privacy is guaranteed) is critical for PHR adoption.

Key principles of PHRs include that they are created upon request and consent of the individuals involved, recognized as the owners, and that information is disclosed only to those authorized by the owners. In addition to height, weight, blood pressure and other quantitative information about a patient's physical body, PHRs can include very sensitive information, including fertility, surgical procedures, emotional and psychological disorders, and diseases. Interestingly, two-thirds of adult consumers are concerned about the privacy and security of their health information, but most of those using PHRs are not worried about privacy implications (CHCF 2010).

Those who are concerned about privacy may change their attitudes with appropriate framing of arguments favoring record use (Angst and Agarwal 2009). The chronically and acutely ill and those who frequently use healthcare services tend to be less concerned about privacy than are health professionals (Hassol et al. 2004). In the context of persons living with HIV/AIDS acceptance of clinicians' need for access to accurate information depends on patients' trust in their primary care providers. A published survey reported that more than 80 % of persons living with HIV/AIDS agreed that the PHR helped them manage their medical problems; however, some users were concerned that their health information was not accurate or secure. In mental health settings, special attention needs to be paid with regards to the handling of sensitive information.

Various threats exist to patient information confidentiality. Disclosure of data can occur due to innocent mistakes made during electronic transfers of data to various entities. Medical or technical personnel may misuse their access to patient information out of curiosity or for profit, revenge or any other purpose. In addition, PHR data collected solely for the purpose of supporting primary care can be exploited for reasons not listed in the contract, such as research, epidemiology or public health. External individuals (former employees, hackers, or others) may access information, damage systems or disrupt operations. No clearly defined architectural requirements and information use policies is available for PHR security and confidentiality. Current security protection mechanisms need to be enhanced for record protection, but to maintain privacy, security levels must not become so tight that health records are unusable (Win et al. 2006; Wiljer et al. 2008; Masys et al. 2002).

Audit trails provide information about identity, times and circumstances of users accessing information. If system users are aware of such a record keeping system, it will discourage them from taking ethically inappropriate actions. In addition, controls such as alerts, reminders and education of users can be implemented. Security policy and its implementation (including the privacy, integrity, and confidentiality of the data, authentication and authorization of users, encryption, firewalls) shall be clearly defined and documented in each PHR project especially with regards to the solutions that have been implemented for emergency access authorization in the absence of patients (Chen and Zhong 2012) and for hiding information. Another critical issue is data credibility. PHRs must identify sources for data received (Rode 2008; Simborg 2009).

13.3.5   Semantic Interoperability

Interoperable EHR systems are the most important enabling tools on the road to patient-centered care, a lifeline for continuity of care and support to mobility of patients. System interoperability is critical to establish linkages between EHRs and PHRs and relies on the development and adoption of interoperability guidelines such as integration and content profiles provided by the international initiative Integrating the Healthcare Enterprise (IHE) and HL7 standards that support information sharing between systems (Stolyar et al. 2005). Some PHRs incorporate clinical documents that conform to standards such as HL7 Clinical Document Architecture (CDA), Patient Continuity of Care (PCC) documents, Continuity of Care Document (CCD) or ASTM Continuity of Care Record (CCR) (Ferranti et al. 2006). CDA-based documents (PCC documents and CCD) rely on an information model based on the HL7 Reference Information Model and are extensible, while CCR is fixed (Benson 2010; Dolin et al. 2001).

Clinical documents may be published in PHRs, considered as document sharing resources, using the cross-enterprise document sharing (XDS) integration profile defined by the IHE information technology infrastructure (ITI) domain. This profile enables a number of healthcare delivery organizations belonging to an XDS affinity domain (e.g. a community of care) to cooperate in the care of a patient by sharing clinical records in the form of documents as they proceed with their patient care delivery activities. Federated document repositories and a document registry create a longitudinal record of information about a patient within a given XDS affinity domain. This profile is based upon ebXML Registry standards, SOAP, HTTP and SMTP. It describes the configuration of an ebXMLregistry in sufficient detail to support cross enterprise document sharing.

With regards to telemonitoring, such as telemonitoring Cardiovascular Implantable Electronic Devices (CIEDs), a key technical challenge is to integrate telemonitoring reports in the PHR. IHE specified a vendor-independent format for telemonitoring CIED report based on the standards specifications from ISO/IEEE 11073 (Health Informatics, Point-of-care Medical Device Communication) and HL7v2.x (Yang et al. 2011).

13.4 Analysis of Health Data Sharing Approaches

Health data sharing approaches can be described using a common description framework. This framework is adapted from (Archer et al. 2011): sponsor; purpose (patient-provider communications, education and lifestyle changes, health self-management); functionalities and record content; system attributes (architecture, privacy and security, semantic interoperability); deployment, adoption and acceptance (adoption and use, acceptance and satisfaction, usability, barriers to pEHR (personal Electronic Healthcare Record) adoption and use, benefits, clinical outcomes and process change).

13.4.1 European Initiatives

13.4.1.1 Austria

Austria is developing the project ELGA "ELektronische GesundheitsAkte", an Electronic Healthcare Record since 2006. ELGA's goals are the sharing of electronic orders, and medical reports using the following modules: (1) "eReport" for hospitalization reports; (2) "eMedikation" for medical orders coupled with pharmacist dispensation (drug interaction detection); (3) "eFindings" for biology and radiology results. Medical data come from various healthcare services providers and from patients.

Interoperability standards are based on IHE profiles (XDS, PIX, PDQ and ATNA) and HL7 CDA. ELGA is based on a national information system which collects data on a patient, retrieving them from the servers of each healthcare provider. ELGA can reconstitute the care pathway of the patient. The service is only maintaining a centralized registry indicating the storage location of all health data related to an individual. The patient is recognized by his/her "E-card".

The organization "Arge ELGA" ("Arbeitsgemeinschaft Elektronische Gesundheitsakte") has been developing the project for 6 years. After a long phase of technical specification, the project is currently in its phase of test in several areas of the country (Vienna, Tyrol, and the north of area in the North of Austria). A first operational version should be deployed starting mid-2013.

13.4.1.2 Belgium

In order to control health expenditure, the Belgian government made a very early choice of a better patient care coordination, organized around the General Practitioner (GP). Indeed, the GP follows his patient for the elementary care, but also for more serious pathologies which require the consultation of a specialist and, if necessary, a hospitalization. This GP must be able access the entirety of the data concerning the health of the person.

The Belgian social protection system (National Institute of Sickness and disablement Insurance, INAMI) has deployed a paper-based health record in 1999 named "Global Medical Record" (DMG). Patients can ask their GP to create a DMG. It costs 28,15 € entirely refunded by the patient's mandatory mutual insurance company. The GP created medical record for the patient, and populates it with any medical data useful for patient's follow-up: medical history, blood analysis, treatments, surgical operations, known allergies and other data domains. The GP can communicate this information to the patient. The specialist, informed by the existence of a DMG, will get in touch with the general practitioner to obtain the information contained in it. For the middle aged patient (45–75), a prevention module called DMG + is specifically designed to detect risks of some major chronic diseases: cancer, cardiovascular pathology and others.

Initially a paper-based solution, the DMG can also be created using a computer since 2003. The electronic version of DMG is called DMI (pEHR). Half of Beligium population already has either a DMG (paper format) or a DMI. This number exceeds 80 % for the elderly people older than 75 years, because they require in priority a follow-up.

At the end of 2004, the Belgian government has created the "BeHealth platform" to organize and coordinate "electronic information exchange". The main mission of the Behealth platform was to allow the rise of several systems of health data exchange. Currently limited to a local level, BeHealth works to interconnect the five regional existing "hubs" (Wallonia, Brussels and three hubs in Flandres) between them, through the project "Hub-Metahub".

13.4.1.3 Denmark

MedCom is a co operative venture between authorities, organizations and private firms for the development, testing, dissemination and quality assurance of electronic communication and information in the healthcare sector. MedCom has built a health secure network which can carry more than 40 types of standardized messages (electronic orders, laboratory results and orders, recommendations, discharge summaries, etc.). Hospital management systems, General Practitioners, home care and pharmacy services have integrated the secured mail system based on this network. Each month, nearly five million messages, including 80 % of all orders, are exchanged. The management of the identity and the data confidentiality are ensured by a system of agreements, an infrastructure with public keys and the production of audit logs.

The health portal, Sundhed (health in Danish), is accessible by the professionals and the citizens using electronic signatures. Citizens have access to general information about health and their personal health data. Professionals can access several services, in particular access to the medical record, under the control of the Danish authorities in charge of the data protection. Today, all General Practitioners and about half of the hospitals use the pEHR. Approximately 15 different Enterprise Resource Planning (ERP) systems are interconnected to support general practitioners and four different softwares are used by the municipalities for home care.

For more than 40 years, all Danish citizens have been receiving a single patient identity number. For access to health data or to the health insurance, electronic signatures by a system of infrastructure to public keys (PKI) are emitted.

13.4.1.4 England

The NHS – National Health Service (1948) – offers a universal and free Social Security covering 50 million people. In 2002, the report "Wanless" recommended a doubling of the informatics expenditures and the government planned the deployment of a national infrastructure by 2010. National infrastructure is divided into five regions linked by a backbone (spine).

These projects called "connecting for health", undertaken by a national agency (NPfIT – National Programs for IT), are divided into five geographical areas to limit the risk of an industrialist or a system failure. These programs, registered in the duration, are divided in four phases between 2003 and 2010: definition of the standards and strategies of purchase (2003), activation of the services of scheduling ("national booking service") and of the orders' exchange (ESP – Electronic Prescription Service) (2005), deployment of the three services of program (appointment, scheduling, EHR), equipment of the hospitals, digital imaging (2007), end of deployment (2010).

Data interoperability is managed by the NPfIT. It is based on the HL7 version 3 RIM and SNOMED CT.

These projects benefit from the largest public IT budget allocated by a government. These projects had many risks and encountered failures of the commercial vendors on several areas, a resistance from hospitals, problems with the pEHR (Greenhalgh et al. 2010) and others. There were also successful components such as imaging network, booking service. Even though, the NHS has now a secure data network which connects every health care institution, with a single identifier for each patient, a national system of booking, a system of transmission of images and an electronic prescription service. The National Audit Office points the lack of clinical functionalities contrasting with the costs estimated to 12,4 Md£. The governance of the projects failed to gain the confidence of doctors. In September 2011, the British government announced an acceleration of the dismantling of the NPfIT and the EHR project, following the conclusions of a new review by the Cabinet Office's Major Projects Authority (MPA). The electronic prescription service, the booking service and the imaging service are now directly managed by the NHS.

13.4.1.5   Estonia

Estonia, one of the Baltic states, is already known for its advanced E-Services, in the business sector (banking), but also in the E-Administration sector (e-voting, e-school, e-ticket, and others). This state has already deployed a nationwide technical infrastructure called "X-road platform". X-road is a platform independent standard interface for secure data processing, connecting all Estonian public sector databases for information exchange.

Digital signatures and Identity card authentication are recent innovations. Patient can use their identity electronic card. The project was launched in 2002 and has now reached a full coverage of the population. These developments are the basis of policies such as mandating a countrywide electronic health record. In 2005, the Estonian Ministry of Social Affairs launched a new eHealth project in four areas: Electronic Health Record, Digital images, Electronic Registration and Electronic Orders. This eHealth project was deployed in December 2008. Three years later, this system is already used by half of the population.

The main goals of the eHealth projects were: (1) decreasing the level of bureaucracy, (2) increasing the efficiency of the health care system, (3) making the information accessible for the attending physician, and (4) developing health care services with higher quality.

Since 2008, the Estonian Parliament has decreed that health care providers are obligated to forward medical data to the PHR. The rules for data usage state that only the health care attending physician or medical assistant currently providing patient treatment has the right to make enquiries about patient's data, i.e., the patient's attending physician or a medical assistant. Since October 2009, using a portal, the patient can exercise their right to set restrictions of access to their health data. In this case the patient will be informed by the information system at the time of setting the restriction that it is dangerous to his/her life and health to provide health care services based on insufficient information.

13.4.1.6   France

Secure Messaging Systems

The secure messaging system appeared because email was not considered as a secure communicating system to exchange medical data. A simple handling error or the general lack of privacy on the Internet can lead to disclosure of personal health information to non-authorized recipients. Health professionals consider the use of secure messaging systems as a good method for exchanging messages and documents while protecting privacy. A secure messaging for inter-personal and/ or inter-system communication manages the transport and the encryption of messages between individuals (health professionals, patients, etc.) or legal persons (institutions, health insurance). ASIP defines the conditions for implementing a secure messaging system in conformance with legal constraints (Informatics, Records and Liberty Law – 1978; Quality and Security of Care – 2002), ease of use and interoperability (compliance with international standards).

Access to Data Hosted by the National Health Insurance

Since July 10, 2007, the national social insurance provides all physicians with an access to claims data of their beneficiaries. This system is called "web doctor". Scope of data viewing is limited to the previous 12 months. The administrative data are: national identifier, last name, first name, date of birth. The "medical" information is related to outpatient procedures and treatments (visit, surgical or dental procedure, lab and radiology procedures, nursing and functional rehabilitation, including preventive procedures and treatments) as well as inpatient ones (date of admission and length of stay, corresponding DRG code, expensive drugs if applicable), information regarding work stoppage (start and end date), mention of possible link with a long-term illness (start date, reason and ICD code). The physician who wants to use this Internet service must inform the patient and obtain his/her authorisation by the means of its insurance card. This authorization relates to the totality of the available data. Authentication is based on the simultaneous use of the health care professional card and to the patient insurance card. Therefore data access outside the office or without the presence of the patient is not possible.

Medication Record

The medication record was created by the law of the 30th January 2007 relating to the organisation of healthcare professions. Its implementation was entrusted to the

National Council of Pharmacists. The medication record is a professional tool intended to secure drugs dispensing. Pharmacists can access the history of all the drugs delivered to the same patient during the last four months, in order to avoid drug interactions. This record facilitates the monitoring of quality and coordination of treatment through shared information. With it, the pharmacist can improve the service given to patients and respond to public health challenges in termes of patient safety. Connection with the national agency of drug safety allows also pharmacists to inform patients of potential adverse events if relevant. This record can be wiewed by the pharmacist only in the presence of the beneficiary, with their health smart card (carte Vitale), to insure patient identification and with the pharmacist's healthcare professional card (CPS, Carte Professionnelle de Santé). This Internet service is led and financed by the National Council of Pharmacists. On the 7th January of 2013, 24,190,413 medication records were created in 22,067 dispensaries. Since 2012, hospital pharmacists can also access this medication record.

Personal Health Record

Based on regional experimentations completed in 2005, the French national EHR (DMP, Dossier Médical Personnel), officially opened in 2010 as a regular service; the patient access is available since 2011. At the date of 6th of January 2013, 265 966 DMP have been created.

The DMP was created by the law of August 13th 2004, based on the law of March 4th 2002 relating to the rights of the patients and the Fieschi's report "patient data sharing: culture of sharing and quality of information to improve quality of care". The DMP project faced difficulties related to its environment. A specific national agency, called ASIP Santé was created in 2009 to restructure the project and facilitate the emergence of eHealth. ASIP Santé defines the DMP as a national and secured pEHR accessible on the Internet by patients and healthcare professionals. The DMP is a public service freely accessible to all the beneficiaries of the national health insurance (52 million recipients out of a total of 66 million French residents). DMP is not mandatory. DMP allows sharing of documents among health care professionals and with the patient, under the control of the patient. The DMP centralizes the health information useful for the coordination and quality of care such as past medical history, allergies, drug orders, hospitalization reports, biological and radiology reports. The DMP can be organized in several sections including medical summary, processing and care, reports, medical imaging, lab results, prevention procedures, certificates and declarations. The patient can also add to his/her DMP any information he/she considers necessary or important. This information is accessible via the Internet through http://www.dmp.gouv.fr website. Access to a DMP is strictly reserved to the patient and to the authorized health care professionals. The management of a DMP is placed under the exclusive control of the patient. The patient constantly keeps the possibility of closing it, removing whole or part of the documents in it or masking selected data. Only the patient can authorize the access of his/her DMP to health care professionals. The patient is the only one to have an "automatic access" to his/her DMP and to be able to determine who else, apart from himself, will be able to view it. Any access to the DMP is traced in a log book and the patient can know who read his/her data. This traceability plays a dissuasive role for the people interested in making an illegal use of the DMP. DMP data are stored by national data hosting providers approved by the Ministry of Health.

13.4.1.7 Germany

The German national project of eHealth tried to connect 2,200 hospitals, 100,000 doctors, 21,000 pharmacies and 200 public insurance companies by the deployment of the "elektronische Gesundheitskarte" (eGK), a smart card compatible with a secure network (an electronic patient record with a lifetime medical history for every patient) (Jha et al. 2008). But the resistance of the medical profession was mounting. In May 2008, the annual Parliament of German doctors ("Deutscher Arztetag") rejected the electronic health card with a majority of 111 over 78 votes. In 2010, the government suspended its national scale project of health smart card. However, local projects continue in regions, via public-private partnerships. After years of trying to roll out a nationwide electronic payments system, the Federal Ministry of Economic Affairs and Work announced on July 18th of 2011 the cancellation of its "elektronischen Entgeltnachweis" (ELENA) national payments records system. German politicians are calling for the elektronische Gesundheitskarte (eGK), to be scrapped following the abandonment of Germany's electronic payments records system amidst concerns over poor systems security and data privacy.

13.4.1.8 Italy

The Italian ministry for Health published on November 11th, 2010 new directives on the electronic medical records and the question of security. The idea is to propose an interactive domestic network in which general practitioners are the most important actors. It is a system of information exchange between experts and health care institutions. Regarding the structures of information (for example, e-precribing and patient's health care summary) the principles are based on HL7 standard and the use of the CDA r2. Some regions (Lombardy, Tuscany) are implementing this system.

13.4.1.9   Spain

Spain is divided in autonomous communities which have full powers regarding their citizens' healthcare. Some areas of Spain have an EHR like in Andalusia and Catalonia.

In Andalusia, Diraya (Arabic word for knowledge) is an integrated, citizen-centered health solution that maintains a unified EHR based on a number of interoperable elements. It is based on four principles: a single health record for each person; unified access to all services; structuring (coding) of all relevant information; and system development by practitioners and providers. As the development of Diraya got underway, a fifth principle was adopted: "customer precedence" in which patients are not considered to be customers or clients, but rather owners. In 2007, Diraya had been implemented in 88 % of the primary healthcare centers which cover 79 % of the Andalucia population.

In Catalonia, there are four main ICT projects: Telemedicine, electronic prescribing, medical imaging and the pEHR.

13.4.1.10   Sweden

Like in many European countries, Sweden has begun the modernization of the health sector, based on new technologies. The objective remains always the same one: take care of an ageing population, use resources efficiently, and adapt care to citizens' needs. With its 9 million inhabitants divided in 21 regions, Sweden launched its own project of National pEHR called "NPÖ" – "Nationell Patientöversikt". After an important period of preparation in 2005, the deployment began in 2009. The Center for eHealth in Sweden has created the conditions necessary for developing and introducing nationwide use of IT in the decentralised health and social care system of Sweden. The Center for eHealth in Sweden is governed by representatives of county councils and regions, the Swedish Association of Local Authorities and Regions (SALAR), municipalities and private care providers. The Swedish pEHR aims at reinforcing:

- The quality of the care – through a better access to the medical data, the pEHR allows the health professionals to reach more reliable diagnoses and prescribe appropriate treatments;
- The efficiency of the health care system – through offering a simplified access to the medical data, the pEHR avoids the multiplication of the acts and the duplication of the same tests several times;
- The safety of the patients – by providing more complete information, the pEHR facilitates the decision making of the health professionals and reduces medical errors.

The electronic card "SITHS" is an electronic identity card for the healthcare professionals, to guarantee that only people authorized have access to information of the pEHR. This smart card is delivered with the professionals and employees of the health care centers of health.

Deployed in half of the Swedish regions, the pEHR should be available on all the territory from by the end of 2012.

The healthcare staff of the municipality of Örebro (southern Sweden) rates highly the National Patient Summary (NPÖ, in Swedish) service. Nine out of ten NPÖ users are 'satisfied' or 'very satisfied' with it, and more than 60 % of them believe that it has been highly or very highly beneficial for their work, according to a recent survey. Örebro is the country's municipality with the longest experience with the National Patient Summary.

Just over 2 years ago, the nursing staff of elderly care services started having access to their patients' data: patient name and contact information, medical records, diagnoses, drugs, patient contacts details at health and healthcare organizations, and results of several tests.

13.4.1.11 The Netherlands

In February 2006, the Netherlands began the generalization of a project named "Landelijk SchakelPunt" which is a National Switch Point (NSP) for family doctors. The project is driven by NICTIZ – the National IT Institute for Healthcare in the Netherlands – which is the national coordination point and knowledge center for IT and innovation in the healthcare sector. The main difference between the Netherlands efforts and the other similar projects is the speed with which the NICTIZ reached determining technical steps.

The national switch point is the core of health electronic communication. Any authorized healthcare practitioner can be connected to the switch point so that he/she can obtain the latest and most relevant information about a patient at any time, from anywhere in the Netherlands and in a simple, secure and reliable way.

The NSP includes a record locator service indicating all the sites with medical data for each patient and a registry of associated clinical systems. The NSP manages the authentication and the authorization of the users and records all their transactions in an audit log.

In consultation with and at the request of the healthcare sector, NICTIZ is continuously developing and refining national standards for electronic communications in healthcare. Furthermore, NICTIZ supports the sector in developing functional IT solutions that can be used nationwide, and contributes to policy making on IT issues as they relate to healthcare on a national and international level.

13.4.2   American Initiatives

13.4.2.1   Canada

Quebec and Alberta have deployed a pEHR for the quasi totality of their inhabitants. On the other hand, Ontario has seen a massive scandal in October 2009: after 10 years and 740 million euros spent for the pEHR, a report denounced the strategic lack of leadership.

13.4.2.2   United States of America

The NwHIN Project

In 2008, a domestic network of health information exchange (Nationwide Health Network Information: NwHIN, recently renamed eHealth Exchange) was started. This is a set of standards, protocols, services and contracts that allow the secure exchange of health information between various entities, geographically close or dispersed throughout the country. The aim is that medical record data "follow" the patient in his displacements and are available to support clinical decisions and beyond, and can be made available to improve public health.

The federal health organizations of which Social Security Administration (SSA), Department of Veterans Affairs (VA), Department of Defense (DoD), Center of Disease Control (CDC), Center for Medicaid and Medicare Services (CMS), and other authorities received a presidential directive which requires them to use standards and the NwHIN for the electronic exchange of medical data.

The specifications of NwHIN are simple:

- A secure platform for medical data exchange between NwHIN members, over the Internet, and based on standards
- No centralized database or patient universal identifier
- The respect of patient preferences to take part or not in this exchange
- Each participant is autonomous and responsible of the veracity of transmitted information;
- A common legal contract that every participant must sign.

A centralized component stores NwHIN participants identities, their URL addresses, and the services they provide (Universal Description Discovery and Integration Registry or UDDI). Under the Office of the National Coordinator for Health Information Technology (ONC), a coordination committee, with volunteer representatives of public and private sectors, is responsible for the planning and the coherence of the deployment.

To join the NwHIN, an organization must fullfil two criteria: (1) sign the contract of confidence (Dated Uses and Reciprocal Support Agreement or DURSA); (2) successfully pass the tests of interoperability and conformance managed by a technical committee. Patient data are mainly exchanged according to a CDA structure. For NwHIN, the CDA was constrained to the Continuity of Care Document (CCD), by imposing certain constraints of optionality and cardinality, and by adding use requirements of medical terminologies, such as SNOMED CT, RxNorm and LOINC. To connect to the NwHIN, an organization needs an adapter and a gateway software package. There exists an "open source" gateway named "CONNECT gateway", but participants can build their own solutions. The adapter is typically developed by each organization to connect the data sources and the gateway. The gateway manages external communication and the messaging with the others gateways connected to NwHIN. The adapter is responsible for the interactions with the internal components, including the gateway, the Master Patient Index (MPI), the clinician interface, the subsystem of management of patient preferences and the organization policy, the patient data sources, and the translation services of medical terminology.

Veterans Personal Health Record

The United States Department of Veterans Affairs (VA) includes over 1,300 hospitals and other facilities. It cares for over six million former military personnel annually. Launched in November 2003, the project "My HealtheVet" provides easy access to the VA EHR (called VistA), to benefits information, personal health journals, vital signs, and prescription refill services. Recently, this project added the "Blue Button", a simple functionality for a Veteran to download a copy of a summary of their medical record and share it with non-VA providers. This summary was initially provided as a text file, but is now also available as a Continuity of Care Document (Hogan 2011). In the future, Veterans will be able to share securely a copy of their medical record with non-VA providers, both in text and a structured CCD format (Bouhaddou et al. 2012). Studies have shown that Veterans were highly satisfied with securely accessing their notes online. Users felt that having ready access to their information improved their self-care and ability to prepare for in-person visits.

OpenNotes

In 2010, more than 100 primary care doctors from three diverse medical institutions across the United States began sharing notes online with their patients (Primary care practices at Beth Israel Deaconess Medical Center (BIDMC) in Boston-Massachusetts, Geisinger Health System (GHS) in Pennsylvania, and Harborview Medical Center (HMC) in Seattle-Washington). These three private health systems offered more than 19,000 patients online access to their doctors' notes, through a secure patient portal (Delbanco et al. 2010). Patients were provided access to clinic notes, and were notified by email when a new document was put into the record. Each site was part of a 12-month study to explore how sharing doctors' notes may affect health care. The OpenNotes study started a movement to enable patients to easily read notes written about their care, and to bring more transparency to medical records. The heart of OpenNotes is to involve patients far more actively in all aspects of care and to improve communication between the doctor and patient. Open notes also encourages patients to share information with others, including those who care for them, and it may help prevent mistakes. At the end of the 1-year study, patients felt more in control of their care, reported greater understanding of their health issues, and were more able to remember their treatment plan (Walker et al. 2011; Delbanco et al. 2012).

13.4.3   Other International Initiatives

13.4.3.1   Australia

NEHTA – National E-Health Transition Authority – is the Australian agency in charge of the deployment of the "Personally controlled electronic health records" (PCEHR) for all Australians. From July 2012, all Australians can choose to register for an electronic health record.

13.4.3.2   New-Zeland

With four million inhabitants, New Zealand family doctors have one of the highest rates of electronic patient record used: almost 95 % of primary care physicians and 100 % of laboratories communicate via secure health data network every day.

13.4.3.3   The epSOS project

Since 2009, the members of the epSOS project (Smart Open Services for European Patients), cofounded by the European Commission, have developed cross-border eHealth services, addressing technical, semantic and legal interoperability challenges across countries. The intention of epSOS is to demonstrate the concept that, with the help of epSOS, medical treatment for citizens residing in other countries can be improved by providing health professionals with the necessary patient data in a secure electronic format. On the 13th of April 2012, the epSOS piloting phase has started, testing the technical, semantic and legal solutions that have been developed in a "real-life environment". European patients and epSOS health professionals who have agreed to take part in the epSOS pilot will be using and evaluating the epSOS pilot cross-border Patient Summary, ePrescription and eDispensation services in real world settings.

13.4.4   Commercial Initiatives

Some companies have developed free or paying systems of personal medical record on the Internet (http://www.phrreviews.com/). Most important are Health Vault (by Microsoft) and Dossia (by AT&T and Intel). Google Health (by Google) is no longer available.

13.4.4.1   Microsoft Health Vault

In October 2007, Microsoft launched its "safe health" which provides to anyone for free of charge an online space for creating and managing a personal electronic health record (pEHR). The initial idea is to empower the patient, in particular when his/her follow-up requires to trace and analyse clinical or biological parameters (blood pressure, capillary glycemia, weight, cardiac monitoring, etc.). The system provides a list of compatible measurement devices; the user chooses one of them, connects it to his/her computer and sends the data in his/her record in order to be able to analyze them over time (trend graph for example). A developer kit is available for suppliers of measuring equipment. There are several ways of adding new medical information in the system: (1) manual entry of information such as medications, vaccinations, weight, personal and family history, etc.; (2) uploading data recorded by a measuring equipment; (3) sending information by a health care institution; (4) importing formated files provided by a health professional (Continuity of Care Document or Continuity of Care Record formats).

13.4.4.2   Google Health

On February 28th, 2008, during the annual edition of HIMSS congress (Health Information and Management Systems Society), the Chief Executive Officer of Google announced the installation of "Google Health" to compete with Microsoft. As the world leader of content hosting, Google proposed an online service allowing any person to collect, store, manage and share his/her medical record and health data by means of a simple Internet browser. With this application, any user can: (1) create his/her profile health; (2) import medical records managed by hospitals, physicians, pharmacists, etc.; (3) obtain information on health problems; (4) access doctors and hospital directories; (5) connect to online tools and services. Google Health did not provide any structured electronic record to physicians or payers. The principle of Google Health is that a patient creates an account for free, then connects to a secure profile stored in the Google system. This profile comprises the following data: diseases, drugs, allergies, results, procedures, vaccinations, age and sex. An option enables a user to import medical records. A list of hospitals that have declared to Google the main features of their structured medical records is available. If the patient finds in this list an organization from whom he/she wishes a copy of his/her record in the Google system then: (1) he/she has to be initially identified by this organization; and (2) then he/she authorizes the establishment to send the data to his/her Google health profile. Thus, the establishment can send to Goggle Health profile messages in free text or Continuity of Care Record format. Technically speaking, Google APIs allow developers to ensure secure connection and data transfer with the establishment. On June 24th, 2011, Google terminated the Google Health project. Many observers wondered about this decision: (1) a lack of connectivity within a project which appeared monolithic and only intended for storage; (2) a difficulty to bring meaning to data both for the patient and the physician: "the Google Health design was elegant and the solution intuitive, but Google seemed to have forgotten that for the patient healthcare is a narrative, not a bunch of lab results, CAT-scans and tests."

13.4.4.3 Dossia

Dossia is a non-profit organization, a consortium of several large American employers that share a common vision to give to their employees the possibility of making the best decisions about their health care. The consortium includes Abraxis Organic-science, AT&T, Applied Materials, BP America, Cardinal Health, Intel, Pitney Bowes, Sanofi-Aventis, Vanguard Health and Walmart. Dossia chose to consolidate the most used industrial standards with regards to the structuring of health data and the use of controlled vocabularies. Each data element refers to a root model specifying the data characteristics. Dossia uses the OpenID identification system to provide users with a SSO (Single Sign On) to access computerized record. The main features of Dossia are: opt-in access to the record only by the patient or his authorized representatives; nobody is obliged to take part; data model are published and APIs are available for data hosting providers and applications developers. The total architecture of the system identifies three data fields: (1) field of personal data self-entered by the patient; (2) a field of imported data from personal health applications (measuring instruments); (3) a field of imported data from medical records managed by health professionals involved in the patient care.

13.5 Use of Personal Health Records

From 2006 to 2009, the number of consumers using online PHRs in the U.S. increased from 2.5–7.3 million A survey was conducted December 18, 2009 through January 15, 2010 among a representative sample of 1,849 adults nationwide, using Knowledge networks. The survey included oversamples among African Americans, Latinos, Asian Americans, and users of online personal health records. About 7 % of those surveyed said they have used a PHR, more than double the proportion identified 2 years earlier in separate research. As a result of their PHR, users cited taking steps to improve their own health, knowing more about their health care, and asking their doctors questions they would not otherwise have asked. Two-thirds of the public remained concerned about the privacy and security of their health information, but the majority of those who are using a PHR are not very worried about the privacy of the information contained in their PHR. These answers must be interpreted according to the culture and of the social protection system of the country. The initiatives of companies such as Google or Microsoft collect only 25 % of favorable opinions, twice less than for organizations directly responsible of care (58 %). Fifty percent of american IT consumers state they would use a system sponsored by their health insurance.

The OpenNotes survey (see Sect. 4.2.2.3), started in 2010, was aimed at evaluating the effect on providers and patients of facilitating patient access to visit notes over secure Internet portals. The participants were 105 primary care physicians and 13,564 of their patients who had at least one completed note available during the intervention period. Patient findings at the three study sites include:

- (87 %) of participants viewed their notes
- Twenty to forty two percent reported sharing notes with others
- Sixty to seventy eight percent reported doing better taking their medications
- Ninety nine percent who viewed notes wanted to continue doing so after the study

Physician findings include:

- Over 90 % of doctors believed that clinic visits did not get longer
- Three to thirty five percent reported changing the way they wrote about mental health, cancer and obesity
- Hundred percent of doctors continued to provide note access after the study ended
- The total volume of secure email messages from patients before and after the study did not change

All three hospitals of the OpenNotes study are working to allow patients who participated to continue to have access to their doctors' notes. Beth Israel Deaconess Medical Center, who already allows all patients to view their test results on a secure patient website, plans to expand the program to allow all patients open access to notes from not only their doctors but also their nurses and other health care providers.

But another debate is rising. A recent survey of 179 HIEs (Adler-Milstein et al. 2011) found only 13 (covering just 3 % of hospitals and 0.9 % of physician practices) capable of meeting Stage 1 Meaningful Use criteria. Of those, only six were reported to be financially viable. More importantly, none of the HIEs surveyed had the capabilities of a comprehensive system as specified by an expert panel, calling "into question whether RHIOs in their current form can be self-sustaining and effective." Multiple HIEs have been shut down, e.g., Washington (DC), Kansas, Tennessee, CalRHIO, and CareSpark (Kingsport, TN, once touted as a national leader). According to Dr. Yasnoff (President of the Health Record Banking

Alliance) these failures are not for lack of funding or of knowing key obstacles, such as privacy, stakeholder cooperation, and financial sustainability. He claims that HIEs are on "the wrong path" by building institution-centric systems that "leave patient information where it's created and retrieve and integrate it in real time only when it's needed." Dr. Yasnoff is currently promoting an innovative patient-controlled approach to lifetime medical records, the patient-centric community health record bank, putting control of healthcare data in the hands of patients themselves. The reality is that the HIE landscape is rich and complex with many innovators attempting a diversity of solutions. The main question is: how and when critical mass can be obtained? In a world where no one seems to have all the answers, it seems better to encourage different paths. Comments about Dr. Yasnoff's claim point out five problems that need to be solved (in order of importance): (1) unique patient ID: social security number or other national identifier; (2) standards of data: HL7, CDA and other structured formats versus plain text, word documents and PDFs; (3) control over who can see the data: patient-provider and provider-provider relationships; (4) getting data into the system: 24/7 or on demand; (5) business model: free of charge, subscription fee, insurance companies or social security sponsorship.

13.6 Conclusion

Health care data sharing is aligned with the digital information landscape of the citizens. Support of microprocessor card, Internet access, electronic financial transactions are now commonly used in healthcare. An evolution is required from the patients to better appreciate the chronicity of pathologies and the involvement and participation of patients in their care. An evolution is also necessary for healthcare systems including harmonization of the practices, reduction of the adverse events, health prevention and education. These evolutions require cultural adaptations: (1) patient implication and reliabilization of the declaratory aspect of personal information (Simborg 2010, Wald 2004); (2) the reliability and the rigor of the traceability by the professional. The objective is to reach and maintain data accuracy, exhaustiveness of uploaded documents in order to facilitate the information flow and care continuity between all professionals (Ross 2005). These evolutions require technical amendments: technical requirements of security and confidentiality of the access and the use of the data (confidence), interoperability between systems components, transparency and ease of use for the end-users. "Adoption requires public awareness. For the healthcare technology and informatics community, this means making the benefits of networked interoperable PHRs apparent to consumers and clinicians." (Ball and Gold 2006). As a conclusion, we can cite Dr. Delbanco, OpenNotes project leader: "We are just giving patients what is already their rights: *Nothing about me without me*."

13.7 For More Information

It is interesting to note that the MeSH expression "Electronic Medical Records" was added in 2010. It is the only instance of the generic term "Medical Records Systems, Computerized" that inherits the concept "Medical Records". The expression "Health Records, Personal" is a direct specification of the concept "Medical Records". Navigating this hierarchy enables to locate the neighboring expressions (entry terms). All the references mentioned in the text of this chapter allow you to deepen the concepts discussed in the text. Try to Google them and see if some non-accessible pages are still available in cache. Using the expression "Electronic Medical Records, Shared Data" with Google engine will also provide you a series of relevant links. An example of Medline query is: ("Health Records, Personal"[Mesh] OR personal health record [Title/Abstract]) AND ((hasabstract [text] AND "loattrfull text"[sb]) AND "humans"[MeSH Terms] AND English [lang]).

The URLs below are related to international or European aspects of health information exchange:

- Bailey A, Jungman A, Sutton T. Google Health's Failure to Bring Meaning to Data. June 28, 2011 (http://designmind.frogdesign.com/blog/google-health-s-failure-to-bring-meaning-to-data.html, last access: 01/10/2013)
- California Healthcare Foundation. Consumers and Health Information Technology: a National Survey. April 2010. (http://www.chcf.org/~/media/MEDIA%20 LIBRARY%20Files/PDF/C/PDF%20ConsumersHealthInfoTechnologyNational Survey.pdf, last access: 01/10/2013)
- Continuity of care document (http://en.wikipedia.org/wiki/Continuity_of_ Care_Document, last access: 01/10/2013)
- Continuity of care record (http://en.wikipedia.org/wiki/Continuity_of_ Care_Record, last access: 01/10/2013)
- Dossia schemas. (http://wiki.dossia.org/index.php/Dossia_Schemas, last access: 01/10/2013)
- European countries on their journey towards national eHealth infrastuctures. eHealth Strategies Report. January 2011. (http://www.ehealth-strategies.eu/ report/eHealth_Strategies_Final_Report_Web.pdf, last access: 12/28/2012)
- European Patients Samrt Open Services. epSOS. (http://www.epsos.eu, last access: 12/28/2012)
- Fieschi M. Les données du patient partagées : la culture du partage et de la qualité des informations pour améliorer la qualité des soins. (http://www. ladocumentationfrancaise.fr/var/storage/rapports-publics/074000714/0000.pdf, last access: 01/22/2013)
- Google Health, a first look. February 2008. (http://googleblog.blogspot.com/ 2008/02/google-health-first-look.html, last access: 01/10/2013)
- International Federation of Information Management Associations. (http://www. ifhima.org/resourcectr.aspx, last access: 12/28/2012)

- International Health Terminology Standards Development Organisation (http://www.ihtsdo.org/, last access: 12/28/2012)
- Microsoft Vault (http://www.microsoft.com/en-us/healthvault/, last access: 01/10/2013)
- Ministero della Salute. Il Fascicolo Sanitario Elettronico Linee guida nazionali. 2010 (http://www.salute.gov.it/imgs/C_17_pubblicazioni_1465_allegato.pdf, last access: 12/28/2012)
- Nazi KM. My HealtheVet Personal Health Record. 2010 (http://www.queri.research.va.gov/meetings/eis/CIPRS-EIS-Nazi-MyHealth*e*Vet.pdf, last access: 01/17/2013)
- OpenNotes. Inviting patients to read their doctor's visit notes. 2012 (http://www.myopennotes.org/wp-content/uploads/2012/08/OpenNotes-Fact-Sheet.pdf, last access: 01/17/2013)
- Shanbhad A. NwHIN Exchange Directories – Status. March 2011. (http://healthit.hhs.gov/portal/server.pt/document/953882/privacy-security-nwhin-exchange-directories-presenation-03-16-2011_ppt, last access: 01/10/2013)
- The Dossia Health Management System. (http://www.dossia.org/, last access: 01/10/2013)
- Yasnoff W. The Sad Truth: HIEs are Falling. Healthcare IT News. 2013 (http://www.nhinwatch.com/perspective/sad-truth-hies-are-failing, last access: 01/19/2013)

A chronicle of the rich diversity of approaches to HIE can be found at the National eHealth Collaborative website http://www.nationalehealth.org. In the free downloadable report entitled Health Information Exchange Roadmap; The Landscape and Path Forward, examples of different approaches to exchanging health information are described including HealthBridge, the Indiana Health Information Exchange, Inland Northwest Health Services, MedVirginia and Surescripts. Each of these approaches is significantly different, and some show how critical mass can be obtained through institutional cooperation.

Exercises

Q1 A French patient asks his general practitioner: "Doctor, why the Government didn't choose the social security number to implement the personal electronic medical record (pEHR)? That would have been easier and especially less expensive?" What is your answer?

Q2 What is the reason to have defined models of clinical document to structure the personal medical record?

Q3 Regarding health data exchange, one basically describes three levels of interoperability. Which ones?

(continued)

Exercises (continued)

Q4 In the French system of personal medical record (DMP), a health care professional can access data according to three ways. Which ones?

Q5 The French national assurance has opened an information system by which any medical doctor can access patient's health data related to health care reimbursements. This system (select all that apply):
A. can be used without patient's consent
B. provides the physician to view patient's data without time limit
C. is restricted to ambulatory care data
D. concerns only the patients with chronic illness
E. provides the physician with a list of all patient's sick leaves

Q6 Which is not a factor in the demand for change to electronic health records?
A. An increase in medical errors
B. Rising health care costs
C. Need for coordination of care
D. Need for saving paper
E. None of the above

Q7 What are the barriers for providers to use HIE? (select all that apply)
A. Time Required to Access and Retrieve
B. Resistance to Change
C. Potential data quality and insufficient volume
D. Inadequate Reimbursement
E. None of the above

Q8 What interfaces are available for providers to interact with HIE data? (select all that apply)
A. Integrated EHR
B. Web Portal
C. Secure messaging system
D. Hybrid approach
E. None of the above

Q9 What are some of the Administrative challenges with HIEs? (select all that apply)
A. Determining a shared population
B. Patient identity matching
C. Managing protected health conditions
D. Defining data interoperability
E. All of the above

Exercises (continued)

Q10 What are the drivers for patients to participate in HIE? (select all that apply)
A. Convenience
B. Safety
C. Quality of Care
D. Health insurance
E. All of the above

Q11 Healthcare information exchange impacts the following except?
A. Reduced medical error
B. Reduced information management labor costs
C. Reduced duplicative tests and procedures
D. Reduced transfer of care
E. Improved service delivery efficiency

Q12 What are the barriers for patients to participate in HIE? (select one answer)
A. Privacy concerns
B. Lack of awareness of the program
C. Enrollment issues, like mismatched signatures
D. All of the above

Q13 On the NwHIN exchange, which service would you use to assert who you are and for what purpose you are requesting a patient health summary?
A. Patient Discovery
B. Query for document
C. Transcription services
D. Retrieve document
E. Authorization framework

Q14 On the NwHIN exchange, which service would you use to find out if your patient is known elsewhere?
A. Patient Discovery
B. Query for document
C. Transcription services
D. Retrieve document
E. Authorization framework

Q15 In order to implement an HIE, what are the problems to be solved? (select all that apply)
A. Unique patient ID
B. Standards of data
C. Control over who can see the data
D. Getting data into the system
E. Business model

(continued)

Exercises (continued)

R1 the French national identifier is not anonymous. It is a composite code and each part of it has a specific meaning (gender, month and year of birth, department of birth, code of birth city...) that can help to discover patient's name. In the French law, the national identifier is considered as a personal and private data and its usage is very constrained.

R2 to make medical record structure homogeneous; to standardize the use of controlled vocabulary; to facilitate data retrieval and applying medical decision algorithms

R3 syntactic interoperability (structure and data arrangement), semantic interoperability (content and context), technical interoperability (service and data transport levels)

R4 personal access (given and registered by patient), undefined access (with patient's consent), emergency access (with tracking and logbook)

R5 E

R6 D

R7 AB

R8 ABCD

R9 ABC

R10 ABC

R11 D

R12 D

R13 E

R14 A

R15 ABCDE

References

Adler-Milstein J, Bates DW, Jha AK (2011) A survey of health information exchange organizations in the United States: implications for meaningful use. Ann Intern Med 154:666–671

Angst CM, Agarwal R (2009) Adoption of electronic health records in the presence of privacy concerns: the elaboration likelihood model and individual persuasion. MIS Quart 33:339–370

Archer N, Fevrier-Thomas U et al (2011) Personal health records: a scoping review. J Am Med Inform Assoc 18(4):515–522

Ball MJ, Gold J (2006) Banking on health: personal records and information exchange. J Healthc Inf Manag 20(2):71–83

Ball MJ, Smith C, Bakalar R (2007) Personal health records: empowering consumers. J Healthc Inf Manag 21(1):76–86

Bates DW, Bitton A (2010) The future of health information technology in the patient-centered medical home. Health Aff (Millwood) 29(4):614–621

Benson T (2010) Continuity of Care Document (CCD). In: Principles of Health Interoperability HL7 and SNOMED. Springer, London, pp 156–160

Bonander J, Gates S (2010) Public health in an era of personal health records: opportunities for innovation and new partnerships. J Med Internet Res 12(3):e33

Bouhaddou O, Bennett J et al (2012) Toward a virtual lifetime electronic record: the department of veterans affairs experience with the nationwide health information network. AMIA Annu Symp Proc 2012:51–60

Chen T, Zhong S (2012) Emergency access authorization for personally controlled online health care data. J Med Syst 36(1):291–300

Chumbler NR, Haggstrom D, Saleem JJ (2011) Implementation of health information technology in Veterans Health Administration to support transformational change: telehealth and personal health records. Med Care 49(Suppl):S36–S42

Cimino JJ, Patel VL, Kushniruk AW (2002) The patient clinical information system (PatCIS): technical solutions for and experience with giving patients access to their electronic medical records. Int J Med Inform 68(1–3):113–127

Consumers and Health Information Technology: A National Survey. California HealthCare Foundation (Lake Research Partners) (2010) http://www.chcf.org/publications/2010/04/consumers-and-health-information-technology-a-national-survey

Delbanco T, Walker J et al (2010) Open notes: doctors and patients signing on. Ann Intern Med 153(2):121–125

Delbanco T, Walker J et al (2012) Inviting patients to read their doctors' notes: a quasi-experimental study and a look ahead. Ann Intern Med 157(7):461–470

Demiris G, Thompson HJ (2012) Mobilizing older adults: harnessing the potential of smart home technologies. Contribution of the IMIA working group on smart homes and ambient assisted living. Yearb Med Inform 7(1):94–99

Demiris G, Afrin LB et al (2008) Patient-centered applications: use of information technology to promote disease management and wellness. J Am Med Inform Assoc 15:8–13

Dolin RH, Alschuler L et al (2001) The HL7 clinical document architecture. J Am Med Inform Assoc 8:552–569

Earnest MA, Ross SE et al (2004) Use of a patient-accessible electronic medical record in a practice for congestive heart failure: patient and physician experiences. J Am Med Inform Assoc 11(5):410–417

Ferranti JM, Musser RC et al (2006) The clinical document architecture and the continuity of care record: a critical analysis. J Am Med Inform Assoc 13:245–252

Frost JH, Massagli MP (2008) Social uses of personal health information within PatientsLikeMe, an online patient community: what can happen when patients have access to one another's data. J Med Internet Res 10(3):e15

Greenhalgh T, Stramer K et al (2010) Adoption and non-adoption of a shared electronic summary record in England: a mixed-method case study. BMJ 340:c3111

Halamka JD, Mandl KD, Tang PC (2008) Early experiences with personal health records. J Am Med Inform Assoc 15(1):1–7

Hassol A, Walker JM et al (2004) Patient experiences and attitudes about access to a patient electronic health care record and linked Web messaging. J Am Med Inform Assoc 11:505–513

Hess R, Bryce CL et al (2007) Exploring challenges and potentials of personal health records in diabetes self-management: implementation and initial assessment. Telemed J E Health 13(5):509–517

Hogan TP, Wakefield B et al (2011) Promoting access through complementary eHealth technologies: recommendations for VA's Home Telehealth and personal health record programs. J Gen Intern Med 26(Suppl 2):628–635

Jha AK, Doolan D et al (2008) The use of health information technology in seven nations. Int J Med Inform 77(12):848–854

Kaelber DC, Jha AK et al (2008) A research agenda for personal health records (PHRs). J Am Med Inform Assoc 15(6):729–736

Luo G (2010) On search guide phrase compilation for recommending home medical products. Conf Proc IEEE Eng Med Biol Soc 2010:2167–2171

Maloney FL, Wright A (2010) USB-based Personal Health Records: an analysis of features and functionality. Int J Med Inform 79(2):97–111

Martin S, Kelly G et al (2008) Smart home technologies for health and social care support. Cochrane Database Syst Rev 4:CD006412

Masys D, Baker D et al (2002) Giving patients access to their medical records via the Internet: the PCASSO experience. J Am Med Inform Assoc 9:181–191

Rode D (2008) PHR debates. The personal record gets political, but there is danger in rushing legislation. J AHIMA 79:18–20

Ross SE, Todd J et al (2005) Expectations of patients and physicians regarding patient-accessible medical records. J Med Internet Res 7(2):e13

Shenkin BN, Warner DC (1973) Sounding board. Giving the patient his medical record: a proposal to improve the system. N Engl J Med 289(13):688–692

Shenkin B, Warner DC (1975) Open information and medical care: a proposal for reform. Conn Med 39(1):33–34

Simborg DW (2009) The limits of free speech: the PHR problem. J Am Med Inform Assoc 16:282–283

Simborg DW (2010) Consumer empowerment versus consumer populism in healthcare IT. J Am Med Inform Assoc 17(4):370–372

Stolyar A, Lober WB et al (2005) Feasibility of data exchange with a patient centered health record. AMIA 2005 Conf Proc 2005:1123

Tang PC, Lee TH (2009) Your doctor's office or the Internet? Two paths to personal health records. N Engl J Med 360(13):1276–1278

Tang PC, Ash JS et al (2006) Personal health records: definitions, benefits, and strategies for overcoming barriers to adoption. J Am Med Inform Assoc 13(2):121–126

Tobacman JK, Kissinger P et al (2004) Implementation of personal health records by case managers in a VAMC general medicine clinic. Patient Educ Couns 54(1):27–33

Turvey CL, Zulman DM et al (2012) Transfer of information from personal health records: a survey of veterans using My HealtheVet. Telemed J E Health 18(2):109–114

Varroud-Vial M (2011) Improving diabetes management with electronic medical records. Diabetes Metab 37(Suppl 4):S48–S52

Vest JR, Gamm LD (2010) Health information exchange: persistent challenges and new strategies. J Am Med Inform Assoc 17:288–294

Wald JS, Middleton B et al (2004) A patient-controlled journal for an electronic medical record: issues and challenges. Stud Health Technol Inform 107(Pt 2):1166–1170

Walker J, Leveille SG et al (2011) Inviting patients to read their doctors' notes: patients and doctors look ahead: patient and physician surveys. Ann Intern Med 155(12):811–819

Wang T, Pizziferri L et al (2004) Implementing patient access to electronic health records under HIPAA: lessons learned. Perspect Health Inf Manag 1:11

Wiljer D, Urowitz S et al (2008) Patient accessible electronic health records: exploring recommendations for successful implementation strategies. J Med Internet Res 10:e34

Win KT, Susilo W, Mu Y (2006) Personal health record systems and their security protection. J Med Syst 30:309–315

Yang M, Chronaki CE et al (2011) Guideline-driven telemonitoring and follow-up of cardiovascular implantable electronic devices using IEEE 11073, HL7 & IHE profiles. Conf Proc IEEE Eng Med Biol Soc 2011:3192–3196

Chapter 14
Computerising the Doctor's Office

A. Venot and M. Cuggia

Abstract In this chapter, we list the main tasks of doctors working in their offices and the reasons for computerising these offices. We describe the architecture of medical office software and the various types of patient data stored in the electronic medical record. The advantages of using computerised provider order entry are explained. We discuss the various types of decision support available to physicians, including diagnostic tools, alerts and dashboard generation. National and regional initiatives for developing medical office computerisation are described, including the design of open-source software in Canada, the certification of drug prescription software in France and incentives and strategies for generalising the use of electronic medical records in the United States.

Keywords Medical office • Primary care • General practitioner • Electronic medical record • Coding systems • ICPC

After reading this chapter you should:

- Be able to identify the main tasks carried out by physicians in their offices that could most benefit from computerisation
- Be aware of the architecture of medical office software

(continued)

A. Venot (✉)
LIM&BIO EA 3969, UFR SMBH Université Paris 13, 74 rue Marcel Cachin, Bobigny Cedex 93017, France
e-mail: alain.venot@univ-paris13.fr

M. Cuggia
Faculté de Médecine, Laboratoire d'Informatique Médicale, INSERM U936, 2, av. Léon Bernard, Rennes 35043, France
e-mail: marc.cuggia@univ-rennes1.fr

A. Venot et al. (eds.), *Medical Informatics, e-Health*, Health Informatics,
DOI 10.1007/978-2-8178-0478-1_14, © Springer-Verlag France 2014

After reading this chapter you should: (continued)
- Know the main functions of medical office software offered to GPs and specialists
- Be able to explain why it is important to use terminological and coding systems to represent patient data in the electronic medical record
- Be aware of the terminological systems developed for primary care
- Be able to cite and describe examples of national or regional initiatives promoting medical office computerisation.

14.1 The Information Systems of General Practicioners and Specialists

14.1.1 The Tasks of General Practitioners and Specialists

In routine practice, General Practitioners (GP) and specialists are repeatedly confronted with the problems of collecting, archiving, recovering, processing and communicating medical and non-medical information.

The principal tasks of these physicians are:

- Recording the data collected during patient interviews and examinations
- Prescribing tests for diagnosis or follow-up
- Storing the results of any laboratory tests, pathology analyses or imaging examinations carried out
- Deciding on the diagnosis and the best way to manage the patient in accordance with the clinical guidelines published by national or international health agencies
- Generating, storing and supplying patients with drug orders (prescriptions) and letters for other physicians if required
- Exchanging information with hospitals, such as storing hospital reports after the discharge of their patients from these institutions
- Exchanging information with the personal records of their patients, which are being developed in many countries
- Exchanging information with health insurance organisations, a process that differs considerably between countries, depending on the specific features of each country's health system.

They also have administrative and financial tasks relating to their medical office activities (revenue management and accounting).

14.1.2   Why Do Physicians Need to Computerise Their Offices?

The electronic medical record (EMR)/electronic health record (EHR) is becoming an integral component of many primary-care outpatient practices. Several countries have implemented successful programs promoting the use of the EMR/EHR in primary care, and governments are increasingly providing support for health information technology.

Doctors now have so much information to deal with that their memory capacity alone is no longer sufficient. They need to work with paper or electronic forms and records. Doctors require support for the memorisation both of data relating to their patients and of the medical knowledge required in daily practice.

It is thought that the EMR/EHR will reduce healthcare costs and improve the quality of healthcare, for the following reasons:

The diagnosis process may be complex when the suspected diseases are not frequent. The best ways of treating patients may change rapidly in certain medical domains (e.g. diabetes and infectious diseases). Physicians therefore need to have rapid access to sources of both diagnostic and therapeutic knowledge. They also need to be able to receive and to send medical information to colleagues from various medical specialities.

Many tasks are repetitive and time-consuming (e.g. the writing of various certificates, accounting, the writing of treatment regimens), but are amenable to partial or total automatisation.

The various exchanges of information required in daily practice can be greatly facilitated by the use of information and communication technologies.

14.2   The Main Functions of Software for the Medical Office

The rapid development of microcomputers in the early 1980s and the steady decrease in their price ever since has led vendors to develop and sell specialist software for the doctor's office (Morgan and Morgan 1984; Bailey 1985). The software now on offer is much more sophisticated than earlier versions. The functions of these programs are very similar in the various countries, although there are some differences due to the specific features of national health systems. The software can be seen as being composed of several modules (Fig. 14.1). The EMR occupies a key position in this architecture.

Further improvements to medical software are required, particularly in terms of ergonomics, but the development of microcomputers, networks and of the Internet has resulted in highly functional software, making it possible for physicians to abandon paper records for most of the tasks carried out in daily practice.

Fig. 14.1 The various modules of medical office software and their links to external systems (laboratories, radiology centres, hospitals, specialists, social security organisations)

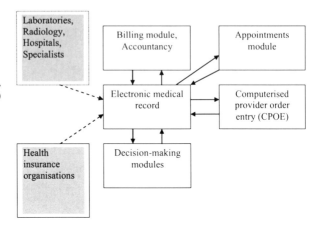

14.2.1   Diaries and the Scheduling of Appointments

This function allows the physician to plan appointments. The secretary selects the doctor and schedules appointments. He or she can view the schedules of all doctors and allocate workloads. The calendar can also alert the physician when the patient arrives and specify the type of appointment (consultation visit, emergency, paediatric consultation, etc.)

Advanced features can be offered, such as the triggering of alerts, the addition of textual or audio annotations for each appointment, printing of the list of appointments and visits with the contact details of the patients and the transmission of patient data directly to the medical record.

Such software has major advantages over a classical paper diary. It is possible to decrease the number of telephone calls and missed appointments if the software allows patients to schedule their own appointments online via the physician's website. The automation of confirmations and reminders also helps to decrease the number of missed appointments

Electronic diaries can also be used retrospectively, to investigate variations in the number of appointments as a function of the day of the week and their links with the number of missed appointments. (Ellis and Jenkins 2012) have shown that the number of missed appointments is greater on some days than on others. Such analyses can help to decrease the time lost by physicians in these circumstances.

14.2.2   The Electronic Record for the Medical Office

14.2.2.1   The Various Types of Patient Data in the Electronic Medical Record

Physicians store various types of data relating to patients in medical records:

- Demographic data (e.g. name, date of birth, sex, address)
- Personal and family medical and surgical histories
- Lifestyle of the patient (nutrition, sport, smoking, alcohol consumption, addictions)
- Data collected at each consultation or visit:

  - Reasons for consulting
  - Symptoms, clinical signs, laboratory test results, imaging data
  - Diagnoses confirmed or ruled out
  - Treatments prescribed; efficacy and tolerance of these treatments; the compliance of the patient with treatment
  - Additional tests prescribed
  - Monitoring elements

- Electronic data transmitted by other doctors or healthcare institutions:

  - Hospital reports
  - Surgical reports
  - Specialist reports

14.2.2.2 The Various Ways of Representing and Coding These Data

General practitioners and specialists often use natural language to record these data. However, over the last 20 years or so, considerable efforts have been made to develop terminology systems appropriate for general practice.

In particular, the International Classification of Primary Care (ICPC) version 2 has been translated into more than 20 languages and has been recognised as an international classification by the WHO (Soler et al. 2008).

http://www.who.int/classifications/icd/adaptations/icpc2/en/index.html

The ICPC has two axes and is composed of 17 chapters, mostly dealing with anatomy and function. Each chapter includes information about complaints and symptoms, diagnosis, screening procedures, test results and drugs. However, this classification has been criticised by some (Botsis et al. 2010), who have suggested that the lack of adherence to this classification on the part of many physicians results from certain deficiencies, such as its incompleteness (i.e. the absence of many clinically important diagnoses).

In the United Kingdom, the coding system developed by James Read (Benson 2011), known as 'Read Codes' has been widely used and is supported by the NHS. Almost all GPs have been using Read Codes since the mid-1990s. This common clinical coding scheme has facilitated many developments in general practice, such as GP quality of care studies.

Some authors have studied the use of general classifications, such as ICD and SNOMED, in primary care (Vikström et al. 2010). The main problem lies in the practical use of these large coding systems. SNOMED-CT will probably be increasingly widely used in primary care in the future. Medical software companies need to develop ways to facilitate the use of classifications for the coding of medical data in primary care. This step cannot currently be considered optimal.

14.2.2.3 The Use of Computerised Provider Order Entry (CPOE) Systems in Primary Care

GPs are increasingly making use of CPOE for the generation of drug orders, which are stored in the patient electronic record (see Chap. 8). Over and above the use of decision support aids, as described below, there are several advantages of storing drug prescriptions in patient records. The prescription is always readable and complete and misspellings are avoided.

The advanced features of some programs can transform drug trade names into international non-proprietary names and pharmacotherapeutic classes, thereby making it possible to construct a meaningful summary of the drug treatment history of the patient. The ergonomic features of CPOE may have a major effect on the likelihood of such tools being adopted by physicians (Khajouei and Jaspers 2010).

14.2.3 Decision Support in the Medical Office

Medical errors may occur frequently in primary care (Abramson et al. 2012; Khoo et al. 2012; Singh et al. 2013). Computerised decision support can decrease the frequency of such errors, thereby making primary care safer.

14.2.3.1 Diagnostic Tools

Diagnostic scores, such as the Mini Mental Score or depression scores, are readily computerised and can therefore be used with ease in primary care consultations (see Chap. 7).

In recent decades, considerable effort has been devoted to the construction of more sophisticated and general diagnostic tools appropriate for use on all patients. However, there is currently no general diagnostic decision support system for primary care integrated into electronic health records.

14.2.3.2 Drug Dose Information

Meta-analyses (Durieux et al. 2008) have indicated that the provision of computerised advice about drug doses could improve drug prescriptions by increasing the initial dose administered, leading to an increase in serum drug concentrations and more rapid therapeutic control, or by reducing the risk of toxic drug levels and the length of time spent in the hospital. Suggestions for drug doses taking into account patient-specific data, such as weight, body mass index, body surface area and renal insufficiency are provided by various medical office software suites.

14.2.3.3   Alerts

The system may generate various electronic record-based alerts designed to improve patient care. Such triggered alerts have been shown to be effective in primary care (Tamblyn et al. 2008), as illustrated by the following examples:

- Alerts reminding doctors about the importance of immunisation (e.g. influenza vaccination in certain categories of patients, such as pregnant women and elderly)

Alerts to prevent medication errors in computerised order entry systems. For example, such systems efficiently detect severe drug-drug interactions (Steele et al. 2005). Clinical conditions contraindicating a particular drug prescription can also be checked (e.g. beta-blockers are contraindicated in patients with asthma). However, it is not possible to issue such alerts if physicians use natural language to enter patient data.. The generation of such alerts requires the patient data in the electronic record to be coded with the same terminology as the information in the drug database used by the CPOE.

14.2.3.4   Dashboards for the Monitoring of Patients

A "dashboard" can be displayed to the physician, to facilitate patient follow-up, particularly for chronic diseases (e.g. patients with type 2 diabetes). Such tools have been shown to be valuable (Koopman et al. 2011) for correct identification of the data required for follow-up, for decreasing the time and effort required to obtain these data with respect to more conventional approaches.

14.2.4   Mail and Word-Processing

Letters can be written to other healthcare professionals with a word-processor, which may be independent (e.g. Word), but is interfaced with the medical office software. Template texts are generally available, for certificates of aptitude for sports or third-party requests for hospitalisation, for example. The software retrieves the relevant information about the patient from the electronic record (e.g. recent history of the disease, current treatment) and incorporates them into the letter. This saves time and overcomes the need for duplicate data entry.

14.2.5 Exchanges with Biological Testing Laboratories

Medical office software generally includes a recovery module for laboratory results. It uses a secure Internet messaging system, in which messages are both signed (the physician is certain of both the sender and the recipients) and encrypted (messages are illegible if intercepted during transmission, protecting confidentiality). This secure messaging service can also be used for other types of exchange (e. g. exchanges of hospitalisation reports between the hospital and the doctor's office).

The messages exchanged between the doctor's office and the laboratory make use of a standard mode of transmission in each country, but standards currently differ between countries. A unique standard (e.g. an international format HL7 CDA) will gradually replace the various existing ones.

On reception, the documents are automatically integrated into the medical record. This integration is based on the approximation of the identity traits (e.g. first name, surname, date of birth) in the received message and the medical record. However, this integration is not entirely reliable, because the patient's name may be misspelled, for example.

14.2.6 Exchanges with the Personal Medical Record

GPs are key contributors to personal medical files. Ideally, medical office software should provide access to the personal medical record, allowing its consultation. It can also send parts of the electronic record maintained by the physician directly to the personal record.

In France, software compatibility is ensured by the application of a set of specifications (known as the personal medical record interoperability framework) provided by the Agency of Patient Information Systems (ASIP). This interoperability framework (see Chap. 13) is based on the exchange standards (in particular HL7 version 3), integration profiles and technical specifications broadcast internationally by the IHE organisation (Health Integrate Enterprise).

14.2.7 Accounting in the Medical Office

Medical office software generally includes an accounting module, dealing particularly with the revenues and expenses of the physician. One of the strong points of such integrated accounting module (as opposed to independent accounting software) is that receipt data are automatically retrieved by the accounting module, avoiding the need for re-entry.

14.3 Examples of National and Regional Initiatives Promoting the Development of Medical Office Computerisation

The computerisation of medical offices, including the systematic use of EMRs, is often encouraged by national authorities and academic institutions. Some of the initiatives developed in various countries are described below.

14.3.1 In Canada: The Open-Source Electronic Medical Record OSCAR

(http://www.oscarcanada.org/)

OSCAR stands for "Open Source Clinical Application Resource". The OSCAR project was launched in 2001 by the Family Medicine Department of McMaster University in the province of Ontario (Canada). Its primary goal was to provide open-source EMR software for use in medical offices (McDonald et al. 2003).

OSCAR has been adopted by many physicians (more than 1,200 in Ontario) and by several family medicine departments, including those at the medical schools of the McGill and Queens Universities, for instance. Two examples of screen captures relating to OSCAR drug prescriptions are provided in Figs. 14.2 and 14.3.

The OSCAR project can be seen as a success story in the domain of open-source software for health.

14.3.2 A Certification Process for the Drug Prescription Modules Included in Medical Office Software in France

The French Health Agency, the *Haute Autorité de Santé* (HAS), introduced, in 2008 a certification process for improving drug prescription software for the medical office. This process had several goals:

- To improve prescription quality;
- To facilitate the work of the prescriber and to promote the compliance of drug orders with regulations and recommendations;
- To decrease the cost of drug treatments without decreasing their quality.

http://www.has-sante.fr/portail/jcms/c_576417/referentiel-de-certification-par-essai-de-type-des-logiciels-daide-a-la-prescription-en-medecine-ambulatoire?xtmc=&xtcr=7

Fig. 14.2 Searching for Sitagliptin with OSCAR (From http://www.oscarmanual.org/oscar-emr_10/clinical-functions/prescription/Rx3Sitagliptin.png/image_view_fullscreen accessed 21st March 2013)

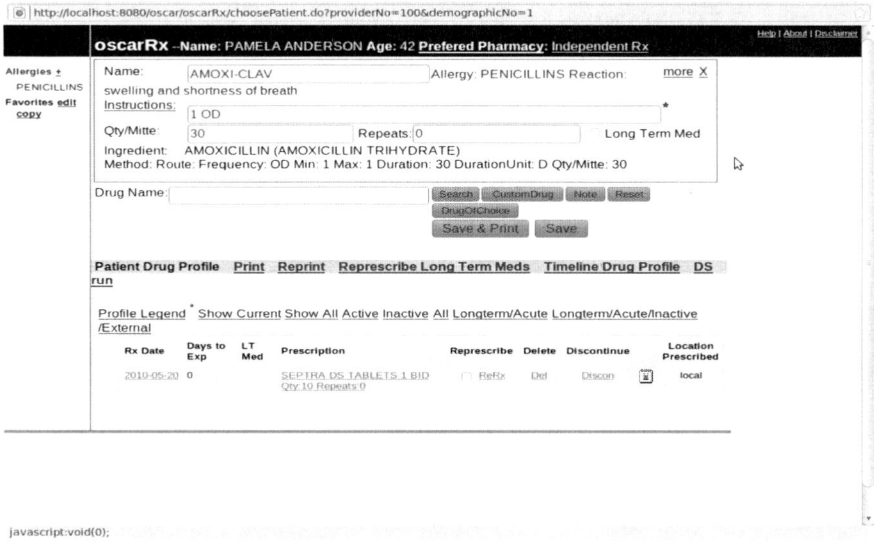

Fig. 14.3 Example of drug prescription and allergy alert with OSCAR (From http://www.new.oscarmanual.org/oscar_emr_12/clinical-functions/prescription/Rx3AllergyAmoxil.png/image_view_fullscreen, accessed 21st March 2013)

The certification process addresses issues such as the choice of drugs (from the common name or brand name), alerts concerning contraindications and interactions and the availability of various types of information about the drug. Safety functions are only truly effective if high-quality information about the drug is made available. The certification process requires the use of drug databases approved by the HAS.

The HAS publishes the tests used for this type of certification. http://www.has-sante.fr/portail/jcms/c_590032/tests-de-certification-des-logiciels-daide-a-la-prescription-pour-la-medecine-ambulatoire

A list of certified software can be obtained from:

http://www.has-sante.fr/portail/jcms/c_672760/logiciels-d-aide-a-la-prescription-pour-la-medecine-ambulatoire-logiciels-certifies-selon-le-referentiel-de-la-has-et-logiciels-ayant-postule-a-la-certification

The existence of this certification process has forced vendors to improve the drug prescription functions of their software.

14.3.3 Standards for the "Meaningful Use" of Medical Records in the United States

http://www.healthit.gov/policy-researchers-implementers/meaningful-use

In 2009, the Obama administration devised the American and Reinvestment Act, which allocated a total of $787 billion to stimulation of the US economy. This act devoted $20 billion to the improvement of health information technology infrastructure. The resulting programme was named Health Information Technology for Economic and Clinical Health (HITECH). It recognises the key role of information technology in improving effective healthcare delivery. Its aims are to improve the control of increasing healthcare costs, to decrease waste and to improve quality, safety and outcomes.

One of the goals of this programme is the universal adoption of EHRs within the space of 5 years, in both hospitals and primary care settings. A qualified EHR has been defined by HITECH as "*an electronic record of health-related information on an individual that includes patient demographic and clinical health information, such as medical history and problem lists; and has the capacity to provide clinical decision support; to support physician order entry; to capture and query information relevant to healthcare quality; and to exchange electronic health information with, and integrate such information from other sources*".

To reach this objective, the "meaningful use" concept has been established and quality criteria have been defined for EHRs. "Meaningful use" is the set of standards defined by the Centers for Medicare & Medicaid Services (CMS). Financial incentives have been established to favour the development and use of certified EHRs. For instance, under Medicare, eligible professionals who display "meaningful use" of certified EHRs are eligible for payments of up to $44,000, whereas eligible professionals who do not do so will be subject to penalties after 2015 (Wright et al. 2013).

Three stages have been defined for the improvement of EHRs. Stage 1 requirements for meaningful use are the use of key EHR functions, including electronic prescribing, drug–drug and drug–allergy checking, and the maintenance of problem, medication and allergy lists. In stage 1, providers must meet 15 core objectives and choose five additional specified objectives from a menu of 10.

In stage 2, additional objectives are introduced, concerning, for example:

- Communication with the patient (one new criterion is "Use secure electronic messaging to communicate with patients concerning relevant health information" and "provide patients with the ability to view online, download and transmit their health information within 36 h of discharge from the hospital")
- Health information exchange between providers, to improve care co-ordination for patients. One of the core objectives is to provide a summary of care records for more than 50 % of patient care transitions and referrals.

Details of these criteria and stages can be obtained from: http://www.cms.gov/Regulations-and-Guidance/Legislation/EHRIncentivePrograms/index.html?redirect=/EHRIncentivePrograms/

14.3.4 Technological Aspects of Medical Office Computerisation

The current trend is to move toward the use of Web-based software rather than software installed on the medical office computer (client–server architecture). The software is installed and maintained by the provider on a server accessed by many users. Web standards are used.

Web-based medical office software has several key advantages:

- The physician can access his data from anywhere, using various devices, such as PCs, tablets and smart phones. This is particularly important for physicians visiting patients at home.
- Data security and back-up are the responsibility of the software provider
- The physician no longer has to worry about installing and managing patches and upgrades.

14.3.5 The Real Impact and Outcome of the EMR Within Primary Care

Several studies exploring the impact of the EMR on care, including primary care, have reported mixed success for decreasing costs and improving the quality of ambulatory services (Holroyd-Leduc et al. 2011). For instance, there is only modest evidence that the use of CPOE decreases prescribing errors. The errors eliminated are mostly minor in nature, and CPOE may actually increase the risks of order duplication and failure to stop treatment.

The EMR clearly results in greater legibility and accessibility than paper-based records and the use of a computer does not seem to impair the doctor-patient relationship.

There are some concerns about losses of privacy and confidentiality relating to the use of EMRs in primary care, due to technical limitations and bad practice. For instance, the data are not protected when GPs replace their computers, and the thousands of patient records on the hard disk of the old computer may become accessible to third parties.

Primary care providers must also take into account the potential decrease in productivity when an EMR system is first implemented, due to an initial increase in the amount of time required for documentation. The installation of an EMR system requires time for learning and training. Costs are initially high, but savings can be made in the long term.

Several other issues, including "alert fatigue", have also been raised concerning clinical decision support.

Many studies have investigated the added value of EMR for primary care, but the methodology of most of these studies has been weak. There is a need to evaluate the real outcomes of EMR in the healthcare process, through randomised control trials.

14.3.6   For More Information

Consult the websites of medical office software providers, to compare the functions of the software provided in different countries. Google can be used. Consult the website for OSCAR software.

Use PubMed to identify and read articles on EMR use in the medical office.

Read the content of US federal websites dealing with "meaningful use", to gain an understanding of the strategy deployed for the generalisation of EMR use.

Exercices

Identify the correct responses

Q1 Medical office software includes modules for

A. Building an electronic record for each patient
B. Prescribing drugs
C. Managing patient appointments
D. Billing and accountancy
E. Receiving and archiving laboratory test results

(continued)

Exercices (continued)

Q2 By comparison to in-office systems, Web-based medical office software is

A. Easier for the physician to maintain
B. Avoids the need to install patches and upgrades in the medical office
C. Requires the installation of expensive software on the physician's PC
D. Allows the physician to access the patient's data anywhere
E. Cannot incorporate a billing and accountancy module

Q3 In the medical office, physicians may benefit from decision support, such as:

A. Facilities for calculating diagnostic scores
B. The generation of alerts in cases of drug-drug interactions
C. Information and alerts about the drug doses to be prescribed
D. Diagnostic advice for all the diseases affecting patients
E. The detection of drug contraindications for diseases expressed in natural language by the physician in the medical record

Q4 Establishing certification procedures for medical office software

A. Complicates the work of the physician
B. Incites providers to improve their medical office software to ensure compliance with the criteria on which the certification is based
C. Requires ways of incorporating imaging data into the electronic record
D. Decreases the health costs to be covered by social security organisations
E. Obliges the providers to incorporate decision support into their software

R1: ABCDE, **R2**: ABD, **R3**: ABC, **R4**: B

References

Abramson EL, Bates DW et al (2012) Ambulatory prescribing errors among community-based providers in two states. J Am Med Inform Assoc 19(4):644–648

Bailey S (1985) Computerizing the medical office. Can Med Assoc J 133(3):221–224

Benson T (2011) The history of the Read Codes: the inaugural James Read Memorial Lecture. Inform Prim Care 19(3):173–182

Botsis T, Bassøe CF, Hartvigsen G (2010) Sixteen years of ICPC use in Norwegian primary care: looking through the facts. BMC Med Inform Decis Mak 10:11

Durieux P, Trinquart L et al (2008) Computerized advice on drug dosage to improve prescribing practice. Cochrane Database Syst Rev Jul 16(3):CD002894

Ellis DA, Jenkins R (2012) Weekday affects attendance rate for medical appointments: large-scale data analysis and implications. PLoS One 7(12):e51365

Holroyd-Leduc JM, Lorenzetti D et al (2011) The impact of the electronic medical record on structure, process, and outcomes within primary care: a systematic review of the evidence. J Am Med Inform Assoc 18(6):732–737

Khajouei R, Jaspers MW (2010) The impact of CPOE medication systems' design aspects on usability, workflow and medication orders: a systematic review. Methods Inf Med 49(1):3–19

Khoo EM, Lee WK, Sararaks S et al (2012) Medical errors in primary care clinics – a cross sectional study. BMC Fam Pract 13(1):127

Koopman RJ, Kochendorfer KM et al (2011) A diabetes dashboard and physician efficiency and accuracy in accessing data needed for high-quality diabetes care. Ann Fam Med 9(5):398–405

McDonald CJ, Schadow G et al (2003) Open Source software in medical informatics–why, how and what. Int J Med Inform 69(2–3):175–184

Morgan RA, Morgan DK (1984) Buying and using an office computer. Can Med Assoc J 130(4): 469–470, 472–4

Singh H, Giardina TD et al (2013) Types and origins of diagnostic errors in primary care settings. JAMA Intern Med 25:1–8

Soler JK, Okkes I et al (2008) The coming of age of ICPC: celebrating the 21st birthday of the International Classification of Primary Care. Fam Pract 25(4):312–317

Steele AW, Eisert S et al (2005) The effect of automated alerts on provider ordering behavior in an outpatient setting. PLoS Med 2(9):e255

Tamblyn R, Huang A et al (2008) A randomized trial of the effectiveness of on-demand versus computer-triggered drug decision support in primary care. J Am Med Inform Assoc 15(4): 430–438

Vikström A, Nyström M et al (2010) Views of diagnosis distribution in primary care in 2.5 million encounters in Stockholm: a comparison between ICD-10 and SNOMED CT. Inform Prim Care 18(1):17–29

Wright A, Henkin S et al (2013) Early results of the meaningful use program for electronic health records. N Engl J Med 368(8):779–780

Chapter 15
Computerizing the Dental Office

V. Bertaud-Gounot, B. Chaumeil, E. Ehrmann, M. Fages, and J. Valcarcel

Abstract Computerizing the dental office aims at improving the dental surgeon's practices and efficiency. The practice management software manages the dentist's activity: patient record, estimates, prescriptions, agenda, accounting... Specific hardware like virtual keyboard is compatible with the high level of hygiene needed at the point of care. The use of an UPS, an antivirus, anti-spyware, daily backups contribute to protect the hardware and software. Digital radiography (using CCD or RLMS sensors) provides the ability to get instantly images and image processing software enable to discover information not visible to the naked eye. Computed image processing can bring help in diagnostic tasks such as early decay detection. They are based on digital radiographs or on Quantitative Light Induced Fluorescence (QLF) techniques. Computer-Aided Design and Computer-Assisted Manufacture (C.A.D.-C.A.M.) allow designing and manufacturing inlays, crowns, orthodontic brackets, surgical guides in implantology... directly in the dental office. Some other software tools are meant whether for patient education or for prevention, post-operative counselling or explanation of the treatment plan. Decision support systems provide relevant care guidelines in order to substantially improve health

V. Bertaud-Gounot (✉)
Prévention-Epidémiologie-Informatique, Faculty of dentistry, Rennes 1 University, Avenue du Professeur Léon Bernard (Bât 15), 35043 Rennes, cedex, France
e-mail: valerie.bertaud@univ-rennes1.fr

B. Chaumeil
Faculty of Dentistry, Clermont-Ferrand University, 11, Boulevard Charles de Gaulle, 63000 Clermont-Ferrand, France

E. Ehrmann
Faculty of Dentistry, Nice Sophia Antipolis University, Pôle universitaire St Jean d'Angely 24, rue des Diables Bleus 06357 Nice Cedex 4, France

M. Fages • J. Valcarcel
Faculty of Dentistry of Montpellier, Montpellier 1 University, 545 avenue du Professeur Viala BP 4305, 34193 Montpellier cedex 5, France

A. Venot et al. (eds.), *Medical Informatics, e-Health*, Health Informatics,
DOI 10.1007/978-2-8178-0478-1_15, © Springer-Verlag France 2014

care quality and potentially reduce errors in practice. The role of the computer in the dental office is no longer confined to administrative and accounting tasks but it takes up more and more important role in the medical field with diagnosis, treatment or prevention tasks.

Keywords Dentistry • Management information systems • Dental digital radiography • Computer-assisted image processing • Cone-beam computed tomography • Computer-aided design • Computer-assisted decision making

After reading this chapter, you should:

- Be able of describing the various functionalities of a typical dental office management software and their usefulness
- Be able of quoting the various necessary actions and protocols for maintaining hardware and software in good working order as well as ensuring data security
- Be able of describing the digital radiography system and identifying its possibilities and limitations
- Know the functionalities in digital imaging (radiography, beam cone computed tomography, intra-oral cameras)
- Know the applications of the CAD-CAM in dentistry
- Know the applications and limitations of computer assisted decision tools

15.1 Introduction

The tasks of the dental surgeon are extensive. They range from diagnostic, preventive, therapeutic, evaluative functions to educational or forensic functions. He also has to design and manufacture custom-made medical devices. He is a manager of his team, an administrator of his own company. He interferes with partners as other doctors in dental medicine, orthodontists, medical doctors, Health Insurances.

Computerizing the dental office aims at improving the practices as well as the efficiency of the dental surgeon in all these functions.

First of all, we will describe the dental office management software as well as some needed equipment in a digital dental office. The following part will be dedicated to dental imaging. Then we will see how the dental surgeon can be helped by computer assisted design and manufacturing systems (CAD/CAM). Information and communication technologies can play other roles which will be at stake in the last part: digital tools for patient education, dental surgeon's web site, computer assisted decision making and finally digital periodontal charting.

15.2   The Dental Practice Management Software

The management software is the environment which allows managing the dentist's activity. Historically speaking, it does not include the digital radiography software nor photographic images, video and medical imaging management software. Nevertheless, over the years, practice management software and radiography software have been bounded by the editors to the point that it could be believed that it is about a single software. Actually this is not the case.

Approximately, the working time of a dental surgeon, can be divided into three parts: a clinical time, an administrative time and an accounting time.

– The clinical time is the actual chairtime; this is the time during which the practitioner treats the patient.
– The administrative time is the time spent in keeping the patient record, writing documents for health insurances, managing drug prescriptions, quotes, delayed payment, dunning letters for unpaid debts, visit reminders, traceability of products, materials, instruments used in health care and medical custom made devices (prosthesis) and managing the stock.
– The accounting time is the time spent in keeping the ledger of income, expenditures and others ledgers, amortization tables, income tax return, etc.... Sound management also requires precise, accurate and immediate knowledge of the dental office accounting elements. It turns out that this is where the computer excels: in fact, it performs repetitive and tedious tasks effortlessly and without qualms. Moreover it does it well, without errors and can display the results at any time.

From a single input (usually in the care form), the recorded data will be processed by all modules where they will be needed.

For example, when you enter a procedure in a patient record, it will be used to: calculate fees, manage traceability, edit documents for health insurances, upload information to health insurances, update statistics, update the ledgers, etc....

Here stands the power of the computer compared to manual entry. We should also mention the clarity and quality of displaying information, making it easier to read.

In the following, we subdivide the software structure into three separate modules: (patient file, accounting and management, utilities), although they are actually nested and in total correlation with each other.

A dental office management software should at least include the information listed further in order the computerization to be efficient and profitable.

In addition, one must be careful about the following points:

– Some software features are spectacular but useless. Their existence is often only justified by purely commercial reasons.
– The "false accounting" which boils down to keeping the income ledger and that requires software for the accounting of the dental office or the use of an accounting firm.

- Radiography and/or digital video imaging can be interfaced to the management but this is not mandatory. This decision should belong to the practitioner and ensues directly from the computerization specification.
- The false teletransmission which requires re-entering data because it is not the management software which is accredited but a utility that controls the chip card reader.
- The software must be able to operate in network.

15.2.1   The Patient Record Module

The software must allow to enter and manage the following data (some variations may occur according to the country and especially its health policy):

- Patient's identity, medical history, current prescriptions, ...
- Patient's health insurance, date of rights opening...
- Procedures: prosthesis, periodontics, surgery, implantology, orthodontics ...
- Treatment plan, comments, therapeutic monitoring (recording estimates, prescription, letters...)
- Accounting...
- Total fees, accounts management of payment on accounts, ...
- Payment's entry
- Document's edition for health insurances
- Traceability of products and management of custom medical devices
- Traceability of sterilization.
- Digital Radiography (optional).
- Digital Video (optional)

In current software, procedures entry is generally made by clicking on the teeth displayed in the dental chart. The dental chart is entered on the first visit and updated at each following visit. It shows the patient's dental condition.

15.2.2   Accounting and Management Module

15.2.2.1   Accounting

It includes the entry, calculation, keeping and edition of account ledgers: income, expenditure, ledgers. It allows calculating the amount of taxes, editing the income declaration (2035 form in France), keeping bank accounts and calculating amortizations (keeping and editing the amortization ledger).

15.2.2.2   The Income Ledger

It automatically records the date (non-falsifiable), the name of the patient, his health insurance, the method of payment, the sum and possible observations. Only the method of payment has to be entered. Actually, it is the only element that is not already present in the other modules.

Edition is done for each bank deposit, thus adding a page to the income ledger and updating business accounts.

15.2.2.3   The Expenditures Ledger

For each expense, the date, recipient, mode of payment, amount, ventilation code and label (reason of the expense, bill number, check number) must be entered.

Editing is generally made in chronological order, for example at the end of each month, thus adding a new page in the expenditure ledger.

15.2.2.4   Counters or Dashboards

They allow to visualize and calculate: the number of procedures realized, working hours, the amount realized, amount paid, (daily, monthly, annual) return, hourly rates, expenses, profit,. . .

15.2.2.5   Management Support

It publishes the revenue and expenditure report, calculates predictions of activity, helps to calculate the fees. Results of activity may be expressed by graphs.

15.2.3   Utility Module

15.2.3.1   Estimates

From the treatment plan and a library of procedures and rates established by the practitioner, the computer can provide extremely well presented and regulatory detailed estimates. The content of the estimate is fully customizable. Computing greatly facilitates the creation and editing of the estimates. They can be stored until the final choice is made.

15.2.3.2   Prescriptions

Using a drug library established by the practitioner, the computer allows the entry and editing of drug prescriptions. Most of software can be interfaced with a drug data base either off-line (CD-ROM) or on-line (Internet). They can provide assistance regarding drug interactions or contraindications.

Standard prescriptions can be registered in advance. They can be edited upon request (for example: oral hygiene instructions, denture maintenance, pre- or post-surgical medication...).

15.2.3.3 Correspondence

Pre-defined and totally editable letters are available, such as letters to colleagues, dunning letters, letters to usual suppliers, various certificates and letters to the authorities.

15.2.3.4 Overdue Management

The software edits fee statements according to a pre-defined minimum overdue threshold and a time periodicity. These letters will be sent to patients. Documents are produced from pre-established models. Several levels of dunning letters are provided.

15.2.3.5 Control Visit Reminders

From the patient file, you can retrieve the patients who agreed to receive reminders and then edit the corresponding letter. The letter will remind the patient that he shall make an appointment to check his oral health, his prostheses...

15.2.3.6 Agenda

This window usually opens at the software launch. It allows managing the appointments. The agenda can be coupled to a phone platform by the way of the internet. One of the advantages is that the platform handles the phone calls outside the dental office opening hours and when you decide you don't want to handle the phone during certain activities (surgery, implantology). It is essential for the platform staff who takes the phone calls and the appointments to be especially trained to dentistry practice.

15.2.3.7 End of Year: Archiving

A series of operations is essential to be able to switch from financial year to the next one: closing of accounting, archiving accounting data of the past year, reset counters before the beginning of a new fiscal year. However, the computer keeps track of the main results in order to be able to present the results of the current year compared to previous years.

15.2.3.8 Inventory Management

A few days before the expiry date of a product, the software shows the need for its renewal. The software also includes an orders management module.

15.2.3.9 Materiovigilance

For each procedure, the used material has to be recorded. Afterwards, multi-criteria searches on a particular lot may be run in order to identify all patients for whom it was used.

15.2.3.10 The Telematic Connection

Now, the computer connection with other machines via a telecommunication network (telephone network, fiber network) is essential. It is needed in order to transmit data regarding the procedures to health insurances. Secure electronic exchange between healthcare professionals and insurance may be implemented, e.g., in France, the SESAM-Vitale system (Electronic System for data entry of Health Insurance).

The dentist can also subscribe to medical intranets where they can access large databases, e-learning and secure e-mail.

15.3 The Specific Computer Equipment

The risk of infection is inherent in dentistry. Protecting patients, practitioners and their staff from infection is therefore a priority. The care instruments must be sterilized but what about the computerized equipment dentists increasingly use?

15.3.1 The Mouse

A wireless mouse or better, a touchpad should be preferably used.

15.3.2 The Monitor

Screens commonly used in dental practice have a size of at least 17 inches (diagonal length). Touchscreens allow control of certain software directly by finger touch on

Fig. 15.1 Soft silicone
membrane keybord

the screen without the intervention of the mouse or keyboard. This feature has a definite advantage in terms of hygiene.

15.3.3   The Keyboard

Basically, it is not designed to be used at the point of care. But some special keyboards are intended to meet the requirements of operating rooms.

Tight keyboards with membrane offer easy cleaning and disinfection but a less good tactile sensation (Fig. 15.1). Other keyboards are tactile, smooth, in toughened shockproof glass, wireless and washable with antiseptics (Fig. 15.2).

There are also virtual keyboards using acoustic signal recognition technology. They are based on the technology of time-reversal acoustics: sound waves propagate through an object when it is touched. It is possible to analyze this signal. Each impact point has a different signal (acoustic signature). Thus, the signal detection enables to recognize very accurately the impact point (Fig. 15.3). This system includes two acoustic sensors and a set of stickers (keyboard). The virtual keyboard is absolutely flat, smooth, wireless, without any mechanical part and can be cleaned with any kind of detergent or disinfectant.

15.3.4   The Smart Card Reader

The practitioner will need a smart card reader if the patients and practitioner identification has to be done with smart cards. This device is similar to that used to read bank cards. It is able to simultaneously read two smart cards: the patient's card and the health professional card. Whatever the model, it requires prior health insurance approval.

Fig 15.2 Tempered glass
Tactile keyboard (Tactys)

Fig. 15.3 Virtual keyboard
(VirtualB keyboard Sensitive
Object)

15.4   The Security, the Hardware and Software Maintenance

15.4.1   The Hardware Security

There are many hazards: fire, water damage, incidents, theft, vibrations, dust, electromagnetic radiation, failure etc.; that have to be considered. Computer systems must be protected first and foremost, placed in areas where the risks are lower.

The vagaries of power supply shall not be neglected: micro or macro cuts, sudden drop or rise in voltage are not appreciated by computers. An Uninterruptible Power Supply (UPS) straightens and regulates the power and provides minimal autonomy in order to shut down the software in case of incident. It is also protects a lightning-protection system.

Networked systems are common in dental offices. Physical Ethernet cabling is more reliable than radio connections (WiFi, Airport™) which are sometimes fragile, considering electromagnetic disturbances present in firms. Furthermore, physical Ethernet cabling is faster (10 Gb/s) than Wifi (100 Mb/s) than Wifi (54 Mb/s) and more secure because it is not accessible from outside.

Finally it is necessary to insure the "hardware" risk.

15.4.2 The Software Security

Software Security starts with the acquisition of licenses for the software whether medical or not. Thus, the practitioner receives updates of these which is extremely important for example when administrative requirements change (Ex: changes in the procedure coding system, changes in data exchanges with health insurances . . ..)

In order to cope with adverse action of hackers, viruses and other malware, it is necessary to use anti spyware and antivirus software and enable protections at your disposal (firewall, etc.) to prevent intrusion from the Internet and electronic mail but also from external storage media (CD, DVD ROM, USB drives, external HDD).

Software security also requires the use of passwords to control access to machines, access to professional software (mandatory) or access to functions restricted to the only computer administrator.

15.4.3 Data Backup

A backup is a data recording on an external or extractible storage medium which has to be placed in a remote and locked location. The backup must be done daily. Saving the data on a single medium is not safe enough. Indeed, the generally admitted consensus is the following one: save the data everyday day of work on a different medium. The rotation is made over a week.

The practitioner who loses all his data is almost professionally dead. The backup is thus the lifebuoy of the office in case of disk crash or other event destroying the equipment and/or the data. In these cases, data will have to be restored. This maneuver will also be used in case of repair or change of disk or machine. It is thus essential to regularly check and test the backup.

Several different storage media can be used, none of them is 100 % reliable: external hard drive, DVD-ROM, Blue Rays, MP3 player, USB keys, memory cards,. . . The choice of the medium depends on the total weight of the data. For the only management data, some hundreds of megabytes are enough and USB keys will be widely sufficient considering their current capacity and considering their excellent value for money. For the imaging (radio, photo or both) several tens of gigabytes will be necessary. The external hard drives (magnetic or better: Solid State Disk=SSD) will then be preferred. All the more if the backups of management and imaging are realized at the same time. SSDs use flash memory (like USB drives). Their lifespan is higher than magnetic hard drives. In addition, the majority of failures on the storage media come from mechanical failures, and SSDs have no moving parts, which obviously protects them from this type of failure.

Professional software integrates all functions for performing data backups.

The full backup of the all the contents of the machine seems not to be relevant as software can be easily reinstalled if necessary.

15.5   The Dental X-ray Imaging

15.5.1   The Hardware

An intraoral (wired or wireless) sensor, a micro-computer (with a specific reading tool for electroluminescent sensors with memory: ELMS) and a printer are needed.

15.5.1.1   The Charge Coupled Device (CCD) Sensor

This is a charge transfer device, which is used instead of the analogue X-ray film. It consists of a scintillator that acts as an amplifier screen, absorbs X-rays and converts them into light. The latter is connected to the central unit of the computer via a cable either directly to an internal digital acquisition card or indirectly via a specific outer unit, or in wireless mode.

Newer systems are constituted by a CCD or CMOS sensor, connected to the computer via a USB Plug and Play connection (Access™ and Ultimate™ by Trophy). These systems may have the advantage that you do not need to invest in several sensors even if you have several treatment rooms.

The Electroluminescent (ELMS) Sensor

It looks like an analogue film and was introduced in the dental community more recently (Digora™) but it was developed by Fuji in the 1970s.

It is a phosphor screen. Once exposed to X-ray it contains a latent image which is "developed" by a device that some have integrated into the computer. The sensor surface is scanned by a laser beam causing an excitation state at the atomic level producing a photonic emission of blue light. These photons are collected by a photomultiplier and then are converted into an analog electrical signal that is digitized by the computer.

The Image Processing System

It has the following functions: the acquisition of the signals supplied by the sensor, the processing of the obtained image (digital image processing software), the temporary storage (RAM memory of the computer), longer-term storage (available mass memory).

The implementation includes sensor set up in patient's oral cavity and X-ray exposure the same way as analog film. The sensor must be equipped with a disposable protection (latex or polyethylene). It is advised to use provided angulators (most sensors are used with RINN™ type angulators). Once the image is obtained, it can be visualized and analyzed on the screen using the possibilities of the software environment. The image processing functions will enable the use of information potentially contained in the raw image, but often not visible to the naked eye. Thus it is assistance in diagnostic radiography. The software functions are generally the followings:

– Enhancement of gray: digital contrast variation from black to white
– Windowing greyscale: grouping greyscale image (every eight gray level for example)
– Positive radio: reverse video (black becomes white and vice versa)
– Image inversion: image return and/or rotate of 90° per image
– Zoom function: magnification of a selected area of the image
– Pseudo color: a color is assigned to each gray level.
– Measures on the image: measure of distances or size of selected elements on the image. (Warning: this measure worth on the image. It may not worth for the actual objects.)
– Density analysis: from profile line determined by the operator, the software produces a histogram showing the image density of gray along this line.
– Pseudo three dimensions: using an algorithm exploiting the image greyscale, the software draws an expression of it in pseudo 3D. This is actually just a histogram.
– Delete, record, archive, backup images.
– Print images: via inkjet, laser, exceptionally sublimation printer.

(See Chap. 5 for more information on medical images treatment).

The digital radiological imaging provides an undeniable comfort of work: indeed, the ability to get instantly images brings a certain speed and convenience to the operative procedures. As regards the ELMS, time required for image acquisition is longer (20 s to 40 s depending on the equipment). As a result, some manufacturers have developed systems allowing to "read" several ELMS simultaneously. The advantage is that it can read sensors of different sizes. Another advantage is the moderate cost of replacement for ELMS. On the other hand the ELMS requires two manipulations, (1) reading it with the laser, (2) repackaging (clear and repackaging) for reuse.

For the CCD, the only manipulation is to wrap them with adequate protection.

Digital radiography in dentistry has reached a level of accuracy equivalent to the analogue radiography. However, a large number of factors intervene which are not necessarily all completely mastered. Indeed, the obtained image will be closely dependent on factors such as the properties of the used X-ray tubes, the performances of the sensors, their sustainability in time, the performance of the operator, the operating conditions, the performance of the IT environment, the possible breakdowns.

Electronic technology and the sensitivity of the sensors lead to a 90 % decrease of the radiation dose for retro alveolar radiography and 70 % for digital panoramic radiography. These are very valuable characteristics of these devices. We must keep in mind that these doses are cumulative and we must not multiply the impact unnecessarily.

The integration of digital radiography in the dental office may be more or less difficult and harmonious depending on the situation: as part of a first installation, imaging should be included in the specifications. As part of an ongoing exercise, the existing situation will guide new investments.

Shall all the images taken during the exercise of a practitioner be stored? Certainly not. All poor quality images will not even be recorded (bad incidence, over and under exposure, bad framing, fuzziness, etc.). But storage has many advantages. It allows to quickly retrieving a radio of a patient in current care or longer-term follow-up for disease course monitoring. In addition, the evolution of technology and lower costs led large capacity hard drives to become common (1TB). In the same time, non-destructive image compression algorithms allow minimizing space requirement on the storage capacity. Thus a large number of images can be stored, especially if the workstation is only used for radiology and the oldest images are archived on external storage capacity.

We shall not forget that the radiographic images, once set, are the patient's property and he can claim them. We must be able to edit them. Images can also be requested for inspection or audit by various organizations (health insurance, expertise . . .). If printed, the image must be as large as possible in order to minimize the loss of quality. Besides, the date of the examination, the identification of the patient and that of the tooth or teeth involved are absolutely indispensable and can be legally mandatory.

15.5.2   The Digital Panoramic Radiography

In these devices the traditional silver photographic film is replaced with a CCD (matrix of about 2,500 × 1,244 pixels) or a CMOS. These systems are grounded on the principles of digital radiography in general. Image processing software are the same. These systems allow some more specific procedures such as Temporo Mandibular Joints sections, maxillary sinus radiography, radiography by sector (anterior, posterior).

The latest generation devices have functions referred to as "3D" only resulting of 2D image software processing. It must be clearly stated that the so called "3D digital panoramic" and cone beam scanners are two different technologies.

Fig. 15.4 3D Kodak 9000

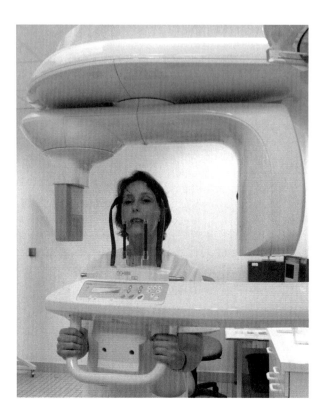

15.5.3   The Cone Beam

(HAS 2009) http://www.has-sante.fr/portail/upload/docs/application/pdf/2009-12/rapport_cone_beam_version_finale_2009-12-28_17-27-28_610.pdf

The cone beam computerized tomography (CBCT) (Fig. 15.4) is a sectional technique that allows 3D imaging of the entire maxillofacial complex and maxillomandibular and dentoalveolar structures. It arises as a result of technological advances in computer science and electronics and has become a key element in the field of digital dentistry.

Devices differ from the traditional scanner which performs multiple superimposing linear sections during multiple rotations of the system. The CBCT, does not work with a thin X-ray beam, but with an open conical beam, which allows to scan in a single revolution the entire volume to be radiographed. The two-dimensional projections once collected by the sensor are then transmitted to a computer for volume reconstruction. The complexity of the system lies in the sensor technology and the three-dimensional reconstruction algorithms. The visualization software allows adjusting the brightness and contrast. The post-processing software can also reconstruct the volumes (Fig. 15.5).

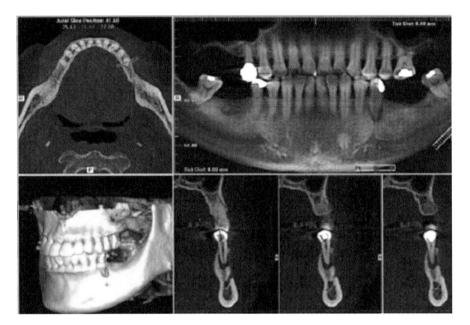

Fig. 15.5 Three-dimensional visualization of structures such as bone and tooth surfaces

This technique is attractive because of its low dosimetry and three-dimensionality of the image. In addition, CBCT scanners have a spatial resolution and reliability comparable to scanner for the exploration of mineralized structures such as bones or teeth and for detection of fractures, cysts or bone lesions, foreign bodies, within its field of view. Indeed, for a dental structures or sinus scanner, the length of one of the voxel's edges (corresponding to the slice thickness) is greater than the other two: the volume is "anisotropic" and sagittal or coronal reconstructions have a lower spatial resolution than the axial (native) sections. On the opposite, the volume of a CBCT examination is isotropic (cubic voxels). Thus, sections obtained have the same spatial resolution regardless of their orientation. On the other hand, only the scanner can measure actual densities and the study of soft tissues, requiring high precision. This constitutes its reserved area.

15.5.3.1 Technical Features

The various CBCT systems are characterized by their technical characteristics:

– Spatial resolution (definition): defined by the voxel size at acquisition time, it depends on the quality and resolution of the plane sensor (or image intensifier on early models), on the field of view (for given number of voxels, the more the field is limited, the better is the resolution) and on the sophistication of the reconstruction algorithm in the software. Finally, the analysis computer system, including among others the operating system and the display system (computer, video card, screen resolution) is crucial in the apparent image definition.

- Density or contrast resolution: the number of gray levels (from white to black) that the system can display. The human eye is able to distinguish only about a dozen shades of gray in the same image. The advantage of having a very high number of gray levels is to allow a wide range of "density windows."
- Automatic exposure control: it allows adjusting the delivered dose to the "opacity" (density, thickness) of the subject (especially useful in children).
- Duration of exposure: kinetic artefacts, due to patient movements during the examination will be limited with a shorter exposure time. All CBCT equipment should include an effective restraint system because of the relatively long duration of exposure compared to the scanner.
- Type of X-ray source: constant or pulsed. Actual exposure is reduced with a pulsed beam, which can limit the delivered radiation dose.
- The volume explored is called the "field of view" or FOV.
- The polyvalence of the equipment (single CBCT or volume tomography modularity "small field" and panoramic radiography).

A variation of the parameters leads to significant differences in the image quality. It is therefore important to select various parameters on a single device aiming at minimizing the X-Ray dose delivered to the patient while having a sufficient image quality for the selected application.

15.5.3.2 Needed Equipment

A computer must be dedicated to this unit. The analysis will take place in the radio room or better, the image will be transmitted over the local area network (LAN) to other computers in the dental office. The needed software to read the cone beam must be installed on each workstation or on a server. A printer able to print photographs is required.

The images can be stored on a server and thus be accessible via the LAN from any workstation. A sufficient storage capacity shall be provided. File size for an exam (or file weight expressed in bytes) will increase with the field of view and the image definition. Such volumes require the use of the latest generation of computers, equipped with large random-access memory (minimum GB) and a powerful graphics card (at least 256 MB video memory).

Data compression is useful for sending data over the Internet. It can reduce the exam to a few MB. But it requires an excellent prior definition. In this area, the scanner is superior with a 40–60 MB file size (DICOM ® data : lossless computer format, standard for the exchange of medical imaging data) easily reduced to 3–10 MB after compression.

Fig. 15.6 The logicon result screen where the decay is outlined and the probability of enamel or dentin decay is displayed

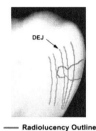

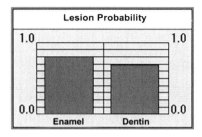

15.5.3.3 Conclusion on the Cone Beam

The cone beam extends the dentist diagnostic capabilities. However, it should be used only when the two-dimensional imaging techniques are insufficient.

The parameters must be selected for each case in order to minimize patient exposure, while maximizing the quality of the images in the area of interest. This requires a good understanding of the phenomena and equipment.

15.6 Diagnostic Means Based on Computer-Assisted Image Analysis

Success of treatment depends, above all on the quality of the diagnosis. Computed image processing can bring invaluable help. Furthermore, image analysis and detection methods have received increased attention because they are non-contact non-destructive techniques.

15.6.1 Digital Radiographs

Logicon Caries Detector software (http://www.kodak.com/dental) is a support tool that can detect proximal decays from radiographs. It extracts features from the radiographic image and correlates them with a database of caries radiographic images (Fig. 15.6). It allows dentists to find twice more early dentinal caries than previously, while not unnecessarily restoring additional healthy teeth. Indeed Initial sensitivity of dentists in detecting early dentinal caries was 30 % and improved to 69 % with the software's density analysis tool. Specificity was 97 % without the tool and 94 % with the tool (Tracy et al. 2011).

15.6.2 Quantitative Light-Induced Fluorescence (QLF) Techniques

(Pretty 2006)

QLF imaging is one of the systems allowing dental caries early detection and longitudinal monitoring of their progression or regression. Using two forms of fluorescence detection (green and red), it may also be able to determine if a lesion is active or not, and predict the likely progression of any given lesion. It opens the way for earlier and non-invasive treatment and true preventive dentistry.

Fluorescence is a phenomenon in which an object is excited by a particular wavelength of light and the fluorescent (reflected) light is of a larger wavelength. When the excitation light is in the visible spectrum, the fluorescence will be of a different color. In the case of the QLF the visible, light has a wavelength (λ) of 370 nm, which is in the blue region of the spectrum. The resultant auto-fluorescence of human enamel is then detected by filtering out the excitation light using a bandpass filter at $\lambda > 540$ nm by a small intra-oral camera. This produces an image that is comprised of only green and red channels (the blue having been filtered out) and the predominate color of the enamel is green. Demineralization of enamel results in a reduction of this auto-fluorescence. This loss can be quantified using proprietary software and has been shown to correlate well with actual mineral loss ($r = 0.73$–0.86).

QLF has been employed to detect a range of lesion types. For occlusal caries sensitivity has been reported at 0.68 and specificity at 0.70, and this compares well with other systems. QLF allows detecting more non-cavitated occlusal lesions and smaller lesions compared to meticulous visual inspection (Kühnisch et al. 2007).

The QLF system offers additional benefits beyond those of very early lesion detection and quantification. The images acquired can be stored and transmitted, perhaps for referral purposes, and the images themselves can be used as patient motivators in preventative practice.

QLF is also used for tooth staining and whitening assessment (Adeyemi et al. 2006), as well as calculus and dental plaque assessment (Komarov et al. 2010).

After scaling the VistaProof camera (http://www.duerr.de/eng) (Fig. 15.7) can detect fissures and occlusal caries tooth by tooth. Powerful LED light with a wavelength of 405 nm stimulates the porphyrins of cariogenic bacteria to emit red light. In combination with a software, the monitor displays a false color image representing active caries in red and healthy enamel in green.

Other innovations allow the visualization of decays already detected on clinical examination and show by example the extent of the attack as a numeric value. This is the case of the KaVo DIAGNODent (DD). It also employs fluorescence to detect the presence of caries. Using a small laser the system produces an excitation wavelength of 655 nm which produces a red light. This is carried to one intra-oral tip which emits the excitation light and collects the resultant fluorescence. The DD does not produce an image of the tooth; instead it displays a numerical value on two LED displays. The first displays the current reading while the second displays the peak reading for that examination. The DD is thought to measure the degree of

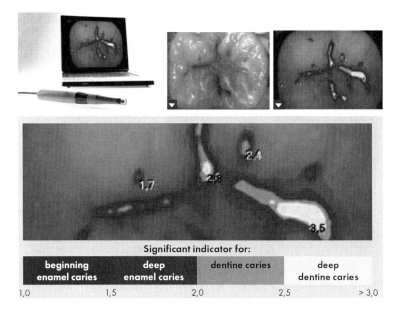

Fig. 15.7 VistaProof system

bacterial activity; and this is supported by the fact that the excitation wavelength is suitable for inducing fluorescence from bacterial porphyrins; a byproduct of metabolism. Initial evaluations of the device showed that; correlation with histological depth of lesions was substantial (0.85) and the sensitivity and specificity for dentinal lesions were 0.75 and 0.96, respectively. Further in vitro studies have found that the area under the ROC was significantly higher for DD (0.96) than that for conventional radiographs (0.66). However, the device requires teeth to be clean and dry. Stain, calculus, plaque have all be shown to have an adverse effect on the DD readings (mostly an increase leading to false positives). The DD clearly is more sensitive than traditional diagnostic methods; however, the increased likelihood of false-positive diagnoses compared with that with visual methods leads to recommend it as an adjunct to visual examinations and radiographs.

15.7 Digital Photography

In dentistry, the most suitable cameras are digital single-lens reflex camera (DSLR) or bridge, for their effective specifications (Bister et al. 2006). Indeed, the interest is based on the high quality of the global intraoral or extraoral images.

Whatever the discipline, and especially in dentofacial orthopedics, patient's initial status' pictures complete clinical and radiological first examination. At this stage, practitioner makes frontal face and profile pictures, with wide and tight smiles. Then with the help of mirrors, he takes intraoral global front, side and occlusal, maxillary and mandibular pictures. The practitioner uses them to elaborate a diagnosis and treatment plan (Goodlin 2011). They can be consulted later during the different phases of treatment. They facilitate communication with the patient and correspondents such as the technician.

It is required to carry out the shooting in the maximum resolution afforded by the device. The camera or the memory card needs to be connected to the computer to transfer pictures. Saving images in native format allows keeping a copy without irreversible loss due to compression. Specific software is required to modify images. They can allow among others to perform simulations of the effects of dental treatments. The modified and compressed pictures are saved in the imaging software.

15.8   Intra-Oral Video Cameras

Intra-oral cameras allow to film or photograph intra-oral views, instantly displayed on a screen to the patient. Whatever the brand, the tool is light, handy and wireless or totally integrated. The captured images can be saved and thus complete the patient's medical record.

15.8.1   Interests

Their main asset is to facilitate practitioner to patient communication. During its first visit, the practitioner may present the patient's oral situation on a large screen. Open discussion about the issues is more easily set. The captured images help the patient to understand and realize his oral condition, and indeed increase his motivation. Regular explanations and various dental prophylactic measures will be easier to establish (Willershausen et al. 1999). Dialogue about prosthetic strategy to adopt is also facilitated thanks to the images. In addition, a patient's progress can easily be tracked, with before and after pictures stored in the patient's digital health record.

Easy handling, powerful lightening and small size of the instrument, allow observing unreachable areas. In addition, high magnification improves vision. For carious lesions detection, as well as for endodontic treatments, it can be an alternative to optical magnifiers and microscopes (Brüllmann et al. 2011; Erten et al. 2006; Forgie et al. 2003; Mirska-Mietek 2010). In restorative as in prosthetic dentistry, a macroscopic picture shot allows a more detailed view of dental surface.

In oral pathology, a lesion is easily shot with intraoral cameras. It helps to follow its evolution and exchange with colleagues or even a pathological anatomy laboratory (Scutariu et al. 2007).

Color choice can be helped with some cameras (Lasserre et al. 2011). Images can be forwarded to the dental technician for improving manufacture of aesthetic dental prosthesis. As its name indicates the intraoral camera is designed to capture intraoral picture and movies. In aesthetic dentistry and for extraoral shooting, digital bridge cameras and single-lens reflex are more appropriate.

In hygiene and motivation research, intraoral cameras seem to be an effective tool to measure plaque thickness in clinical and epidemiological studies (Smith et al. 2006; Splieth and Nourallah 2006). Finally, from a legal point of view, non-subject to change captured images can be used in forensics matters.

15.8.2 Managing the Images

All cameras are equipped with a USB 2.0 or S.video connection. They are combined to the digital imaging software, which edits archives and classifies all digital data such as radiographs and photographs. The digital imaging software can be used independently or integrated to the dental practice management software. Images are saved in a proprietary format but can be exported in universal formats such as JPEG and TIFF.

15.9 Computer-Aided Design and Computer-Assisted Manufacture (C.A.D.-C.A.M.)

15.9.1 Principles

Computer-aided design and computer-assisted manufacture (C.A.D.-C.A.M.) arose from the meeting between the progress of the computing, the optics sciences, the materials sciences and the robotization of machine tools (Sarava 1965; Pryputniewicz et al. 1978; Leith and Upatnieks 1964).

They match systems of digital readings and digitally operated machine tools for the realization of diverse products, intended for the world of the industry (aeronautical, cars, etc.) as well as in the health sector. The odontology uses these technologies in many areas (implantology, orthodontics and prosthodontics).

All the process of C.A.D.-C.A.M. works with the same principles and equipments :

- An image acquisition unit allowing the data capture: a scanner for the optical impression (Fig. 15.8).
- A processing unit for modelling the recorded image and the virtual therapeutic project: it is the computer-aided design (C.A.D.) (Fig. 15.9).
- A computer-controlled manufacturing, machining the virtual model in the chosen material: it is the computer-assisted manufacture (C.A.M.) (Fig. 15.10).

Fig. 15.8 Scanner et software C.A.M. (Nobel Biocare®)

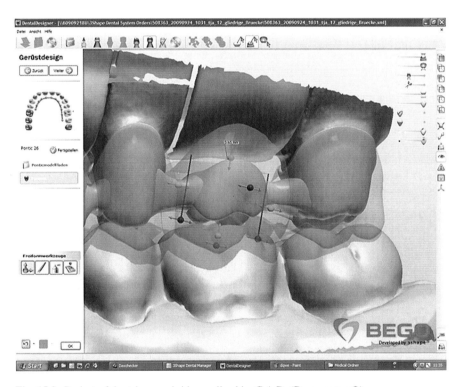

Fig. 15.9 Project of dental crown bridge realized by C.A.D. (Bego system®)

The concept of C.A.D.-C.A.M. applied here to dentistry appeared in 1973 (Duret 1973) and has grown in an exponential way since François Duret's founding works. It now allows accurate therapeutic realizations in a lot of disciplines in dentistry. Three stages can be distinguished:

– Data capture with a digital impression realized directly in mouth (intra-oral scanner) or on a plaster cast stemming from a classical impression. Various

Fig. 15.10 Milling Unit
Cerec Mc-XL (Sirona®)

technologies can be used, so-called by triangulation or by flight time. The profilometry is gradually phased out. The recorded files are said "open" (STL) when they are exploitable by different process of C.A.D.-C.A.M. or "proprietary" or "closed" if they can be exploited only within a same brand of C.A.D.-C.A.M. process.
- The exploitation of the data by C.A.D. creates the 3D image. On this virtual reality, the prosthodontic project(increased reality) is built in a prosthodontic laboratory and more and more, directly in the dentist office.
- The exploitation of the data modeling the planned prosthesis by C.A.M. allows piloting a machine tool. As required by the clinical case, the prosthodontic project can be carried out in the material chosen by the practitioner.

The fabrication methods are numerous: the subtractive method (manufacturing by milling is the most spread), the additive method (stéreolithography, compaction or electro-mechanical assembly).

15.9.2   In Orthodontics

15.9.2.1   Lingual Orthodontics

In orthodontics, C.A.D.-C.A.M. is mainly applied in lingual orthodontics with the design of individual orthodontic ties brackets and personalized bows from a personal highly accurate impression (Figs. 15.11 and 15.12). Individual effective orthodontic systems for shorter treatments and more precise dental teeth movements are obtains by this way (Wiechmann and Wiechmann 2003).

From an initial imprint realized with a high precision dental material, the procedure begins with the realization of a set-up representing the end of the treatment simulating the final therapeutic result. On this set-up, the dental lingual

Fig. 15.11 Lingual
orthodontics devices realized
with C.A.D.-C.A.M.

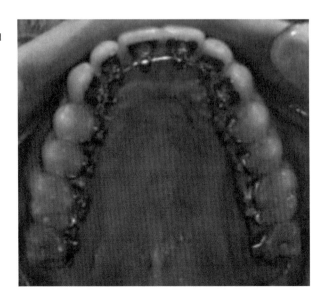

Fig. 15.12 Lingual
orthodontics devices realized
with C.A.D.-C.A.M.

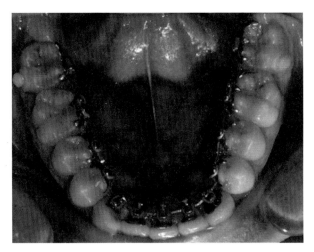

surfaces are digitalized with a high-resolution optical scanner in order to provide a
3D set- up. According to the treatment, this stage allows the conception of brackets
adapted in shape and in ideal position on the dental surfaces with a computer-aided
design software (C.A.D.). In a secondly C.A.M. stage, a wax printer makes the wax
models of the brackets which will then be poured in gold by the process of the lost
wax process (310 kg/mm^2 Vickers) (Wiechmann et al. 2003).

The shape of the morphological bows is made by a robot which accurately folds
various types of threads (SS, TMA, CoCr, NiTi) from the digital data according to
the position of brackets (height and thickness) previously transferred on a model
reproducing the initial clinic situation. Several systems are proposed in

orthodontics using this process (Incognito® of 3M, Lingualjet® of Baron and Gualano.Harmony® of Curiel).

15.9.2.2   Others Orthodontic Applications

Orthodontic corrective appliances usually called "aligners" like Invisalign ® can also be designed by a C.A.D.-C.A.M. process. Removable, transparent, thin, flexible polyurethane patches including the entire dental arch are elaborated from a 3D digitalization scan by scanning the dental surfaces of a whole arch on a dental cast. This 3D digital model can be obtained by several 3D scanners (laser, white or structured light, photogrammetric or CT scan processes) with a level of accuracy around 20 μm. Digital adjustments (between the initial state and the wished final result) allow to design several dental devices adapted to every stage of treatment.

Finally, simulation processes using combinations of imaging techniques (Scan CT, MRI, C.A.D.-C.A.M.) able to recreate 3D images will be applied to orthodontic, orthognatic and orthopaedic rehabilitations in oral and maxillofacial surgery.

15.9.3   In Implantology

Computer-aided design in implantology was introduced in the 1990s for implants manufacturing (Priest 2005). Then, it was employed for the clinical planning (diagnosis and treatments) with the numerical simulation of the surgical implants position and the associated prosthodontics rehabilitation.

The manufacturing of the attachments systems and the implant devices by C.A.D.-C.A.M. uses numerous types of materials (titanium alloys, oxide of alumina ceramic or zirconium oxide) with a an important reliability and clinical sustainability (Kapos et al. 2009). Several kind of implant systems, with or without abutments, conceived by C.A.D.-C.A.M. are available (Procera®, Atlantis®, Encode®, C.A.M. StructSURE®, CARES®, Biocad®).

In dental clinic, two surgical techniques use the C.A.D.-C.A.M. systems:

– Computer-aided surgery
– Computer-aided surgical navigation technology with a real time guide during the intervention.

15.9.3.1   Computer-Aided Surgery

The computer-aided surgery virtually restores the volume of the osseous anatomy in 3D and simulates the implants position in accordance with the anatomy and the prosthodontic project. The radiological reproduction of the anatomical osseous structures with their details in 3D images (3D DICOM format) by the tomography

(CT scan or MRI) allows to optimize a series of digital information on the conception or the choice of implants, their surgical guide on the operating site and the ideal surgical preparation phases adapted to the prosthodontic project (Azari and Nikzad 2008).

This stage uses preoperative and operative guides (bone-supported, tooth-supported or gingiva-supported guides) previously designed. Indeed, in addition to the 3D radiological imaging, the intra-oral reliefs must be studied. It allows realizing a forward-looking assembly of the prosthodontic rehabilitation on articulator and to establish radiological guides (to study the implant positioning) and surgical guides (to guide their positioning during the surgical intervention). The capture of the digital data of the oral anatomical reliefs with an intraoral 3D optical system (scanner or intraoral 3D camera) or more often on a model (scan laser) can be used for that purpose.

The construction of the surgical guides is realized on study models by rapid 3D prototyping process by addition or subtraction of material (stereolithography rapid prototyping, 3D wax or plastic prototyping, laser additive construction). These guides have radiopaque position indicators (gutta percha or small diameter balls) necessary for the diagnostic analysis by imaging CT or CBCT or MRI. With these digitized data, we can draw and simulate the implants position in a 3D anatomical context adapted to the prosthodontic project. Many softwares allow these conceptions (Implametric®, SimPlant®, Nobel Guide®, med3D®, Litorim®, RoBoDent®, Surgiguide®, Stent Cad®).

15.9.3.2 Computer-Aided Surgical Navigation Technology

The computer-aided surgical navigation technology (Watzinger et al. 1999) is based on a monitoring of the surgical procedure. The system indicates in real time the position of the surgeon's instruments in connection with the digital image data in order to guide the drilling. Bone-secured optical marks (balls or diodes) allow to follow and to guide with a camera, by triangulation, the movement of the rotary drills on the site by referring to the images used for operative and surgical planning.

The practitioner must "separate his eyes of his hand". Eyes look at the screen and, at same time, the hand manages the instruments. Every stage is in real time controlled on the screen (positioning of drills, angulation and nearness of anatomical structures). Several systems of computer-aided surgical navigation exist (Artma Medical Technologies®, Robodent®, IVS Solutions AG®, Vector Vision 2®).

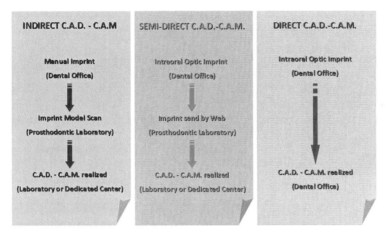

Fig. 15.13 Various prosthodontics approaches with C.A.D. – C.A.M.

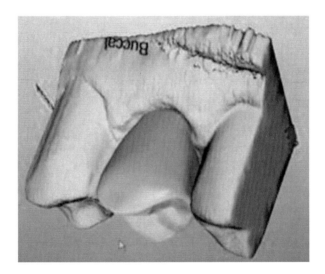

Fig. 15.14 Modelling of a prosthodontic project with direct C.A.D. – C.A.M. (Cerec 3D® Sirona)

15.9.4 In Prosthodontic

C.A.D.-C.A.M. systems in prosthodontics are realized according to the general principles expressed previously. Direct, semi-direct or indirect C.A.D.-C.A.M. methods are used (Fig. 15.13).

With monoblock ceramic, direct C.A.D. – C.A.M. (Cerec® Sirona) allows the manufacturing and the positioning of single denture construction (inlays, onlays, facets or crowns) in a one clinic session (Fig. 15.14). An experimented operator realizes and places theses prosthodontics elements in approximately 1 h. Manufactured ceramic is strengthened vitreous ceramics or lithium disilicates

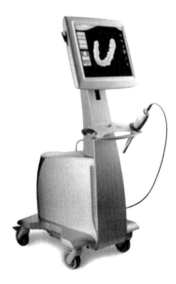

Fig. 15.15 The Lava C.O.S®. Intraoral Imprint unit for semi-direct C.A.D. – C.A.M.

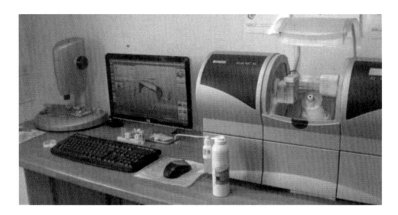

Fig. 15.16 Indirect Prosthodontic optic Process. From *left* to *right* : scan, C.A.D. unit, milling unit

ceramics. These devices also allow the realization of temporary resin bridge (3–4 elements).

Semi-direct or indirect C.A.D. – C.A.M. performs prosthodontics restorations from the single fixed denture element to the complete bridge for dental or implant rehabilitations in several steps. Firstly, an optical imprint has to be taken directly in mouth (C.O.S Lava system.® of 3M) (Fig. 15.15). Imprints are sent to the dealing laboratory by the internet.

In the indirect C.A.D. – C.A.M. (Fig. 15.16), a traditional work model in plaster will be scanned. Among the implemented materials, we can cite Ceramics (vitreous

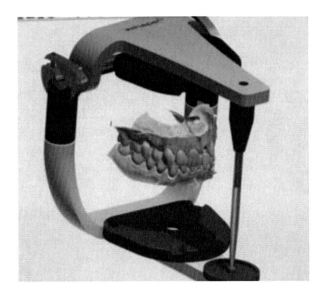

Fig. 15.17 Virtual Articulator

ceramics, aluminas, zirconia ceramics), metals (titanium, Cobalt-Chrome), carcinable or not carcinable resins or wax. Indirect C.A.D. – C.A.M. allows treating fixed and removable partial prosthodontics. On the scanned models C.A.D. – C.A.M. process realizes virtual frames. Then, these are conceived by C.A.M. additive processes (stereolithography). Carcinable materials models (wax-resin) are made and poured according to the conventional processes of foundry. If the C.A.D. – C.A.M. allows to obtain an excellent accuracy and quality, the practitioner must adapt his therapeutic gesture to the requirements of these new prosthodontic processes as to those of the dental materials used for rehabilitations and implemented assembly.

15.9.5 In Occlusion Approach (Virtual Articulator)

The virtual articulator (Fig. 15.17) allows to record and to modulate digital data reproducing the parameters of the dental occlusion (condylar guidance, Bennett angle, protusion, retrusion and distraction movements). The reconstruction of the dental anatomical reliefs in touch with their antagonists is exactly and rapidly realized. Procedures of transfers of these data by a software interface allow realizing the execution of prosthodontic project according to similar processes used in the prosthodontic C.A.D. – C.A.M. methods.

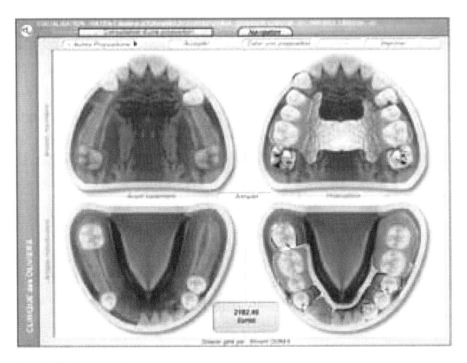

Fig. 15.18 Dentalvista®

15.10   Other Applications

15.10.1   The Patient Education

Communication with the patient, whether for prevention or education purpose, postoperative counseling or explanation of the treatment plan is an important facet of the profession of dentist. One says "a picture is worth a thousand words".

Communication tools and interactive multimedia are available on DVD or in the form of patient education software including images and movies which can be modified to be adapted to each patient. Some software can be integrated with the practice management software. More educational for the patient, they also promote transparency and allow him to understand better treatment plans and specifications.

For example, Dentalvista ® software (Fig. 15.18) allows previewing prosthetics for communication with the patient. It allows to study the possible configurations for prosthetic and implant and perform automatic chassis layouts adapted to the specific case of the patient. The animation of the treatment plan can show the process of care from the first consultation to the final solution.

Consult-Pro ® (Fig. 15.19) and CAESY ® provide 3D animations on prosthetic, periodontic or endodontic therapies. Dental Master ® (Fig. 15.20) can create its own 3D animations to customize treatment plans (Fig. 15.21).

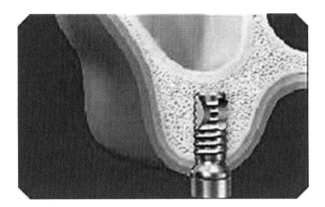

Fig. 15.19 Consult-Pro®

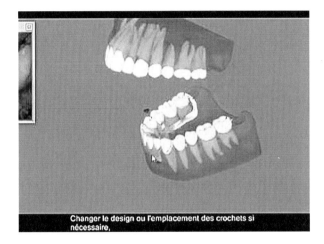

Changer le design ou l'emplacement des crochets si
nécessaire,

Fig. 15.20 Dentalmaster®

Nowadays, the internet is also a resource for patient education. It may be helpful to recommend high quality health web site, such as those certified by HON Code Certification (Health on the Net), to the patients; for example, dentalespace patients (www.dentalespace.com) or dentalhealthonline (http://dentalhealthonline.net/).

15.10.2   Decision Support Systems

Clinical Decision Support Systems (CDSS) are not intended to replace the practitioner who is solely responsible for its diagnosis, treatment options and prescriptions whatever aid he used. These systems use embedded clinical knowledge to help health professionals analyze patient's data and make decisions regarding diagnosis, prevention, and treatment of health problems.

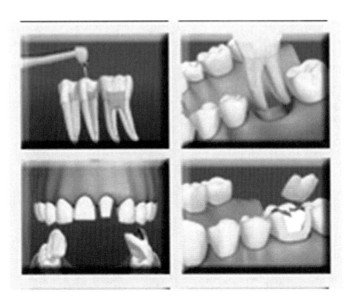

Fig. 15.21 QuickDental®

We will refer to Chap. 7 for the theoretical and methodological approaches. In the following, we focus on applications within the field of dentistry.

There has been research and development of CDSSs in dentistry for over two decades. These systems have utilized different types of knowledge representation and have addressed several major areas of dental practice (Mendonca 2004):

– Dental emergencies and trauma, orofacial pain (differential diagnosis),
– Oral medicine and surgery: oral screening and dental management of patients with head and neck cancer; Brickley and Sheperd 1996 developed a neural networks based system to provide reliable decision support for lower-third-molar treatment planning,
– Oral radiology (interpretation of radiographic lesions and automated interpretation of dental radiographs),
– Orthodontics (analysis of facial growth, landmark identification of cephalometric radiographs, and treatment planning),
– Caries management and pulpal diagnosis and restorative dentistry (removable partial denture design)
– Intelligent agents for dental treatment planning (Finkeissen et al. 2003).

For example, the Oral Radiographic Differential Diagnosis (ORAD) software program was designed to evaluate radiographic and clinical features of patients with intrabony lesions in order to assist in their identification (White 1989). It is a probabilistic system based on disease prevalence, sensitivity and specificity of signs and the use of Bayes' theorem. Ninety-eight jaw lesions were described by their prevalence and distribution by age, sex, race, presence of pain, number, size and

Patient Information

CLINICAL FEATURES
What is the sex of your patient? Female ·
What is the race of your patient? Nonblack ·
What is the age of your patient? 26 50 ·
Does your patient have pain or paresthesia? No pain ·

RADIOGRAPHIC FEATURES
Location
Which jaw contains the lesion? Mandible only ·
The lesion center is in what region ? Canine/Premolar Region ·
The relationship of the lesion to teeth is: Not tooth associated ·
Please estimate the number of lesions: One ·
What is the maximum size of the lesion? 2 to 3 cm ·
Where is the origin of the lesion? Unselected ·
Periphery
The borders of the lesion are: Corticated ·
The loculation of the lesion is: Unilocular ·
Internal Structure
The contents of the lesions are: Radiolucent ·
☜ Does the lesion contain one or more teeth ? No ·
Effects on Surrounding Structures
Does the lesion expand the bony cortex? No ·
Does the lesion cause root resorption ? No ·
Does the lesion cause tooth displacement or impaction? No ·

Differential Diagnosis (lesion data as of May 6, 2013 15:55:52)

Top Ten Significant Lesions (greater than 0.5%)

☐ 35% Simple (traumatic, solitary) bone cyst
☐ 23% Lateral periodontal cyst
☐ 11% Bone marrow cavity
☐   6% Residual cyst
☐   5% Anterior salivary gland defect
☐   4% Osteitis fibrosa cystica (Brown tumors from hyperparathyroidism)
☐   2% Surgical defect
☐   2% Keratocystic odontogenic tumor (odontogenic keratocyst)
☐   2% Cemento-ossifying fibroma
☐   2% Central odontogenic fibroma-mandible

Total percent is 93.36%

Fig. 15.22 ORAD®

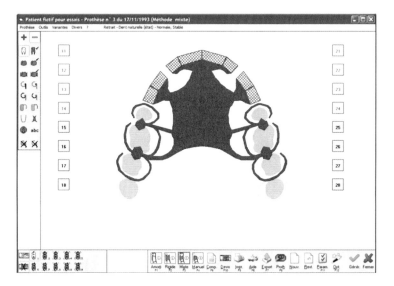

Fig. 15.23 Stelligraphe Software®

location of lesions, association with teeth, extension, locularity, borders, content and impact on adjacent teeth. Patient specific information is entered in a menu of 16 questions in order to characterize a specific lesion. The program output (Fig. 15.22) is a list of the lesions ranked by their estimated probability for that described condition. It also computes a pattern match estimating how closely the set of entered characteristics match the typical presentation of each of the considered program's conditions. An online version of this program is available at http://www. orad.org.

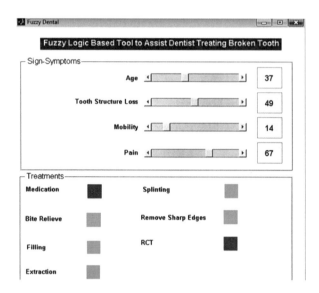

Fig. 15.24 The graphical user interface of the system designed for clinical decision-making in the treatment of cracked or broken tooth (Mago et al. 2012)

In prosthetic dentistry, Stelligraphe® software (Gaillard and Jourda 1991) (Fig. 15.23) is an aid in designing removable metal partial denture. Stelligraphe ® can design the optimal prosthetic solution and realize the automated tracing from a base of 400,000 prosthetic solutions that meet the traceability and supplemented with a detailed specific quote. The software allows also a manual method offering complete freedom in the conception.

More recently, Suebnukarn et al. 2008 presented a decision support model for predicting endodontic treatment outcome. Mago et al. 2012 developed a system to assist dentists in clinical decision-making process of choosing treatment for a cracked or broken tooth (Fig. 15.24).

CDSSs are designed to provide relevant care guidelines to providers to substantially improve health care quality and potentially reduce errors in practice. Despite the recognized need for CDSSs, the implementation of these systems has been limited because of the cost and difficulties involving the generation of knowledge bases, practitioner skepticism about the value and feasibility of decision support systems, among others. Another barrier is the lack of interaction with the patient's oral health records. This can be explained because of the uncomputability of a significant portion of relevant clinical facts in the electronic record (free text = unstructured and non standardized data entry). The adoption of controlled vocabularies (Systematized Nomenclature of dentistry: SNODENT) and the use of ontologies allowing more interoperability will most probably positively affect the development of CDSS.

In the last decade, the evidence based practice has gained strength. The growing number of practice guidelines provides reliable evidences to ground knowledge

Fig. 15.25 Florida Probe
System®

bases on. At the same time health care providers are encouraged to use the best current evidence and to act in conformity with patient care guidelines in clinical practice and health services. This will in turn, hopefully increase practitioner's willingness to use decision support tools

In computer aided prescription, we can refer to Chap. 8. Drug databases enable consultation of drug monographs and prescription analysis to automatically indicate possible drug interactions.

15.10.3 *Periodontal Computerized Electronic Probing*

The periodontal charting is a routine examination that measures periodontal pocket depth and gingival recession tooth-by-tooth and site-by-site, thus allowing calculating the loss of attachment. Further details have to be noted down in this examination such as the amount of plaque, tooth mobility, bleeding or suppuration. These huge amounts of measurements and observations have to be recorded simultaneously in the patient record. Furthermore, the classic periodontal probing using manual probes can be invasive if the pressure applied in the gingival sulcus is too strong. To address this issue, a second generation of probes, electronic, has been developed to apply a constant pressure less than 200 N/cm^2. Now a third generation of probe, with the same principle of controlled pressure probing is integrated into a computerized system. They are linked to software that records data during the exam. Several of these systems have been developed such as the Florida Probe System® (Fig. 15.25). The purpose of this device was also able to obtain a reproducible method of periodontal probing for clinical or research use (Niederman 2009).

15.11   Conclusion

The future of information technology and communication is great. The role of the computer in the dental office is no longer confined to the only administrative and accounting tasks but it takes up more and more important in the medical field with diagnosis, treatment or prevention tasks. Computer at the chair side, "paperless" dental office, digital images, CAD-CAM and many others new technological innovations will further improve practices. These advances are unavoidable and will forever change the dental practice as dental turbine has forever changed the clinical dentistry.

Technological advances can make the dentist more efficient. Training of the dental team (including dentists) to effectively utilize advanced technology is at least as important as the hardware and software components themselves.

In the longer term, the integration of information systems will grow as we are already seeing it in medicine. This means that the information collected in the dental office will be possibly linked with information from electronic medical records (hospital, GP or specialists practitioners), with large genomic banks at the national and even international level (see European Project TRANSFoRm) for epidemiologic and research purposes. This raises of course many ethical problems but opens up prospects for improving both the care of patients and research resources.

15.12   For More Information

Among the many references on which this chapter is based, we can particularly recommend:

- van der Veen MH, de Josselin de Jong E (2000) Application of quantitative light-induced fluorescence for early caries lesions. In: Faller RV (ed) Assessment of oral health: diagnostic techniques and validation criteria. Monogr Oral Sci. Basel, Kager 17: 144–162
- Azari A, Nikzad S (2008) Computer-assited implantology : historical background and potential outcomes – a review. Int J Med Robotics Comput Assist Surg 4: 95–104
- de Almeida EO, Pellizzer EP, Goiatto MC et al. (2010) Computer-guided surgery in implantology: review of basic concepts. Craniofac J Surg 21 (6): 1917–21
- Mendonca EA (2004) Clinical Decision Support Systems: Perspectives in Dentistry. J Dent Educ 68(6): 589–597

Exercises

Q1 What are the various functionalities of a typical dental office management software?

Q2 On what does the management software base quotes or prescriptions?

Q3 What are the necessary actions for maintaining hardware and software in good working order as well as ensuring data security?

Q4 List the different types of sensors used in digital radiology and their characteristics.

Q5 Cite some diagnostic means based on computer assisted image analysis.

Q6 what are the applications of the CAD-CAM in dentistry?

Q7 what are the applications of computer assisted diagnostic tools?

R1 Patient record, accounting, estimates, prescriptions, correspondance, control visit reminders, agenda, inventory management, materio vigilance and telematic connection.

R2 Quote: treatment plan and a library of actions and rates established by the practitioner. Prescription: a drug library established by the practitioner or connection to a drug database.

R3 The use of an UPS, a cable network, an antivirus, anti-spyware, daily backups placed in a remote location

R4 CCD (charged Coupled device) : immediate image, high sensor cost sensors and electroluminescent sensors: a few minutes needed before getting the image; moderate sensor cost

R5 Logicon Caries Detector software can detect proximal decays from radiographs. QLF imaging provides dental caries early detection and longitu dinal monitoring of their progression or regression. According to the various systems, it provides a color image (Vistaproof) or only a numerical value (Diagnodent).

R6 C.A.D.-C.A.M. is used in dentistry for design and manufacture in orthodontics (brackets, bows, aligners), in implantology (surgical guide), in prosthodontics(ceramic crowns, resin bridges, removable prostheses), computer-aided surgery, computer-aided surgical navigation and virtual articulator.

R7 CDSS have been used in numerous areas of dentistry: diagnosis of dental emergencies and trauma (treatment choosing for a cracked or broken tooth), orofacial pain, oral medicine and surgery(oral screening and dental manage ment of patients with head and neck cancer; decision support for lower-third-molar treatment planning, ORAD for bone lesion diagnosis), oral radiology (interpretation of radiographic lesions and automated interpretation of dental radiographs), orthodontics (analysis of facial growth, landmark identification of cephalometric radiographs, and treatment planning), caries management and pulpal diagnosis and restorative dentistry (removable partial denture design), intelligent agents for dental treatment planning.

References

Adeyemi AA, Jarad FD, Pender N, Higham SM (2006) Comparison of quantitative light-induced fluorescence (QJF) and digital imaging applied for the detection and quantification of staining and stain removal on teeth. J Dent 34(7):460–466

Azari A, Nikzad S (2008) Computer-assited implantology : historical background and potential outcomes – a review. Int J Med Robot Comput Assist Surg 4:95–104

Bister D, Mordarai F, Aveling RM (2006) Comparison of 10 digital SLR cameras for orthodontic photography. J Orthod 33(3):223–230

Brickley MR, Sheperd JP (1996) Performance of a neural network trained to make third molar treatment planning decisions. Med Decis Making 16(2):153–160

Brüllmann D, Schmidtmann I, Warzecha K et al (2011) Recognition of root canal orifices at a distance – a preliminary study of teledentistry. J Telemed Telecare 17(3):154–157

Duret F (1973) Empreinte optique. Master's thesis – Claude Bernard – Lyon University

Erten H, Uçtasli MB, Akarslan ZZ et al (2006) Restorative treatment decision making with unaided visual examination, intraoral camera and operating microscope. Oper Dent 31(1):55–59

Finkeissen E, Stamm I, Müssig M et al (2003) AIDA: web agents in dental treatment planning. Adv Dent Res 17:74–76

Forgie AH, Pine CM, Pitts NB (2003) The assessment of an intra-oral video camera as an aid to occlusal caries detection. Int Dent J 53(1):3–6

Gaillard J, Jourda G (1991) Computer-assisted design in removable partial dentures. Expert system and software for framework tracing. Rev Odontostomatol 20(3):223–229

Goodlin R (2011) Photographic-assisted diagnosis and treatment planning. Dent Clin North Am 55 (2):211–227, vii

Kapos T, Ashy LM, Gallucci GO et al (2009) Computer–aided design and computer-assisted manufacturing in prosthetic implant dentistry. Int J Oral Maxillofac Implants 24:110–117

Komarov GN, Anderson C, Barr R et al (2010) The Effect of QLF-D in improving toothcleaning. In: Caries research. 45(2):188. Proceedings of the ORCA congress, Kaunas, 2010

Kühnisch J, Ifland S, Tranaeus S et al (2007) In vivo detection of non-cavitated caries lesions on occlusal surfaces by visual inspection and quantitative light-induced fluorescence. Acta Odontol Scand 65(3):183–188

Lasserre JF, Pop-Ciutrila IS, Colosi HA (2011) A comparison between a new visual method of colour matching by intraoral camera and conventional visual and spectrometric methods. J Dent 39(3):e29–e36

Leith E, Upatnieks J (1964) Wavefront Reconstruction with diffused illumination and three-dimentional objects. J Opt Soc Am 54(11):1295–1301

Mago VK, Bhatia N, Bhatia A et al (2012) Clinical decision support system for dental treatment. J Comput Sci 3(5):254–261

Mendonca EA (2004) Clinical decision support systems: perspectives in dentistry. J Dent Educ 68 (6):589–597

Mirska-Mietek M (2010) Diagnosis of caries on approximal surfaces of permanent teeth. Ann Acad Med Stetin 56(2):70–79

Niederman R (2009) Manual and electronic probes have similar reliability in the measurement of untreated periodontitis. Evid Based Dent 10(2):39

Pretty IA (2006) Caries detection and diagnosis: novel technologies. J Dent 34:727–739

Priest G (2005) Virtual-designed and computer-milled implant abutments. J Oral Maxillofac Surg 63:22–32

Pryputniewicz C, Burston J, Bowley W (1978) Determination of arbitrary tooth's deplacements. J Dent Res 57(5):663–674

Sarava B (1965) Application of photogrammetry for quantitative study of tooth and face morphology. J Phys Antrhopol 23(4):427–434

Scutariu MM, Voroneanu M, Ursache M (2007) Detection strategies and risk factors in oral cancers. Rev Med Chir Soc Med Nat lasi 111(1):221–227

Smith RN, Rawlinson A, Lath DL et al (2006) A digital SLR or intra-oral camera: preference for acquisition within an image analysis system for measurement of disclosed dental plaque area within clinical trials. J Periodontal Res 41(1):55–61

Splieth CH, Nourallah AW (2006) An occlusal plaque index. Measurements of repeatability, reproducibility, and sensitivity. Am J Dent 19(3):135–137

Suebnukarn S, Rungcharoenporn N, Sangsuratham S (2008) A Bayesian decision support model for assessment of endodontic treatment outcome. Oral Surg Oral Med Oral Pathol Oral Radiol Endod 106:e48–e58

Tracy KD, Dykstra BA, Gakenheimer DC et al (2011) Utility and effectiveness of computer-aided diagnosis of dental caries. Gen Dent 59(2):136–144

Watzinger W, Birkfellner F, Wanschitz F (1999) Computer-aided positioning of dental implants using an optical tracking system-case report and presentation of a new method. J Craniomaxfac Surg 17:77–81

White SC (1989) Computer-aided differential diagnosis of oral radiographic lesions. Dentomaxillofac Radiol 18(2):53–59

Wiechmann D, Wiechmann L (2003) Les finitions occlusales assistées par ordinateur. Orthod Fr 74:15–28

Wiechmann D, Rummel V, Thalheim A et al (2003) Customized brackets and archwires for lingual orthodontic treatment. Am J Orthod Dentofacial Orthop 124:593–599

Willershausen B, Schlösser E, Ernst CP (1999) The intra-oral camera, dental health communication and oral hygiene. Int Dent J 49(2):95–100

Chapter 16
E-health

R. Beuscart, E. Chazard, J. Duchêne, G. Ficheur, J.M. Renard, V. Rialle, and N. Souf

Abstract E-health is a large domain of research and applications of Information and Communication Technologies (ICT), not only in Medicine, but in the broad field of healthcare, including homecare and personalised health. The history of e-health started as soon as the 1960s, but e-health continues to extend its range of innovation and applications, particularly in developing countries and in the homecare domain. E-Health scientific background is based upon the theories of "Computer-Supported Cooperative Work" theorised by Schmidt, Ellis, and Johansen, in the 1990s. In this chapter, we present different fields of development of telemedicine, and Home-based tele-health. We present also how e-health contributes to the constitution of large networked data warehouses to be now exploited with the relevant methods.

Keywords E-health • Medical informatics • Telemedicine • Tele-health • Electronic health records • Computer supported cooperative work • Personalised health

R. Beuscart (✉)
Faculté de Médecine – Pôle Recherche, CERIM, Lille 2 University, E.A. 2694, 1, place de Verdun, Lille, Cedex 59045, France
e-mail: regis.beuscart@univ-lille2.fr

E. Chazard • G. Ficheur • J.M. Renard
CERIM, Lille 2 University, E.A. 2694, 1, place de Verdun, Lille, Cedex 59045, France

J. Duchêne
Université de Technologie de Troyes, Troyes, France

V. Rialle
Université Pierre Mendes-France, Grenoble, Grenoble, France

N. Souf
IUT Castres, Castres, France

A. Venot et al. (eds.), *Medical Informatics, e-Health*, Health Informatics, DOI 10.1007/978-2-8178-0478-1_16, © Springer-Verlag France 2014

After reading this chapter, you should be able to:

- Define e-health.
- Chart the major steps in e-health development since the 1960s.
- Explain the anticipated impacts of e-health.
- Define cooperative work.
- Define telehealth and its different components (telemedicine, telecare).
- Explain the impact of new information and communication technologies on personal autonomy.

16.1   Introduction

16.1.1   Definitions of E-health

E-health can be broadly defined as the application of information and communication technology (ICT) to health and healthcare (Telemedicine 2010, report from the WHO).

The World Health Organisation (WHO) stated recently that "health" is not merely limited to the absence of disease (whether acute or chronic) but also corresponds to a "state of complete physical, mental and social well-being". So the definition of e-health by WHO is "a cost-effective and secure use of information and communication technologies in support of health and health-related fields, including health-care services, health surveillance, health literature, and health education, knowledge and research" (WHO). The term "health" is also related to activity limitations and participation restrictions in life in society (i.e. disabilities).

Another definition of e-health has been proposed by the The European Union which defines e-health as comprising four interrelated categories of applications:

- Clinical information systems
- Telemedicine and homecare
- Integrated health information networks and distributed, shared-access health databases
- NCSC Non-clinical systems

This last field encompasses education for healthcare professionals, health education, health promotion for patients and the general public (such as information portals) and decision support. It also includes care management systems.

Currently, the definition of E-Health encompasses much of medical informatics (Blum and Duncan 1990) but tends to prioritize the delivery of clinical information, care and services rather the function of technologies. Unfortunately, no single consensus, all-encompassing definition of e-health exists (Fatehi 2012). The field of e-health spills over from informatics by integrating all aspects of telecommunications, video and the Internet and by taking account of organisational and human aspects. Hence, the development of e-health services is tightly coupled to the particular environment; for example, teleconsultation has grown more quickly in the Australian desert and in the north of Norway than in Northern France, where nobody lives more than 20 km from a hospital.

In this chapter, we will consider that e-health comprises two complementary fields:

(a) **Tele-health, which includes:**

- Telemedicine, which covers the set of techniques and applications that enable physicians and healthcare professionals to remotely establish diagnoses, initiate therapies, provide follow-up and support and monitor coordinated care.
- Telecare in everyday life and for social welfare, through the provision of support and monitoring for elderly, disabled and dependent people and compensation for the loss of personal autonomy. These services extend beyond medical care and so encompass social care." This growing field also includes what some people call p-health (personal health), in which the patient uses ICT directly. P-health encompasses all the health information available on the Internet (whether generated by companies, governments, charities or individuals). The web is a huge source of information for patients and their families. The impact of these websites on public health merits careful analysis. Moreover, p-health also includes teleconsultations, telepharmacy for drug prescriptions and some aspects of home assistance (i.e. the whole field of ICT in relation to the general public, patients and their family and friends). This field is likely to have a major commercial impact in the near future.

(b) **The use of information systems** in health and healthcare (and not limited to medicine,) including methods and technologies for exploiting and analysing data.

16.1.2   A Brief History

The first telemedicine experiments were performed in the 1960s, with a teleconsultation organized by the Nebraska Psychiatric Institute (Wittson and Benschoter 1972). The need for teleconsultation was driven by the lack of local skills and the fact that patients were often isolated in this sparsely populated state of the USA. In 1965, the first ever tele-assistance videoconference on open heart surgery was broadcast from the United States to Switzerland. The first congress on telemedicine was held in 1973 in Michigan. At this time, telemedicine essentially relied on telephony and video broadcasting.

The 1970s witnessed the emergence of a large number of structured projects, although the latter were handicapped by poor technical performance, high telecommunications costs, the absence of suitable medical and social organizations and, of course, the absence of business models.

The military further developed telemedicine in the 1980, with a view to use on the battlefield and in space. The main driver for these research projects and applications was the limited access to soldiers or astronauts. Hence, in the Vietnam War, the US Army developed telemedicine projects for deployment in the field. At the end of the 1980s, the US Navy developed a system for teleconsultation within a fleet of ships. The NASA (Bashshur 1980) tested various means of providing medical assistance to astronauts. At the same time, very similar civil applications were being developed for providing remote care and monitoring to people in scientific field stations in Antarctica and on North Sea oil platforms.

Telehealth really took off in the 1990s. The European Commission co-funded a number of major projects and provided fostering measures for developing e-health applications throughout Europe: remote imaging, teleconsultation, e-learning, surgery-hospital communication, telesurgery, virtual staff meetings, emergency telemedicine, and so on. Scientists, healthcare professionals and institutions generated a large number of projects...very few of which have survived to this day. Nevertheless, the foundations of e-health had been laid.

E-health matured in the first decade of the new millennium. Some applications are now used routinely: emergency telemedicine, remote imaging, electronic health records, interhospital communication, laboratory-physician communication, secure messaging, etc. Today's projects are often mature, large-scale projects that are designed to have a strong impact on professional practice: teleradiology, teledialysis, telehealth in prisons, telemedicine in rural zones, personal medical records, pharmacy records, etc (Bashshur and Shannon 2009).

From 2010 onwards, all the factors required to accelerate the deployment of e-health solutions were in place: an increase in the prevalence of chronic disease in the general population, the need to find alternatives to hospitalization or repeat consultations and the need to develop telecare for everyday life and social well-being. Furthermore, greater IT skill levels helped to change mentalities and make e-health more acceptable. Organizations are now changing by adopting the latest available technologies (tablet computers, wireless Internet, smartphones, etc.). In many countries, the regulatory situation has also evolved to match the technological changes. However, aspects such as data protection, funding and reimbursement have yet to be comprehensively addressed.

16.2 Scientific Theories of Computer-Supported Cooperative Work

16.2.1 Introduction

E-health activities involve the participants (whether healthcare professionals, patients, family members or helpers) in collaborative work activities. In this section, we shall present a few theories from the field of Computer-Supported

Cooperative Work (CSCW) because they are useful for modelling cooperative activities in e-health (Schmidt and Bannon 1992)

16.2.2 Cooperative Work

Communication and cooperation are critical points for improving and maintain high-quality patient management over time. Cooperation is essential for patient care and medical practice:

– Cooperative activities are extremely *frequent in medical practice*. Modern healthcare prompts the various healthcare professionals (e.g. nurses, general practitioners, specialist practitioners, physiotherapists, etc.) to coordinate their activities on a regular basis. Even in situations where these people rarely meet, they never work alone and their care activities must be coordinated with those of other healthcare professionals.
– These activities have an extremely *significant impact on care quality*. Many studies have shown that as soon as the quality of communication falls, errors in care management will appear (e.g. a change in care that is not noted or inability for a third party to check information). These coordination problems can have serious and even potentially life-threating repercussions for the patient.

In order to introduce effective tools, it is advisable to have a good understanding of the concept of **Cooperation**. The dictionary defines cooperation as a situation or behaviour characterized by the achievement of a mutually beneficial outcome for each of the participants. In fact, this concept is a continuum that depends on the activities in question.

When the work is subdivided into several sub-tasks to be performed by different participants, coordination of the sub-tasks becomes an important issue in its own right. The challenge is then to synchronise the sub-tasks and ensure that resources are forecast and allocated accordingly.

This is why e-health is based on fundamental research in the field of CSCW, the goal of which is to study how cooperative activities and coordination can be performed by computer systems by looking at how people work together and how the introduction of ICT-based tools will affect their behaviour as a group (Schmidt and Bannon 1992; Ellis et al. 1991)

16.2.3 Classification of Cooperative Tools (Groupware) as a Function of Space and Time

In order to better establish which cooperative tools might be useful for a given activity, it is useful to define when and where the cooperation is being performed.

Fig. 16.1 The time-space matrix (According to Johansen)

	Same Time	Different Times
Same Place	Face-to-Face Interaction	Asynchronous Interaction
Different Places	Synchronous distributed interaction	Asynchronous distributed interaction

Johansen (1988) is suggesting using a time-space matrix to classify groupware tools according to the following parameters: same place vs. different places and same time vs. different times. Hence, there are four main types of interaction: face-to-face interactions, asynchronous interactions, synchronous distributed interactions and asynchronous distributed interactions (Fig. 16.1).

Computerized systems for e-health activities must therefore be designed to take account of these different aspects of cooperative work, with a view to offering tools that best meet user needs. This analytical step is crucial for obtaining well-accepted, efficient tools (Fig. 16.2).

These four types of interaction correspond respectively to the following examples of cooperative applications:

– Same time, same place: tools for facilitating face-to-face meetings, such as interactive whiteboards, conference rooms and decision support tools.
– Same place, different times: a series of people successively attending to the same patient (e.g. in a hospital ward or examination room) can benefit from tools for classifying, filtering and displaying changes in the patient's status, displaying messages and managing shared documents.
– Same time, different places: teleconsultations, such as remote staff meetings and teleconferences.
– Different places and times: coordination activities are aided by workflow tools, e-mail and version management tools.

16.3 Tele-health: Telemedicine and Telecare

16.3.1 Definition

Telehealth (also known as health telematics) covers the activities, services and systems performed remotely using ICT with respect to worldwide needs in health promotion, disease management and control, health management and health-related research.

Fig. 16.2 The time-space
matrix applied to tele-health
applications

	Same Time	**Different Times**
Same Place	Interactive whiteboards, group working	e-mail
Different Places	Teleconsultation	Teleradiology

16.3.2   Telemedicine

Telemedicine (as defined by the WHO, enables "the delivery of health care services, where distance is a critical factor, by all health care professionals using information and communication technologies for the exchange of valid information for diagnosis, treatment and prevention of disease and injuries, research and evaluation, and for the continuing education of health care providers". In fact, telemedicine is telehealth performed by the medical profession. It includes four fields of application. The first three fields are performed by one or more physicians. In contrast, the term "tele-assistance" is more used in social care:

- **Telemonitoring**: a device is used to remotely monitor vital signs in a patient at home or, more broadly, outside a hospital setting. The data are transmitted for analysis by a physician.
- **Teleconsultation**: a patient consults a physician who is in a different place. The patient may or may not be accompanied by another healthcare professional.
- **Tele-expertise**: is a specialised form of teleconsultation where a physician requests the opinion of a colleague, in the patient's absence.
- **Tele-assistance**: a remote healthcare professional helps a person (generally another healthcare professional) to perform medical act.

Other terms are often used (Fatehi and Wootton 2012) and may overlap with those given above. For example, telediagnosis may correspond to the outcome of a teleconsultation or a tele-expertise, whereas tele-surveillance is a kind of telemonitoring. These various fields can be subdivided by clinical speciality: telecardiology, teledermatology, telesurgery and so on.

The non-medical applications of telehealth also include data dissemination (care networks), teaching and training (e-learning) and healthcare systems management.

In the following sections, we review the four fields of telehealth, i.e. telemonitoring, teleconsultation, tele-expertise and tele-assistance. We shall define each of these fields, explain their general principles and give some typical examples. We shall then discuss new constraints that are associated with these medical activities.

16.3.2.1 Telemonitoring

Telemonitoring consists of the transmission of clinical, radiological or biological data. The data may be collected by the patient or by healthcare professionals. Interpretation of these data may prompt a treatment decision. Telemonitoring is a synchronous medical activity as it is based on a stream of patient information that is collected and (in most cases) transmitted immediately.

Telemonitoring is based on the regular, automated transmission of health-related patient data during activities of daily living. Data capture and transmission can be performed by an implantable medical device (an implantable cardioverter-defibrillator (ICD), for example), an external medical device (an electrocardiograph, for example) or an external non-medical device (a mobile phone or a smartphone, for example). In terms of the time point and timescale, patient data can be collected either continuously, at pre-programmed times of day or during particular events (arrhythmia detected by an implantable defibrillator, for example). The signal may be analyzed continuously or only following the occurrence of an alert.

Pacemakers and Telemonitoring

The first clinical implantation of a pacemaker took place in 1958 and defibrillators have been available since 1994. These devices were able to adapt their stimulation to the myocardium's electrical activity but were not able to export the information analyzed. In 2001, a wireless pacemaker transmitted information to an external receiver for the first time. Currently, the implanted pacemaker communicates wirelessly with a logger in the patient's home. The logger then transmits data on the device itself (e.g. the battery level and self-test results) and on the patient's cardiac activity (e.g. episodes of arrhythmia detected and perhaps even resolved by the device) over the telephone network on a regular basis. The information is uploaded to a secure server and, if necessary, alerts can be sent by e-mail or text message to the physician, as required. The physician can then consult detailed information via a secure web platform. Studies have shown that the quality of the transmitted information ranges from 92 % to 100 %. The clinical studies have clearly demonstrated that this type of device is associated with a reduction in the number of consultations, the number of adverse events and the duration of post-operative hospitalization. The cost-efficacy balance seems clearly beneficial.

The monitoring of at-risk pregnancies is a good example of the use of external devices. Some hospitals in Paris (France) notably use foetal telemonitoring in pregnant women suffering from diabetes or kidney failure. When applied as part of homecare, this technique is less cosltly than traditional hospitalisation and guarantees the same level of monitoring. More generally, the use of external devices for telemonitoring of chronic conditions is a fast-expanding field and includes peritoneal dialysis at home, the monitoring of patients with hypertension or chronic respiratory deficiencies, glycaemia monitoring for diabetics, the adjustment of anticoagulant treatments, etc. These devices enable interventions to be performed intelligently and earlier than in a physician's office or in hospital. They also decrease the number of inappropriate consultations and hospitalizations

The systematic production of high volumes of information poses a new set of problems [9]. The information throughput is very high and requires the use of analytical and alerting devices that enable an appropriate response by healthcare professionals – sometimes in an emergency context. Healthcare professionals must be vigilant at all times, including during stand-by and on-call periods. The lack of a response to a significant event will raise new professional liability issues in the field of medicine. Moreover, in order to make this type of vigilance practical and keep costs to a reasonable level, abnormal signals have to be filtered out of normal signals. Decision support systems must thus be developed (Koutkias et al. 2011). An additional problem then arises: integration of the different types of data available (the nature of the alert, the health professional's level of knowledge and the patient's characteristics, on-going treatments and medical history) in order to avoid the occurrence of false positives (i.e. alerts in situations that do not justify intervention, i.e. over-alerting) and false negatives (i.e. the absence of an alert when an intervention is necessary) (Beuscart et al. 2009).

Conversely, these devices will create huge data warehouses – the retrospective exploitation of which (i.e. data mining) will generate extremely valuable knowledge. This point will be addressed in Sect. 16.5.

16.3.2.2   Teleconsultation

Teleconsultation enables a patient to consult a physician located elsewhere. Another healthcare professional may be present with the patient, to provide assistance. If the teleconsultation involves two physicians, the physician with the patient is referred to as the "requesting physician", whereas the remote physician is referred to as the "requested physician". Hence, teleconsultation is a synchronous medical act performed in real time and with the patient present.

Teleconsultation is essentially based on videoconferencing. This overcomes problems related to distance and, to a certain extent, enables the patient to obtain a consultation more rapidly. The patient may be an outpatient at home, receiving homecare or living in a retirement home or healthcare establishment. He/she consults the requested physician alone or in the presence of an accompanying physician or nurse. The latter might use a handheld camera or perform a clinical examination when invited to do so by the requested physician. In all cases, teleconsultation requires the requested physician to have access to the patient's medical records.

Fig. 16.3 Teleconsultation

Teleconsultation also provides benefits to the patient (as has been demonstrated in clinical gerontology and in chronic diseases management) and his/her carers and family circle (with improvements in their psychological status) (Mair and Whitten 2000).

The Emergence of Teleconsultation

Examples of teleconsultation include multidisciplinary team meetings in oncology and, more recently, teleconsultations for patients in mountainous regions or care homes.

In the USA, teleconsultation has been developed for psychiatric care and prison health in particular. Use of teleconsultations in this context avoids the need for a sometimes reluctant physician to visit the prison and reduces the otherwise high cost of taking a prisoner to an external medical centre.

In developing countries, teleconsultations provide a new access-to-care to distant rural areas, where access to physicians and healthcare professionals is difficult, complex or too expensive. (Wootton 2008; Piette et al. 2012) (Fig. 16.3).

16.3.2.3 Tele-expertise

A Tele-expertise is a diagnostic or therapeutic activity performed by two healthcare professionals in the absence of the patient. It is based on a discussion between two or more physicians. Diagnostic or therapeutic decisions are based on clinical and other data present in the patient's medical records. In many cases, this corresponds to a second opinion on a complex case. In all cases, tele-expertise in an asynchronous medical act – the patient is not present.

Fig. 16.4 Tele-expertise

Tele-expertise is justified by the extreme specialization of medical knowledge and the need for multidisciplinary discussions. Perin@t networks first developed tele-expertise on the basis of ultrasound datasets recorded during pregnancy. These networks have improved screening for foetal malformations, enabled local care to be maintained (with non-specialists receiving remote support) and, above all, reduced the cost and discomfort of travel for the pregnant women concerned. It should be noted that this is not a teleconsultation because the distant physician does not intervene at the same time as the diagnostic procedure is performed; he/she interprets the data asynchronously.

Tele-expertise: A True Medical Act

A tele-expertise is more than just a discussion between colleagues: it is a true medical act that generates a report signed by all practitioners involved. In the event of major disagreement, the patient has to been informed of the various options suggested by the physicians involved in the appraisal. Tele-expertises are subject to the same regulatory constraints as multidisciplinary staff meetings performed in the patient's absence (Fig. 16.4).

16.3.2.4   Tele-assistance

In tele-assistance, a healthcare professional helps another healthcare professional to perform a diagnostic or therapeutic act on a patient. It is a synchronous activity and the healthcare professional present with the patient may not necessarily be a physician. Although the best-known example of tele-assistance is telesurgery, this type of activity also applies to remote imaging and can be adapted to suit applications in emergency situations and remote places.

A Few Examples of Tele-Assistance

The first telesurgical operation (the so-called "Lindbergh operation") was performed in 2001: it consisted of a cholecystectomy and lasted 45 min. The patient was located in Strasbourg (France) and the surgeon (located in New York) operated a robotic arm via a high-speed fibre-optic link.

Tele-assistance is especially important in medical emergencies in which a physician does not have access to the patient. Several emergency medical services perform around a 100 tele-assistance procedures a year for in-flight medical problems on board Air France airliners. Similar systems have been set up for mountain refuges and islands.

All the usual regulatory obligations (e.g. professional liability, patient consent and the obligation to provide a result) still apply to tele-assistance. If the health professional is found to be liable though through no fault of his/her own and if possible under the circumstances, he/she must be able to prosecute third parties involved in the technical processes, i.e. the companies that make or maintain the equipment or telecommunication networks involved. Hence, if the patient invokes the physician's or the healthcare establishment's professional liability, the latter can file a recourse action against the technology providers.

16.3.3   Telecare: E-health and Personal Autonomy

16.3.3.1   Introduction

In the previous section, we reviewed the benefits of e-health in the field of telehealth in general and telemedicine in particular. In general, telehealth enables care provision for patients, the elderly, people with disabilities or people who have lost some degree of personal independence. Patient management via e-health goes beyond the framework of telemedicine *per se* and involves carers, helpers, social workers and sometimes the patient him/herself, in addition to physicians. The involvement of all these stakeholders means that the boundaries between medical care, home care and social support have become blurred. This is why the following section will consider the different applications of e-health in maintaining personal autonomy in general (i.e. not just the management of medical care) (DelliFraine and Dansky 2008)

The sudden or progressive loss of personal autonomy can affect elderly people, people with activity limitation and/or participation restrictions and people hospitalized at home. Although these people have similar needs, we shall focus here on the particular case of loss of independence in the elderly.

In some countries, a distinction is made between "medical technical aids" (for treatment and/or prevention) and "social technical aids" (intended to compensate for the elderly person's loss of independence and help him/her to remain "socially integrated").

16.3.3.2   Technology and Ageing ("Gerontechnology")

On one hand, the number of people suffering from loss of personal independence is growing rapidly. On the other hand, technological innovation is producing ever more products and services. The integration of these technologies into the everyday environment of frail or dependent people (or those at risk of losing personal independence) is a major challenge.

The increase in the number of elderly people is accompanied by the fragmentation of the nuclear family that characterises modern society, which inevitably increases the number of elderly people living alone. Population aging is increasing the prevalence of chronic diseases (such as diabetes and its complications). More effective treatment of acute medical conditions also increases the number of patients with permanent sequelae of various degrees of severity, such as myocardial infarction survivors with rhythm disorders. Lastly, for psychological and economic reasons, health authorities are seeking to avoid traditional hospitalization whenever possible; this translates in greater numbers of patients being treated in their own home or in a retirement home.

Technology and Ageing: The Main Fields of Activity

Devices are commonly classified into three main classes as a function of the type of impairment:

- Devices acting on physical impairments: these provide more detailed information on physical problems, loss of mobility and activity disorders – all of which have an impact on the risk of falls.
- Devices acting on social impairments: these devices act on social isolation by recreating social links that the person no longer has the physical or psychological ability to maintain.
- Devices acting on cognitive impairments: this type of device provides support (and, in some cases, treatment) for people with cognitive disorders in general and dementia in particular.

Some social technical aids are home-based and are closely integrated into the building, whereas others are worn or carried by the patient. Lastly, communication technologies (video calls and teleservices in particular) can improve monitoring, follow-up and socialization by promoting the maintenance of social links.

16.3.3.3   Medical Home Automation

Domotics refers to the set of information technologies used in buildings and habitations. Many research programmes are seeking to promote homecare for elderly or dependent people. This new type of habitat comprises a complete information processing chain, from the sensing of information to its analysis in terms of medical parameters (diagnosis, functional assessments, telemonitoring, etc.), decision-making (data fusion, alarm detection, etc.), statistical analyses (day/ night actimetry) and management (access rights, archiving, retrospective analyses, etc.).

The information processing chain starts with sensors that can be placed in everyday household objects (domestic appliances, the shower, the bath tub, sets of scales, beds, window blinds, shutters, etc.) and networked. The sensors are connected to a local information system that centralizes, aggregates and filters the signals. This local information system is coupled to an analysis system (for detecting abnormal situations) and a telecommunications network (the telephone system or the internet) so that e-mail or text message alerts can be sent automatically to carers, helpers, families, neighbours, etc.

Some sensors are used to quantify the person's activity: an elderly person can be monitored in the room by infrared movement sensors or through detectors in doorways. Some domestic objects (like a kettle) can also be equipped with activity sensors and thus testified to activity in the home. Lastly, real-time monitoring of water or electricity consumption can also provide this type of information. The sensors used for monitoring include radio-frequency identification chips, microphones (acoustic sensors) and cameras (optical sensors).

As mentioned above in the section on telehealth, sensors can be integrated to medical devices (blood pressure monitors, oximeters, non-invasive ventilation devices, etc.) so that vital signs can be monitored; this is part of telemonitoring.

To conclude this section on Medical Home Automation, we shall give two examples of domestic applications of robotics:

- Domestic robots that provide assistance with housework.
- Robot pets used to reassure an elderly person with dementia.

The use of robots (and, more generally, monitoring a patient to ensure his/her safety) poses some obvious ethical problems, which must always be taken into account when these technologies are applied.

16.3.3.4   Technical Aids for the Patient

The technical aids mentioned here are portable, "on-board" technologies. Some are already in everyday use, whereas others are in the research and development phase. Some devices merely monitor the patient's activity, whereas others intervene directly on the patient's actions.

Actimetry

An actimeter is a device that uses accelerometers or magnetometers to sense and then quantify movement. By extrapolation, actimetry corresponds to measurement of a person's physical activity. When used to tackle the loss of personal autonomy, actimetry seeks to answer three main questions: "Is the person walking around?", "Has the person fallen over?" and "Is the person wearing his/her monitor?". These actimeters are generally watch- or belt-type devices that notably detect falls, violent impacts and suspicious periods of total lack of movement. When a suspicious event occurs, the device sends an alert (by telephone or via the Internet) to a telemonitoring centre. In some cases, the device also sends an alert when the owner forgets to wear it.

Location Monitoring

Going missing is a significant source of morbimortality in elderly people suffering from dementia. The available technical aids range from door contact sensors (mentioned in the section on domotics) that inform the monitoring centre whenever the door is opened to more sophisticated geotracking solutions (such as GPS-enabled devices worn on the wrist or integrated into a belt or an item of clothing) that transmit the wearer's position on a regular basis.

Mobility Aids

Many devices are designed to serve as both a mobility aid and a communicative monitoring system. For example, Zimmer frames are widely used by people with reduced mobility. The simplest versions have 4 ft but Zimmer frames can also be fitted with wheels. Indeed, more sophisticated versions under development can avoid collisions, choose the optimal itinerary and help the user to stand up. Some of these systems also monitor falls and track physiological parameters. The final example corresponds to sat-nav devices fitted with an alarm button and that can send out an alert if the wearer leaves a specified zone.

Cognitive Aids

Cognitive aids are devices and software that help people to compensate for cognitive impairments, such as memory and executive disorders. A simple example of a cognitive aid is a portable diary with a simplified interface and that reminds the wearer to take his/her medications, keep appointments and so on.

16.3.3.5 Communication and Teleservices

The above-mentioned remote alarm systems can be coupled with telephone or videophone communication and thus become tele-assistance systems.

Tele-assistance and the remote monitoring of information collected on an elderly person (and within his/her home in particular) is a growing market. Service companies collecting domestic data are expanding in the same way as in medical telemonitoring.

A growing number of personal services are now available as teleservices and are being progressively implemented on appropriate devices; these include home meal delivery and home help management.

Lastly, communication technologies can help elderly people to maintain social contact with their family and friends. Nevertheless, use of modern web-based communication and information tools by the elderly requires the provision of initial support and training.

16.4 E-health Enables the Constitution and the Exploitation of Networked Data Warehouses

16.4.1 The Available Data

E-health and new ICTs have enabled the rapid transfer of nominative or non-nominative medical data. The transmitted data now constitute real "treasure troves", i.e. data warehouses that can be exploited for scientific, clinical, epidemiological and economic purposes and can reveal new potential for medical knowledge.

Data warehouses are used to store data from many heterogeneous sources. They constitute decisional databases that can be queried to extract information in batches. The information constitutes new knowledge and can be used to make decisions. Production of this information is based on online analytical processing and dashboards for decision support.

Several types of information are now collected routinely and fed into data warehouses. The information is heterogeneous and can be aggregated around a patient's unique identifier.

The database is always accessible. Hospitals are obliged to collect information on patients and patient care (diagnoses, treatments, demographic information, etc.). In physician's offices, many other items of information are systematically collected e.g. on medications, devices, services and consultations that can be reimbursed. Moreover, some medical establishments and physician's offices can produce computerised versions of laboratory test results and lists of medications administered to their patients.

This basic information can be enriched with information collected by the telemonitoring devices (home oximeters, pacemakers, sets of scales).

These sources of information enable the constitution and exploitation of data warehouses. Information of value in invoicing and reimbursing medical goods and services constitutes gigantic data warehouses with a homogeneous format. From these repositories, millions of medical records are available and constitute a huge source of information to be exploited for medical, epidemiological, economical purposes.

16.4.2 Problems Related to the Constitution of Data Warehouses

The constitution of data warehouses poses a number of problems in terms of confidentiality, interoperability and analytical methodologies.

When the warehouse contains medical data (and notably diagnostic data), steps must be taken to prevent a patient from being identified directly or indirectly. This need for confidentiality conflicts with the ability to crosscheck all the information from a given patient so that a comprehensive clinical picture emerges.

A second problem relates to interoperability, i.e. the need to merge heterogeneous data from different databases. This aspect is detailed in Chap. 13.

A third problem relates to the complexity and volume of the data; there are huge numbers of patients, variables and relationships between variables, making it impossible to apply traditional statistical methods. Furthermore, many data items (e.g. biological parameters) are repeatedly measured over time and require very specific analytical methods.

16.4.3 Methods and Perspectives for the Analysis of Data Warehouses

The challenge in the healthcare sector is similar to those in other sectors (banking, insurance, retail, etc.): exploiting the billions of items stored in the data warehouse and extracting "market knowledge" and "business intelligence".

Data Warehouse Analysis

Large-scale data warehouses can be analyzed for various end purposes:

- For a given patient, in order to monitor changes over time in parameters and thus provide a more comprehensive medical picture
- In a population, to calculate the incidence and prevalence of morbid events.
- In a population, to identify the complex time-domain motifs or signatures that result in or predispose to the occurrence of a harmful event.
- For a given patient, to alert health care professionals if a harmful event is likely to occur with a sufficiently high probability (based on the analysis of previous events).

The identification of motifs that enable the prediction of a morbid event is based on data mining and knowledge discovery. Data mining combines techniques derived from statistics, artificial intelligence and data management to extract new knowledge from large-scale databases. These methods are also used in research projects.

The European PSIP and AKENATON Projects

The European project entitled "Patient Safety through Intelligent Procedures in Medication" (PSIP: http://www.psip-project.eu) analyses hospital stays in order to identify treatment-related adverse events and the circumstances that are responsible for the latter's occurrence. This knowledge helps to prevent the occurrence of adverse events when the circumstances are met at the time of drug prescription (i.e. decision support) (Beuscart et al. 2009; Koutkias et al. 2011).

Given that patients and their treatments are becoming increasingly complex, taking into account the full context should help to prevent drug-related adverse effects [13].

The "Automated Knowledge Extraction from Medical Records in Association with a Telecardiology Observation Network" (AKENATON) project (http://www.u936.univ-rennes1.fr/index.php) analyses the signals produced by ICDs. The many alerts that the device sends to cardiologists are thus filtered and ranked in order of importance.

By limiting and ranking the importance of information streams, the project is aimed at preventing care providers from being "submerged" by data and thus encourages wider use of telemonitoring.

16.5 Other Fields of E-health

The information provided to the physician during continuous medical education, the health information available to the patient on the web, electronic health records and the issue of data security are other aspects of e-health. They have been dealt with in other chapters and so are briefly listed here merely by way of a reminder that they form an integral part e-health or p-health.

16.5.1 Electronic Health Records

Shared-access electronic health records (see Chap. 13) enable healthcare professionals to consult relevant information on the patient (when authorized by the latter) over the Internet. Electronic health records have been authorized in many countries since the 2000s. They are commonly managed by national or regional agencies so that patients are able to view and (if required) correct their personal medical data.

16.5.2 Provision of Information to Physicians and Patients

The Internet and the web act are major vectors for e-health by promoting the dissemination and sharing of skills and knowledge, e-learning and the exchange of information and ideas.

In this sense, the provision of information to physicians and patients represents an important aspect of e-health (covered in Chap. 3).

16.5.3 Health Data Security

E-health services involve Internet technologies in general and web technologies in particular and thus constitute a truly distributed information system. Data security is a very important issue for all users (whether patients or healthcare professionals) who contribute to and use this system. Security is essential for creating a high level of confidence and thus enabling high-quality care. The main properties required are reliability, integrity, confidentiality, and interoperability. These different aspects are covered in Chaps. 11 and 13.

16.6 The Anticipated Impacts of E-health

The criteria for the development and application of e-health are now met, by virtue of the widespread availability of ICT infrastructure (the Internet, in particular) and changing mentalities. E-health applications can have major repercussions on the person receiving care and on professional practice (Ekeland et al. 2010; Eleand-de Kok et al. 2011; Wootton 2012)

- Providing medical decision support and improving the quality of care. The objectives are to have access to sufficient information on the patient whenever and wherever required, adapt the treatment to suit the patient (p-health) and limit treatment-related adverse effects (Beuscart et al. 2009).
- Optimizing the use of information systems in health research, notably by aggregating data from databases and population cohorts. The challenges in clinical and epidemiological research are considerable.
- Promoting information sharing and the dissemination of know-how among professionals (Wootton 2012).

- Facilitating a patient's access to diagnosis and care – above all in geographical areas with a low density of healthcare services. Being able to receive medical advice regardless of the time or place represents a considerable gain for the elderly, carers, patients and families (Wootton 2008; Piette et al. 2012).
- Offering economically viable solutions to the issues related by population ageing, dependence and maintenance or restoration of personal autonomy.
- Rebalancing the patient-professional relationship (Wootton 2012). Information on the Internet, assistive devices and home monitoring can make people more independent and more responsible for taking care of their health; this will progressively modify the way patients interact with healthcare professionals.

16.7   Conclusion and Perspectives

Telemedicine has now become a routine part of patient care. Until recently, medical acts could not be performed remotely (other than as part of a few feasibility trials). In many countries, the legal framework for telehealth has now been clearly defined. However, telemedicine does not exonerate the physician from any of the liabilities and precautions associated with conventional medical acts. The role and liability of third-party technology providers are also now well defined.

The particular case of telemonitoring has prompted the emergence of new constraints and liabilities. Previously, home-based medical acts were only performed when explicitly requested by the patient; the healthcare professional could thus not be accused of not acting spontaneously in the patient's interest. Telemonitoring reverses this chain of command: a constant stream of information is sent to healthcare professionals, who have to take the initiative if an abnormal event occurs. In order to deal with these new responsibilities and liabilities, healthcare organisations must carefully organize 24/7 monitoring, filter alerts that threaten to submerge staff and impose full traceability for actions implemented after each alert because "best efforts" must still be made. In contrast, failure to make "best efforts" could conceivably be interpreted as an establishment's failure to offer appropriate telemonitoring services to a patient.

E-health raises problems that are not purely technical or professional in nature. E-health is shaking up care management practices, clarifying the relationship between the various healthcare professionals and driving the debate on new professional liabilities. These various organisational, regulatory and legal aspects must be clarified when e-health applications are implemented (Wootton 2012; Saliba et al. 2012).

The future of e-health is very open. Of course, healthcare professionals and healthcare organisations are interested in technologies and applications that can improve practice, promote recruitment and boost their reputation.

However, the future of e-health truly lies with patients, their families and the general public. Now that people buy goods, read newspapers, book holidays and travel tickets over the Internet, it hard to understand why they prefer going to an overcrowded accident and emergency unit , rather than having a teleconsultation.

It has already been estimated that 70 % of patients with Internet access check their prescriptions on health websites, in order to understand why the medication was prescribed and learn about the likelihood of adverse effects, drug interactions and so on. The development of personal teleservices in general (whether for patients or not) is certainly the key challenge for the decade 2010–2020.

16.8 Further Reading

Telemedicine – The Cochrane Library. http://www.thecochranelibrary.com/details/collection/806797/Telemedicine.html

Telemedicine – Opportunities and developments in Member States. Report on the second global survey on eHealth. http://www.who.int/goe/publications/goe_telemedicine_2010.pdf

Journal of Telemedicine and Telecare. www.jtt.rsmjournals.com

Telemedicine and e-health (Journal) www.liebertpub.com/TMJ

International Journal of Telemedicine and applications www.hindawi.com/journals/ijta

Telemedicine Journals – American Telemedicine Association www.americantelemed.org

---

Exercises

Q1 Give a list of six devices or sensors that are used in the field of "gerontechnology" and that could be used to telemonitor a person with dementia living alone.

Q2 Present the four main types of cooperative interaction defined by Johansen and give an example of each.

Q3 Use an internet search engine to find and view a demonstration of teleconsultation.

R1 The following six devices or detectors can be used to telemonitor a person with dementia living (the list is not exhaustive):

- A water consumption sensor.
- An infrared movement sensor.
- A domestic robot
- A geotracking device (e.g. a GPS-enabled wrist band)
- A cognitive aid (e.g. a portable diary)
- A fall detector (e.g. a belt actimeter)

(continued)

Exercises (continued)

R2 The four main types of interaction in cooperative work are:

- Face-to-face interactions (same time, same place, e.g. an interactive whiteboard).
- Asynchronous interactions (same place, different times, e.g. teleconsultation)
- Distributed synchronous interactions (different places, same time) e.g. e-mail,
- Distributed asynchronous interactions (different times, different places) e.g. teleradiology.

R3 Many websites offer teleconsultation demonstrations. Here are two examples:

- The website: "http://www.askthedoctor.com" is a free service where you can ask a network of physicians any health related questions.
- The website: "http://www.consultdrs.com" proposes: response to any medical question or problem; to understand routine screening tests; to know a second opinion on the medical condition or treatment, etc.
- Many other websites are available. Do not hesitate to test them.

Acknowledgments The authors wish to thank Nicolas Leroy for his assistance.

References

Bashshur RL (1980) Technology serves the people: the story of a co-operative telemedicine project by NASA, the Indian Health Service and the Papago people. National Aeronautics and Space Administration, Washington, DC, 115 p. Available from: US GPO, Washington, DC, 017-028-00009-0

Bashshur RL, Shannon GW (2009) History of Telemedicine: evolution, context and transformation. Marie-Anne Libert, New Rochelle

Beuscart R, Hackl W, Nohr C (eds) (2009) Detection and prevention of adverse drug events: information technologies and human factors. IOS Press, Amsterdam

Blum BI, Duncan K (1990) A history of medical informatics. ACM Press, New York

DelliFraine J, Dansky K (2008) Home-based Telehealth: a review and meta-analysis. J Telemed Telecare 14:62–66

Ekeland AG, Bowes A, Flottorp S (2010) Effectiveness of telemedicine: a systematic review of reviews. Int J Med Inform 79:736–71

Eleand-de Kok P, van Os-Medendorp H, Vergouwe-Meijer A, Bruijnzeel-Koomen C, Ros W (2011) A systematic review of the effects of the e-health on chronically ill patients. J Clin Nurs 20(21–22):2997–3010

Ellis C, Gibbs J, Rein G (1991) Groupware: Some issues and experiences. Commun ACM 34:38–58

European Union definition of E-health. http://ec.europa.eu/health-eu/care_for_me/e-health/index_en.htm

Fatehi F, Wootton R (2012) Telemedicine, telehealth or e-health? A bibliometric analysis of the trends in the use of these terms. J Telemed Telecare 18:460–464

Johansen R (1988) Groupware. Computer support for business teams. Free Press, New York/London

Koutkias V, Niès J et al (eds) (2011) Safety informatics: adverse drug events, human factors and IT tools for patient medication safety. IOS Press, Amsterdam

Mair F, Whitten P (2000) Systematic review of studies of patient satisfaction with telemedicine. BMJ 320:1517–1520

Piette JD, Lun KC et al (2012) Impacts of e-health on the outcomes of care in low and middle-income countries: where do we go from here? Bull World Health Organ 90(5):365–372

Saliba V, Legido-Quigley H et al (2012) Telemedicine across borders: a systematic review of factors that hinder or support implementation. Int J Med Inform 81:793–809

Schmidt K, Bannon L (1992) Taking CSCW seriously: supporting articulation work. CSCW 1:7–40

Telemedicine (2010) Opportunities and developments in Member States. Report on the second global survey on eHealth, World Health organisation. http://www.who.int/goe/publications/goe_telemedicine_2010.pdf

WHO Global Observatory on e-health. http://www.who.int/goe

Wittson CL, Benschoter R (1972) Two-way television: helping the Medical Center reach out. Am J Psychiatry 129:624–627

Wootton R (2008) Telemedicine support for the developing world. J Telemed Telecare 14:109–114

Wootton R (2012) Twenty years of telemedicine in chronic disease management – an evidence synthesis. J Telemed Telecare 18:211–220

Chapter 17
Translational Bioinformatics and Clinical Research Informatics

C. Daniel, E. Albuisson, T. Dart, P. Avillach, M. Cuggia, and Y. Guo

Abstract Two new emerging sub-domains in biomedical informatics – translational bioinformatics and clinical research informatics – provide information technology (IT) solutions supporting research – whether basic, clinical or translational research. The aim of the so-called "translational research" is to improve the continuum between research and care and to facilitate personalized medicine. Translational research requires better cooperation between basic research centers, healthcare facilities, clinical research and public health organizations and therefore better integration of information systems of these sectors.

This chapter presents the main characteristics of information systems used in the biomedical research domain and especially focuses on the information technology (IT) infrastructures developed to better integrate clinical care and research activities. The opportunities for the different stakeholders and the main challenges faced while developing such infrastructures are presented. The technical challenges

C. Daniel (✉)
Assistance Publique-Hôpitaux de Paris, INSERM UMRS 872 équipe 20, CCS Domaine Patient
AP-HP, 05 rue Santerre, 75012 Paris, France
e-mail: christel.daniel@crc.jussieu.fr

E. Albuisson
Université de LorraineFaculté de Médecine de Nancy 9, avenue de la Forêt de Haye BP
184 - 54505 Vandoeuvre-les-Nancy cedex

T. Dart
INSERM UMRS 872 équipe 20 15, rue de l'école de médecine, 75006 Paris, France

P. Avillach
INSERM UMRS 872 équipe 22 15, rue de l'école de médecine, 75006 Paris, France

M. Cuggia
INSERM U936 Faculté de Médecine, DIM - CHU Pontchaillou, rue H. Le Guilloux,
35033, Rennes

Y. Guo
Department of Computing, Imperial College London, 180 Queen's Gate,
London SW7 2BZ, UK

A. Venot et al. (eds.), *Medical Informatics, e-Health*, Health Informatics,
DOI 10.1007/978-2-8178-0478-1_17, © Springer-Verlag France 2014

are especially addressed (semantic interoperability, data integration, solutions ensuring data quality, data security and patient privacy, data mining). The potential of EHRs and PHRs to improve patient recruitment, conduct feasibility studies, refine inclusion/exclusion criteria, enhance safety data and, in general, to inform basic and clinical research is addressed. Examples of key national or international information technology (IT) infrastructures dedicated to translational research are shortly described.

Keywords Translational bioinformatics • Clinical research informatics • Biomedical research • Clinical research • Translational research • Epidemiologic study characteristics as topic • Evaluation studies as topic • Patient selection

After reading this chapter you should:

- Be able to give the main features of information systems (IS) used in biomedical research
- Be able to define translational bioinformatics and clinical research informatics
- Understand the opportunities for the different stakeholders and the challenges of information technology (IT) infrastructures dedicated to translational research
- Be able to identify the current challenges faced by researchers for efficiently identifying new knowledge in biomedicine and conducting biomedical studies
- Be able to identify the need of integrating multiple clinical resources to assemble data sets describing ever-larger patient cohorts in modern biomedical studies
- Be able to identify the need of repurposing in the context of biomedical research clinical data routinely collected into electronic healthcare records (EHRs) during patient care activities
- Understand specific challenging obstacles of cross-health enterprise studies design such as heterogeneity of data sources, data privacy control and performance of clinical data warehouses or distributed data platforms
- Know the most important national or international information technology (IT) infrastructures dedicated to translational research

17.1 Introduction

The previous chapters have introduced the role of medical informatics for patient care coordination or public health. Medical informatics also facilitates research – whether basic, clinical or translational research.

17.1.1 Information Systems for Biomedical Research

Basic medical research provides a better understanding of the patho-physiological mechanisms of complex diseases – such as hypertension, diabetes or cancer, for example – and especially of the associations between genetic variations and individual phenotypic characteristics in order to finally invent new biomarkers and targeted therapies. In this context, **bioinformatics** refers to the study of information processes in biological systems and proposes mathematical and computational methods for managing biological data and knowledge and finally supporting researchers in the identification and validation of physiological models.

Clinical research is based on the results of basic research and aims at demonstrating, in controlled experimental trials, often in highly constrained conditions, the efficacy of new strategies to prevent, diagnose, treat, and monitor health conditions. **Clinical research informatics** develops methods for managing clinical data and knowledge in the context of studies in therapeutic or non therapeutic human experimentation.

Comparative effectiveness research is based on the results of clinical research and aims at informing patients, providers, and decision-makers about which interventions are most effective for which patients under specific circumstances. Effective treatment provides positive results in a usual or routine care condition that may be controlled in the sense of specific activities are undertaken to increase the likelihood of positive results. **Clinical and public health informatics** aims to facilitate comparative effectiveness research.

17.1.2 Information Management Challenges in Clinical and Translational Research

The aim of the so-called "translational research" is to improve the continuum between research and care. It is on the one hand to facilitate patient access to diagnostic and therapeutic innovations from research ("from bench to bedside") and on the other hand to allow researchers to better exploit routinely collected clinical data and specimen to develop and/or validate research hypotheses ("and back again to the bench").

The development and execution of multidisciplinary, clinical or translational studies are significantly limited by the propagation of "silos" of both data and expertise. Translational research requires better cooperation between basic research centers, healthcare facilities, clinical research and public health organizations and therefore better integration of information systems of these sectors (Bernstam 2009; Gaughan 2006; Sarkar 2010).

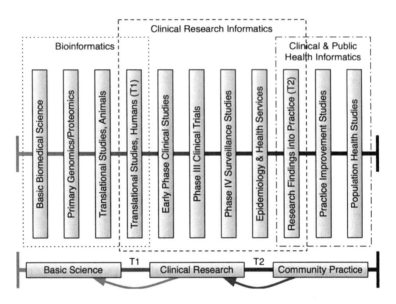

Fig. 17.1 Illustration of types of research across which the sub-domains of bioinformatics, clinical research informatics and public health informatics are focused. These sub-domains parallel the focus areas associated with the translational science paradigm, including the commonly referred to T1 and T2 blocks in translational capacity (where the T1 block is concerned with impediments to the translation of basic science discoveries into clinical studies, and the T2 block with the translation of clinical research findings into community practice). (Embi and Payne 2009)

17.2 Information Systems for Biomedical Research

17.2.1 Information Systems for Basic Science

17.2.1.1 Computing Approaches for Basic Research and Translational Bioinformatics

Bioinformatics is a branch of biological science which refers to the study of information processes in biological systems and proposes mathematical and computing approaches for managing biological data and knowledge (Fig. 17.1).

Bioinformatics applies informatics methods and technologies – such as algorithms, databases and information systems, web technologies, artificial intelligence, information and computation theory, software engineering, data mining, image processing, modeling and simulation, discrete mathematics and statistics – to the biological domain.

The scope of bioinformatics includes the development and implementation of tools that enable efficient access to, use and management of, various types of data (nucleotide and amino acid sequences, protein domains and structures, pathways, etc.) and the development of new algorithms and statistics supporting the process of analyzing and interpreting these data. This includes the development of complex interfaces whereby researchers can access existing data, submit new or revised data and assess relationships among members of large data sets in order to get a comprehensive picture of how normal cellular activities are altered in different disease states.

Major research efforts in the field include sequence alignment, gene finding, genome assembly, drug design, drug discovery, protein structure alignment, protein structure and/or function prediction, prediction of gene expression and protein–protein interactions, genome-wide association studies and the modeling of evolution.

Recently, the American Medical Informatics Association (AMIA) defined **translational bioinformatics**,[1] a major new domain, as: …*the development of storage, analytic, and interpretive methods to optimize the transformation of increasingly voluminous biomedical data into proactive, predictive, preventative, and participatory health. Translational bioinformatics includes research on the development of novel techniques for the integration of biological and clinical data and the evolution of clinical informatics methodology to encompass biological observations. The end product of translational bioinformatics is newly found knowledge from these integrative efforts that can be disseminated to a variety of stakeholders, including biomedical scientists, clinicians, and patients.*

17.2.1.2 Standards in Translational Bioinformatics

With the increasing volume of bioinformatics investigations and publications there have been concerns, within the scientific community about disappearing databases, lack of interoperability and general quality and integrity issues. Recent efforts have been coordinated by the project on Minimum Information about a Biological or Biomedical Investigation (MIBBI) (Taylor et al. 2008) to promote checklists of minimum information reporting. As part of this effort, the Minimum Information about a Bioinformatics investigation (MIABi) initiative specifies, through a series of documentation modules, the minimum information that should be provided for a bioinformatics investigation. The scope of the MIABi initiative includes algorithm, analysis, database or resource, software, and Web server. In addition, many efforts have been made to implement increasingly controlled vocabularies, standardized terminologies, nomenclatures or ontologies (Ashburner et al. 2000) and infrastructural and informational interoperability, such as use of international computational grids and cloud computing as backend computing resources for the maintenance and sustainability of knowledge resources of ever-increasing sophistication. The ultimate goal would be for the community to achieve a systematically organized, universally adopted approach in building and organizing the corpus of biological knowledge that is accurate, reliable, trustworthy, consistent and persistent (Tan et al. 2010).

---

[1] http://www.amia.org/applications-informatics/translational-bioinformatics

17.2.2   Information Systems for Clinical Research

17.2.2.1   Clinical Trial Data Management Systems

The pharmaceutical industry as well as the institutional research invests in tools to improve the efficiency and speed of clinical research. Using paper-based format for data collection and management as well as for their submission to health agencies (e.g. the "Food and Drug Administration" (FDA) or the "European Medicines Evaluation Agency "(EMEA)), if required, results in unnecessarily tedious and expensive processes. There is an increasing use of computerized systems in clinical trials to generate and maintain source documentation on clinical trial protocols and source data about enrolled subjects. Electronic Case Report Form (eCRF) are replacing paper case report forms and clinical trial data are more and more directly captured from the patient's home and/or from the treating physician.

Such electronic source documentation and data must meet the same fundamental criteria of data quality (e.g., attributable, legible, contemporaneous, original, and accurate) that are expected of paper records and must comply with all applicable statutory and regulatory requirements. Health agencies's acceptance of data from clinical trials for decision-making purpose depends on the ability to verify the quality and integrity of the data during on-site inspections and audits.

A guide called "Computerized Systems Used in Clinical Investigations" (CSUCI)[2] provides to sponsors, Contract Research Organizations (CROs), data management centers, clinical investigators, and Institutional Review Boards (IRBs), recommendations regarding the use of computerized systems to create, modify, archive, maintain, retrieve or transmit electronic records during clinical investigations (FDA 2013). The objective of the recommendations is to ensure the reliability, quality and integrity of data from electronic sources in establishing control procedures that prevent errors during data management operations.

A clear policy should be defined with regards to (1) the role of computerized systems in study protocols, (2) specific procedures and controls (Standard Operating Procedures [SOPs]), (3) source documentation and retention, (4) internal security safeguards (limited access, audit trails, etc.), (5) external security safeguards, and (6) the training of personal.

Such guidance does not establish legally enforceable responsibilities and should be viewed only as recommendations, unless specific regulatory or statutory requirements are cited.

---

[2] This guide, edited in May 2007 by the "Food and Drug Administration" (FDA) and the U.S. Department of Health and Human Services supplements the guidance for industry entitled "21 CFR Part 11" (*Electronic Records; Electronic Signatures – Scope and Application*), dated August 2003 and the Agency's international harmonization efforts *(E6 Good Clinical Practice)* when applying these guidances to source data generated at clinical study sites.

17.2.2.2   Standards in Clinical Research Informatics

In the field of clinical research international, efforts has been made to facilitate data sharing and exchange through the definition of interoperability standards consisting in standard formats for data structure and coding rules according to standard international controlled vocabularies (standard healthcare terminologies or ontologies).

The Clinical Data Interchange Standards Consortium (CDISC), created in 1997 at the initiative of the clinical research industry with the approval of the FDA, is a global, open, multidisciplinary, non-profit organization that has established standards to support the acquisition, exchange, submission and archive of clinical research data and metadata. CDISC develops and supports standard formats dedicated to biological (LAB for "Laboratory Standards") and clinical (ODM for "Operational Data Model") data. ODM defines the data structure of an electronic case report form (eCRF). CDISC also defines formats for the submission of data to regulatory health agencies (SDTM for "Study Data Tabulation Model" and ADaM for "Analysis Dataset Model") (CDISC 2013).

Recently, CDISC specified the unambiguous semantics of a number of common data elements that are deemed "common" to all trials (CDASH for "Clinical Data Acquisition Standards Harmonization" initiative). As such, CDASH represents a significant first-step in achieving cross-trial semantic interoperability. The further step is to define a global electronic library of standardized core common data element definitions and metadata that can be used in software applications in both biomedical research and healthcare domains (SHARE for "Shared Health And Clinical Research Electronic Library"). At last, CDISC recently published a standard format for protocol definition (PRM for "Protocol Representation Model"). The PRM was developed to support the generation of a protocol document, research study (clinical trial) registration and tracking as well as regulatory needs, and also to facilitate single-sourced, downstream electronic consumption of the protocol content. The global CDISC data standards are platform-independent, vendor-neutral and are freely available via the CDISC website.

17.2.3   Role of Electronic Healthcare Records in the Research Area

The modern clinical environment provides an ideal medium by which human subjects or populations can be characterized using existing or relatively low-cost data sources. Recently, a number of investigators have examined the various roles that Electronic Health Records (EHRs) – generated and maintained within an institution, such as a hospital, integrated delivery network, clinic, or physician office – or personal health records (PHRs) that the individual patient controls – might assume in the research context. Access to routinely collected clinical data is valuable for both basic research and clinical research. Therefore, data repositories for the repurposing of clinical care data for research are increasingly being developed, especially in academic medical centers (Murphy et al. 2012).

Clinical data can contribute to enrich biobanks and support the validatation of in-silico models and the systematic evaluation of data-centric patterns can inform the modeling of the etiology or course of normal and disease states (Butte 2008a).

Vast clinical data stores have also the potential to inform clinical research (Ohmann and Kuchinke 2009; Prokosch and Ganslandt 2009). Some authors have pointed out that EHR data may be useful during clinical trial design by providing trial planners with a better understanding of the available cohorts and to support them in conducting feasibility studies, refining inclusion/exclusion criteria (Dugas et al. 2010; Weng et al. 2010; Murphy et al. 2007; Kohane et al. 2012). Clinical Trial Recruitment Support Systems (CTRSS) can efficiently improve patient recruitment (Cuggia et al. 2011). Others (Kush et al. 2007; El Fadly et al. 2011) have specifically addressed the lessening of the burden of clinical trial data collection through the targeted re-purposing of EHR data during a trial's execution phase. If part of the clinical trial data entered into dedicated electronic clinical trial data management systems is specific to the research protocol, for certain types of trials, an important amount of the clinical trial data is also entered into the EHR. Whether the patient record or the clinical trial data management system is updated first depends on local workflows but usually clinical data is manually (re-)entered into the clinical trial data management system.

Not surprisingly, many healthcare professionals and researchers wonder why the different electronic sources cannot be linked. This would avoid data transcription and the resulting errors and allow busy healthcare professionals to use a single system designed to fit their patient care workflow. Today, many efforts focus on leveraging routinely collected clinical data for enhancing clinical trial data collection and also for supporting adverse event detection and reporting. Many initiatives aim at developing integrated "single-data-entry" systems for clinical research. In this context of increasing variability and complexity of the clinical data management systems, the 'source' of the data used for clinical research may no longer be obvious and verifying clinical research data becomes more difficult. At last, the integrity of the healthcare systems holding the source data may not always meet the rigorous rules for clinical research systems laid out in regulations and guidelines.

17.2.4 Translational Informatics Infrastructures

The ever-increasing computerization of both patient care and biomedical research sectors results in the availability of high-value patient-centric phenotypic data sources as well as biomolecular measurements such as "omics" (genotypic, proteomic, metabolomic, etc.) expression profiles produced by a growing suite of instrumentation platforms. The size and complexity of data sets that researchers can collect, store, retrieve and process on a regular basis are growing at an exponential rate (Butte 2008b; Chen et al. 2012; Masys et al. 2012).

At the same time, the data management and processing practices currently used in many basic science and clinical research settings rely on the use of conventional databases or individual file-based approaches that are ill-suited to enabling interaction with large-scale data sets (Kush et al. 2008; Fridsma et al. 2008).

Within the broad context of clinical and translational research, there is a reoccurring demand for tools that can reduce the data management burden and assist investigators as they seek to aggregate and reason upon multi-dimensional and multi-scale data sets. Such information needs exist at multiple points in the research lifecycle, including hypothesis discovery and study planning, as well as both retrospective and prospective data analyses.

In the context of the development of translational research, the challenge is not so much the development of information system in each area of activity (research or patient care) that the sharing of information between heterogeneous information systems that were developed in silos.

Therefore, many initiatives focus on the development, dissemination and adoption of advanced **information technology (IT) infrastructures** that allow researchers and their staff to focus on fundamental scientific problems rather than on practical informatics needs (Embi and Payne 2009; Murphy et al. 2012).

Biomedical informatics provides solutions for translational research by facilitating data and knowledge acquisition from heterogeneous sources and distributed organizations, their representation, integration and storage in a format and in a manner appropriate to the desired processing to be achieved in analysis platforms (Fig. 17.2).

17.3   Organizational, Social, Economical and Ethical Challenges of Translational Bioinformatics and Clinical Research Informatics

The activities conducted in both new domains – translational bioinformatics and clinical research informatics – are of great importance, not only to data managers and statisticians, but to individuals across the realm of biomedical research: from the biologist, the study sponsor, coordinator and investigator, federal agencies, policy makers in public health, to the clinician, and eventually and most importantly, to the patient.

The increasing development of **IT infrastructures for translational research** represents a great opportunity (Murphy et al. 2012) but at the same time many challenging issues that are not only technical (discussed in Sect. 17.4) but also organizational, regulatory and ethical need to be carefully identified and addressed in collaboration with all the stakeholders in order to meet their expectations.

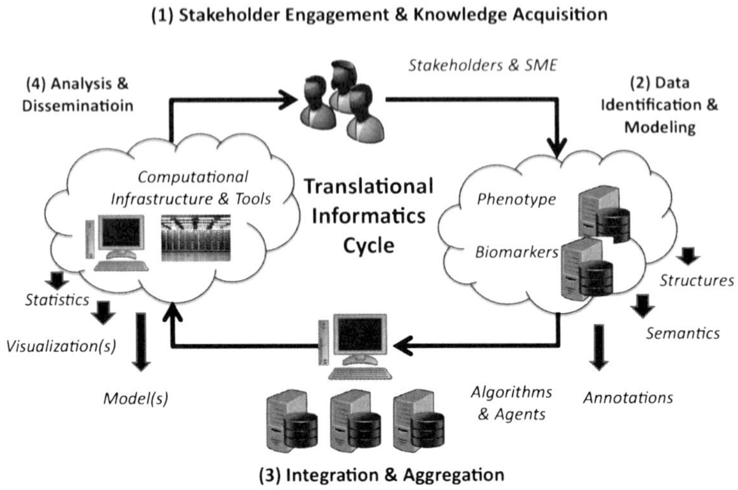

Fig. 17.2 Information cycle in translational research and major phases of the design and execution of translational informatics projects (Embi & Payne, Physiol Genomics 2009)

17.3.1 Opportunities for Researchers, Healthcare Professionals and Patients

Information technology infrastructures developed for translational research directly benefits to **data managers and statisticians** by providing them an easier access to the data and advanced services for its management and analysis.

Such infrastructures also benefit to **researchers** in general.

From a **biomedical scientist** perspective, information technology shall optimize the integration of increasingly voluminous biomedical data in order to generate new hypothesis, to design new models, new biomarkers and therapeutic targets.

From a **clinical research** perspective, in the current context of increasing time and costs of research and development, information technology shall facilitate clinical trial design (feasibility studies) as well as its execution (patient recruitment, data collection, monitoring, analysis and submission, adverse event reporting). Through international multi-centre trials involving the required number of patients, new drugs and treatments would be developed in a shorter time frame.

From a **public health** or **epidemiology** perspective, information technology shall similarly facilitate re-use of EHRs/PHRs data, enhance data collection and processing various indicators in relation to populations/groups, settings/facilities and regions/geographic units. The long-term impact would be to reconcile environmental variables to improve the understanding of health determinants.

From a **health policy maker** perspective, information technology shall support the identification and dissemination of the most efficient preventive or therapeutic strategies. The long-term impact would be to identify public health signals and emerging trends, and to manage public health issues more effectively.

From a **healthcare professional** perspective, the quick and robust translation of basic discoveries into diagnostic and therapeutic applications provides the opportunity of better management of the disease suffered by a patient, or a predisposition to this disease. Information technology shall support the clinician in choosing among the various diagnostic or therapeutic options those that are likely to give the best results according to the specific clinical, environmental and possibly "omics" profile of the patient. Information technology solutions consist in decision support systems integrating individual clinical data to different sources of up-to-date knowledge (e.g. recommendations or results of clinical trials, specific alerts or real-time access to population-level data) that are relevant in the context of the patient and his environment.

Most importantly, from a **citizen** perspective, the expected outcome of translational research is to provide an earlier access to new preventive, diagnostic, therapeutic strategies that are best suited to their personal characteristics. Information technology shall support the reporting of recent scientific discoveries to patients (development and validation of new therapies, methods for health promotion and prevention, new diagnostic tools and medical technologies) and support their involvement in biomedical research. In practice, patients shall be appropriately informed in order to have the opportunity to consent to be subject of experiments – enrolled in studies – and/or being subjects of discovery by making their clinical data and/or biological samples available to the scientific community. Information technology is an important lever towards innovative medicines and will ultimately lead to a more holistic healthcare approach through a closer coordination between care providers and citizens resulting in safer and more evidence-based diagnosis and treatment.

17.3.2 Organizational Challenges and Institutional Support

The need of increasing collaborations between research and healthcare institutions, including the re-use for research purposes of routinely collected clinical data, is supported by institutional national and international research programs (Ash and Anderson 2008).

In Europe, the objective of the health research program is to improve the health of European citizens, and increase and strengthen the competitiveness and innovative capacity of European health-related industries and businesses. One of the main focuses of European-funded health research is translational research in major diseases (brain-related diseases, infectious diseases, cancer, cardiovascular disease, diabetes/obesity, rare diseases, other chronic diseases including rheumatoid diseases, arthritis and muscoskeletal diseases).

As part of the Seventh Framework Programme (FP7) many research projects were funded in order to address key challenges of translational informatics such as semantic interoperability issues among distributed EHR/EDC systems, the development of integrated environment that enables the reuse of EHR data for clinical research such as the early detection of adverse event reactions.

Since 2007, the European Commission launched Joint Technology Initiatives that are long-term Public-Private Partnerships involving industry, the research community and public authorities proposed to pursue ambitious common research objectives at European level in areas of major interest to European industrial competitiveness and issues of high societal relevance. As part of the Joint Technology Initiatives, the Innovative Medicines Initiative Joint

Undertaking (IMI JU) addresses the bottlenecks currently limiting the efficiency, effectiveness and quality of the drug development activities needed to bring innovative medicines to the market. Some IMI calls were specifically dedicated to Knowledge Management (Drug/disease modeling: library & framework, Open pharmacological space and Electronic Health Record).

In US, Clinical and Translational Science Awards (CTSA) program, launched in 2006 by the US National Institutes of Health (NIH), aims to speed up the process of transforming laboratory discoveries into therapeutic treatments by restructuring the translational research enterprise in the US and seeks to strengthen the full spectrum of translational research. Institutional CTSA are the centerpiece of the program, providing academic homes for translational sciences and supporting research resources needed by local and national research communities to improve the quality and efficiency of all phases of translational research. CTSAs support infrastructure and resources for clinical research, including clinical trials as well as the training of clinical and translational scientists and the development of all disciplines needed for a robust workforce for translational research.

17.3.3 Regulatory and Ethical Challenges

Advanced information technologies currently developed to address the increasing need of collaborations between research and healthcare institutions shall be designed and certified to protect patient rights and interests, to avoid conflicts of interest and shall be adequaltely reported to Institutional Review Board (IRB). An important challenge is the mitigation of complex regulatory frameworks that collectively impede or prevent the secondary use of routinely collected clinical data for research purposes,

In Europe, research initiatives addressing ethical and personal data issues raised in the domain of e-health, clinical and translational research are funded as part of the *Seventh Framework Programme* (FP7). Regulatory and ethical challenges of information technologies used in the biomedical research area are specifically addressed in Chap. 11.

17.3.4 Economical and Industry Challenges

Research-based small and medium enterprises are the main economic drivers of healthcare, biotechnology and medical technologies. Strong biomedical research will enhance competitiveness of the pharmaceutical and healthcare industries. It is therefore imperative to create an environment conducive to innovation in the public and private sectors.

Information technology (IT) to support biomedical research has steadily grown over the past 10 years. Many new applications at the enterprise level are available to assist with the numerous tasks necessary in performing research. Despite this significant technological progress, important gaps currently limit the use of technology solutions, such as the regional diversity, the emergence of multiple stand-alone sponsor-hospital approaches, and the lack of a systematic and robust business model that would provide an operational framework to enable the proper deployment of innovative scalable and cost-effective services for translational research at national or international levels.

It is still not clear how rapidly this technology is being adopted and there are still few studies demonstrating whether it is making an impact upon how research is being performed. On-going projects will develop metrics to evaluate how far IT platforms may speed up translational research, qualitatively and quantitatively improve patient recruitment and data collection and finally improve the development of innovative diagnostic and therapeutic strategies. These projects do not only demonstrate, through pilots, the viability and scalability of innovative health IT platforms (systems, organizational structure, data interoperability, governance model, etc.) they also contribute to define a business model. They define how IT platforms and complementary services shall be funded and sustained, ethically, legally; how they are professionally accepted, how to ensure that value can be delivered to all stakeholders and how return on investment by using its platform could be expected.

The business model shall provide a practical and systematic framework for the deployment and consistent delivery of sustainable, innovative, high-quality and cost-effective services to patients, health care providers, the pharmaceutical industry and society. Especially in the context of re-purposing routinely collected clinical data for research clear guidance is still needed on how to establish sustainable and scalable solutions to connect EHRs providers (e.g. Hospitals), service developers (e.g. trusted third part), and service receivers (e.g. pharmaceutical industry, academic research) in an ethical, safe and efficient way. In practice, the business model shall specify in detail the product and service offering, shall define appropriate governance arrangements for the platform services and for international networks as well as operating procedures, third party service requirements and a roadmap for international adoption and for funding future developments.

The international adoption of such platforms will be promoted and strengthened by the certification of products and the accreditation of clinical trials units. A rich understanding of the economic and political eco-system in which systems certification and investments occur in the fragmented eHealth marketplace is needed to construct a realistic strategy.

17.4   Technological Challenges of Translational Bioinformatics and Clinica Research Informatics

The challenge of biomedical informatics in the context of biomedical research – and in particular of translational research – consists in developing, beyond the state of the art, the methods, tools and services allowing a better collaboration between researchers and to support the link between research and healthcare. The adoption rate for IT-supported research recently increased remarkably, specifically electronic data capture for clinical trials, data repositories for secondary use of clinical data, integrative platforms and infrastructures for supporting collaboration (Murphy et al. 2012) and brought with it a number of notable challenges such as heterogeneity of data sources, performance of distributed data manipulation engines, data quality and privacy control (Butte 2008b; Embi and Payne 2009; Payne et al. 2009).

17.4.1   Semantic Integration of Heterogeneous Data and Knowledge

Semantic interoperability is an important objective to achieve in order to support large scale multi-national research. In order to enable semantic interoperability, it is necessary to univocally identify the meaning of clinical information in the different systems and that the meaning of data imported or queried from heterogeneous sources can be interpreted by computers as well as by humans (Burgun and Bodenreider 2008).

However, although multiple initiatives have been provided by different Standards Development Organizations (SDOs), major challenges remain unsolved. Due to the still emerging adoption of health informatics standards, clinical data are, in fact, often structured and encoded differently within health information systems. Table 17.1 gives an example of different ways of recording the same clinical information. Although different, these expressions are semantically equivalent.

Even when the same syntax is used to represent the clinical information, the controlled vocabulary used to encode the different terms of the expression (semantic annotation) may be different since different coding systems may be used depending on the context. For example, "mesothelioma" can be coded R-00162 in the patient record using SNOMED Clinical Terms ® (SNOMED CT ®), C3762 using ICD-O-3, C7865 in a database of clinical research using the thesaurus NCI-T, D-008654 in bibliographic databases using the MeSH thesaurus and C0206675 in UMLS.

Several ongoing research efforts address the definition and widespread adoption of robust, scalable and sufficiently expressive research-centric data modeling and exchange standards, including reference information models, controlled terminologies and ontologies [Kush08].

Table 17.1 Five different expressions used to record the fact that "malignant mesothelialcells were found in a pleural fluid aspirate"

Attribute (question)	Value (answer)
Pleural fluid finding	Malignant mesothelial cells
Site of malignant mesothelial cells	Pleural fluid
Lab test result	Malignant mesothelial cells in pleural fluid
Type of mesothelial cells in pleural fluid	Malignant
Type of malignant cells in pleural fluid	Mesothélial

In Europe, the Network of Excellence, Semantic Health Net (http://www.semantichealthnet.eu/), develops a scalable and sustainable pan-European organisational and governance process for the semantic interoperability of clinical and biomedical knowledge. The aim is to ensure that electronic health record systems are optimised for patient care, public health and clinical research across healthcare systems and institutions. The goal of semantic interoperability is to be able to recognize and process semantically equivalent information homogeneously, even if instances are heterogeneously represented with great variety by using different combinations of information models and terminologies/ontologies.

In addition, many efforts focus on clinical and molecular knowledge representation. Gene summarization systems are developed in order to discover relationships between genes, variants, and proteins and phenotypes (diseases or drugs) using different methods such as knowledge extraction from semantic resources (biological ontologies such as Cell Ontology, Unified Medical Language System, the Mammalian Phenotype Ontology, etc.) or natural language processing of the medical literature.

At last, in the domain of clinical research, recent efforts have focused on bridging the various international efforts in both healthcare and research domains into closer alignment. In 2004, CDISC and HL7 – along with the National Cancer Institute and the FDA – began collaboration on the development of a domain analysis model which defined the static/informational semantics of "the domain of protocol-driven research involving human, animal, or device subjects as well as all relevant, associated regulatory and post-marketing data". The effort has produced the Biomedical Research Integrated Domain Group (BRIDG) model which, on one side, contains representations of clinical research data with underlying mappings to the HL7 RIM and, on the other side, covers a superset of the scope defined by CDASH (Fridsma et al. 2008). The most recent version of the BRIDG model also contains – in addition to its original UML representation – a semantically equivalent representation using the Web Ontology Language (OWL). CDISC has committed to migrating all of its standards to be expressed using BRIDG semantics, and the HL7 Regulated Clinical Research Information Management (RCRIM) Work Group within HL7 is committed to developing all of its message specifications in the context of BRIDG compliance.

In 2007, the Joint Initiative Council was formed as a partnership between HL7, CDISC, ISO TC 215, IHTSDO, and CEN TC 251 with the stated goal of increasing collaboration between standards organizations based on the recognition of a common goal of computable semantic interoperability.

In 2011, ISO defined a set of data type definitions (ISO 21090:2011) for representing and exchanging basic concepts that are commonly encountered in healthcare environments in support of information exchange in both the clinical care and clinical research contexts.

Recent efforts in both patient care and clinical research consist in defining metadata and vocabulary standards for clinical information and thereby in building **Common Data Elements (CDEs,** also called metadata repositories or item banks) (Nadkarni and Brandt 2006). The CDEs are structured data elements, consisting of precisely defined questions and answers (e.g. NCI's caDSR (Covitz et al. 2003; Warzel et al. 2003), caMatch approach in the ASPIRE project, CDISC SHARE CDISC 2013).

17.4.2   Data Integration

The trend to integrate multiple clinical resources arises from the necessity to assemble data sets describing ever-larger patient cohorts in modern biomedical studies. In practice, client software installed in healthcare centres can connect to both local and remote resources to select medical records of interest for biomedical research (cohorts, clinical trials, biobanks, etc.) and researchers, with authorized access, can extract clinical data to continue investigation.

Different methodologies or generic tools are currently being developed to set up clinical datawarehouses or to achieve heterogeneous data stores mediation and federation.

17.4.2.1   Clinical Data Warehouses

Biomolecular databases are becoming increasingly large, complex and interconnected. This increase of data scale, complexity and the need of interoperability means that there are many fundamental challenges in the database development, deployment and distribution.

For decision support in health care and biomedical research especially with a translational approach, data warehousing is a solution for the integration of multiple data sources and the proper use of data mining strategies. Data warehousing can support the re-purposing of routinely collected clinical data in the context of biomedical research and therefore avoid or limit duplications and errors and enhance the quality and avaibility of clinical data for basic research and clinical research.

We can define a data warehouse as a collection of structured data, supplied by operational databases of electronic healthcare records, and oriented to data mining for decision support. Clinical data warehouses are key components of an infrastructure for translational research.

The operational databases within health institutions are traditionally built to enable the processing of personal data, but insufficient to perform transversal queries (e.g. number of patients aged over 45, diabetics with a glycosylated hemoglobin greater than 6 %) because of a non-integrated structuration and of an inappropriate exploitation strategy. A clinical data warehouse is specifically developed to perform transversal queries and is a preliminary step highly recommended (although sometimes unrealized) for the use of data mining. In clinical datawarehouses, data are typically ready to use (full quality), available (wait time, navigation), clearly identified and defined (consistency), detailed rather than processed. They must be modified without disturbances (continuity and consistency access).

Deploying a clinical data warehouse in a healthcare organization is long and methodical process, requires a solid sponsoring from the decision-makers, the motivation of staff of various professions, a strong and durable cooperation between information departments and operational services, a strong interest coming from the operational services for the value of information and its analysis. The data shall be available and business rules shall be precisely defined.

17.4.2.2   Data Federation and Mediation

Alternatively to clinical data warehouses, where clinical data are centrally loaded and stored, data federation solutions mediate legacy heterogeneous distributed sources and enable cross-federation querying. Current iniatives explore the ability to transform relational data stores into semantic repositories to enable semantic querying (using SPARQL).

A key challenge is to adopt a systematic methodology to develop ontological models for clinical data, leveraging existing information model standards (e.g. 13606, openEHR, HL7 and CDISC) and biomedical terminologies and ontologies in order to implement semantic interoperability between heterogeneous databases. Models built on repositories of Common Data Elements can enable local repositories and software to interface faster with larger global infrastructure such as the National Cancer Institute's cancer Biomedical Informatics Grid (caBIG). Ontologies and Semantic Web technologies are more and more developed and used in Translational Bioinformatics and Clinical Research Informatics. Designed using Semantic Web technologies, prototype systems currently integrate in real time, structured and unstructured data from clinical information systems across different hospitals and enable the seamless integration of both relational and semantic data sources. Prototypes built with ontology-based models (e.g. of anatomy) can reason about normal and diseased body structures. Other prototypes perform text-parsing on clinical documents and compare these to local or global profiles (e.g. for biosurveillance, pharmacovigilance, patient selection, etc.).

17.4.2.3 Standard-Based Integration Profiles

The collective efforts of the multiple organizations currently focused on defining the various standards required to achieve computable semantic interoperability. However, given the reality that achieving broad-based, scalable computable semantic interoperability across multiple domains will require the integration of multiple standards, it became apparent that an "integration organization" involving multiple stakeholders (including both vendor and provider organizations) could serve a valuable role by defining – based on stakeholder dialogue – "real-world usage scenarios" – that, in turn, could then be instantiated using existing standards. The organization **Integrating the Healthcare Enterprise (IHE** – http://www.ihe.net) has, in fact, emerged as that organization. The "real-world usage scenarios" that are published by IHE are referred as Integration Profiles. Each profile defines a series of "transactions" which specify how existing standards should be applied to meet the overarching business goal described by the profile.

Unlike preliminary pilot studies demonstrating the use of EHRs for clinical research that were one-time-only solutions limited to a single vendor-single pharmaceutical company context, and not using standards for data representation, IHE defines standard-based integration profiles dedicated to research.

Two IHE profiles dedicated to research and public health – as proposed by both the IHE Quality Research and Public Health (QRPH) and Information Technology Infrastructure (ITI) domains – address the issue of multi-vendor, scalable interoperability required for multicenter trials (IHE QRPH 2013).

The **Retrieve Form for Data-capture (RFD)** integration profile combined with the Clinical Research Document (CRD) content profile collectively provide a conceptual framework for implementing the "single-point-of-data-collection" approach to EHR/CDMS integration. Specifically, the case report form (eCRF) is centrally implemented in the CDM system of the clinical research organization acting as a "Form Manager." The EHR system, acting as "Form Filler," retrieves the eCRF. Source data are entered in this interface to the CDM displayed in the EHR. Clinical trial data are stored in the CDMS acting as "Form Receiver".

Alternatively, the **Clinical Research Document (CRD)** content profile describes how EHRs (regardless of the underlying EHR semantics) can through generation of a Continuity of Care Document (CCD) provide clinical data for pre-populating eCRFS. The rationale of the IHE the Clinical Research Document (CRD) content profile is based on the mapping between CDISC CDASH data elements – a minimal core dataset that is valid across all regulated clinical trials – and the data elements of the Continuity of Care Document (CCD) templates.

17.4.3 Data Mining

Data mining (machine learning, intelligent data analysis) was defined in 2001 by the Massachusetts Institute of Technology, as one of the ten emerging technologies that "will change the world" during the twenty-first century.

This is a non-elementary process, applying to a large dataset, to help the decision making by the use of appropriate analytical methods of varying complexity (clustering, classificataion, association discovery, causality analysis, etc.).

Data mining allows the discovery of knowledge and the structuring of this knowledge. In clinical research informatics, data mining is widely adopted in all aspects ranging from bioinformatics study (e.g. omics data analysis in biomarker discovery) to public health (e.g. epidimilogy study). Data mining technology is also combined with the application of vasualisation technology where large scale complex data can be displayed with graphic representations where the structure and relationships among the data can be revealed intuitively.

17.4.4 Ensuring Data Quality

The use of clinical data from electronic healthcare records, which have not been initially created in a research perspective, represents specific problems. For example, the coding quality performed in a medico-economic aim can be unreliable when the data is used in research many paper states about validated methods for identifying special diagnosis, conditions and adverse events using administrative and claims data (Ramsey et al. 2009; Zhan et al. 2009). Therefore, it remains challenging to rely on qualitative data and is necessary to assess data quality within electronic healthcare records prior to its storage and use within datawarehouses or IT platforms dedicated to translational research. At the beginning of the 1990s, computer scientists considered the issue of defining, measuring and improving data quality. Wang introduced a Total Data Quality Management (TDQM) methodology based on Deming wheel that defines a quality improvement process (Wang 1998). Likewise, the data warehouse communities have introduced various methods to measure, enhance and monitor the quality of data in CDWs into six quality dimensions: accuracy, timeliness, comparability, usability, relevance and privacy & security. The healthcare domain has other resources built over years that could help in data quality assessment, namely, the standardized information models and domain terminologies or on-tologies (Brown and Sönksen 2000).

17.4.5 Implementing Solutions for Maintaining Data Security and Patient Privacy

Due to the sensitive nature of healthcare data, an additional key issue to be addressed is to ensure data confidentiality and security and to elaborate data protection mechanisms.

The re-use of patient data from electronic healthcare record systems can provide tremendous benefits for clinical research, but measures to protect patient privacy while utilizing these records have many challenges. Some of these challenges arise from the fact that the privacy of the patient should be considered on three fronts: technical de-identification of the data, trust in the researcher and the research, and the security of the underlying technical platforms (Murphy et al. 2011). Recent publications focus on technical issues (anonymisation tool, cryptographic solutions, etc.) (Aryanto et al. 2012; Canim et al. 2012) as well as on more general issues about for example intellectual property management for data sharing and the impact of development of population-based study in the biomedical field on laws and regulations (Te et al. 2012). These issues are more specifically addressed in Chap. 11.

17.5 Information Technology (IT) Platforms for Translational Research

17.5.1 Translational Bioinformatics

Advances in knowledge of the human genome have brought biomedical informatics in an era of linking fundamental biological discoveries with medical knowledge and patient data.

In this context, a comprehensive approach is needed to develop information systems and methods that incorporate genomic, phenotypic and environmental knowledge.

Ultimately, these efforts will contribute to personalized medicine that is the diagnosis, and treatment proposals from specialized clinical, pathological, but also genomic, transcriptomic, proteomic and metabolomic on the patient.

Pharmacogenomics example takes into account genetic factors that determine how a person will respond to a drug and seek to adapt the treatment to the patient. This means being able to adapt a generic drug metabolism based on such patient data, with the long-term perspective to relate this model to a model of operation or malfunction of the organ studied.

The concepts of translational research and personalized medicine are developing around the center point of the analysis from genotype to phenotype (G2P) and analysis from phenotype to genotype (P2G).

This work mobilizes mathematical and computational methods and algorithms for sequence analysis (e.g., prediction of function or location); structural analysis (detection, prediction and structure comparison), analysis of data expression (e.g., microarrays, regulatory networks, annotation), systems biology (systems modeling cellular signaling cascades, interaction networks, regulatory networks, metabolic pathways); methods of data mining and text data on patients, scientific articles, and biological data and databases and ontologies in the medical and biological field.

This requires a major effort of integration and therefore a suitable infrastructure.

17.5.1.1   ELIXIR (Europe)

At European level, the FP7 project ELIXIR (European Life Sciences Infrastructure for Biological Information) was tasked to create an infrastructure to support European-wide research in life sciences, and its translation to the health and environmental infrastructure that provides a central hub that is at the EMBL-EBI, and ensures the maintenance, dissemination and expansion of databases essential to biomedical research.

Also ensuring the link between biological resources and medical infrastructure ELIXIR will facilitate the understanding of disease mechanisms, and support strategies for prevention, diagnosis and treatment for the populations concerned.

17.5.1.2   TransMART (US, Europe)

The project tranSMARTt (Translational Medicine Mart), aims to provide an infrastructure for translational data warehouse, incorporating both patient data from clinical trials, aspects of "omics" and concerns of clinical research.

Developed by J & J and Recombinant since June 2009, tranSMART aims to help investigators to identify therapeutic targets, to operate a transversal data "omic" and clinical data to discover new diagnostic biomarkers or prognostic especially in complex areas such as immunology and oncology.

Combining the technologies used in the "omics" methods of decision support developed in the field of medical informatics, Transmart performs data integration and knowledge as a warehouse and offers specific bases "tailored" to meet specific scientific issues (marts).

17.5.1.3   eTRIKS (Europe)

eTRIKS (European Translational Information & Knowledge Management Services) (http://www.etriks.org/) is an EU Innovative Medicines Initiative (IMI) project aiming to provide a cloud based knowledge management (KM) platform and service infrastructure capable of the efficient storage and effective analysis of experimental data from studies in man, in animals and in pre-clinical models.

The main eTRIKS objectives are:

- **Service**: Deploy and host the eTRIKS platform based on the tranSMART technology and provide training, support and consultation activities to all IMI project partners on using the platform.
- **Platform**: Develop and maintain a sustainable, interoperable, collaborative, re-usable, open source and scalable translational research (TR) KM platform, as well as conduct research & development into effective analytics methods and tools to support TR and computational molecular biology research in general, thus evolving and extending the platform with tools for omics, imaging data and text analysis that can leverage cloud-based operations for system biology research.
- **Content**: Establish eTRIKS as a unique European TR data resource supporting cross-organisation TR studies, including clinical studies and pre-clinical studies, omics data analysis for biomarker discovery and validation, genetics and NGS studies and populate eTRIKS with existing and active data from TR studies and supporting the integration of standardised legacy TR study data.
- **Community**: Promote and lead an active international TR analytics & informatics community, centred around eTRIKS, through active stakeholder engagement and by disseminating tools and expertise worldwide and engage in, and influence, international standardisation activities in areas relating to TR informatics.

17.5.2   Clinical Research Informatics

17.5.2.1   Feasibility Studies, Patient Recruitment and Protocol Execution

Many on going projects develop IT platforms dedicated to Clinical Research to enhance the process of feasibility studies, patient recruitment as well as clinical trial data collection through the targeted re-purposing of EHR data during a trial's execution phase.

Retrospective analysis of a large amount of clinical data in EHRs/PHRs or clinical data warehouses can contribute to the design of a new biomedical study by generating new research hypotheses. In addition, access to clinical data of patients multiple health facilities allows developers to test patient enrollment trends and anticipate recruitment problems even before the design of the study protocol is finalized. During feasibility studies, information technology is used to distribute formalized queries based on eligibility criteria over heterogeneous EHRs/PHRs or clinical data warehouses and to determine the sensitivity or effectiveness of particular inclusion/exclusion (I/E) criteria to ensure that when the protocol is actually published, a suitable subject base will be able to be identified.

In practice, the variation in the number of patients meeting the criteria is based on the change of threshold (e.g. number of diabetic men older than 45 years, 50 years or 55 years).

Another challenge addressed by ongoing projects is to develop efficient Clinical Trial Recruitment Support Systems (CTRSS) in order to increase the inclusion rate in clinical trials and to support investigator in recruiting patient in time and especially special populations (women, children, minorities) that are usually under-represented in trials. Patient recruitment is a complex process, and a successful strategy requires several complementary approaches to improve participation rates. CTRSS are implemented according to two possible approaches: patient-centered approach and the sponsor/investigator-centric approach. In the patient-centered approach individual patients (or their providers) may submit individual demographic, phenotypic and/or genotypic data against a repository of protocols, the specific I/E criteria of which are available in a computable form, to determine which protocols the patient could be eligible for and should choose to participate. In the protocol-centric approach, for a specific protocol, the computable protocol's I/E criteria are compared to patient demographic, phenotypic and/or genotypic characteristics of patients in order to identify from a repository a cohort of potential trial subjects. At last, some projects have also specifically addressed the re-purposing of EHR data during a trial's execution phase and developed mechanisms to prepopulate electronic case report forms (eCRFs).

- I2b2/SHRINE

    The i2b2 (Informatics for Integrating Biology and the Bedside) project (https://www.i2b2.org/) was launched by the NIH-funded National Center for Biomedical Computing based at Partners HealthCare System and is coordinated by Isaac Kohane, a pediatrician and bioinformatics researcher who co-directs the Harvard Medical School Center for Biomedical Informatics in Boston (Fig. 17.3). The i2b2 Center is developing a scalable informatics framework that enables clinical researchers to use existing clinical data for discovery research and, when combined with IRB-approved genomic data, facilitates the design of targeted therapies for individual patients with diseases having genetic origins.

    The I2b2 open-source platform enjoys wide international adoption by the CTSA network, by more than 40 academic health centers, and industry.

    Data on millions of patients at hospitals and clinics – collected with consent, and used in anonymous databases – can help researchers spot adverse drug effects and find people to enroll in clinical trials, among other uses. Search criteria can include patient demographics, ICD coded diagnoses, coded procedures, laboratory test results, inpatient pharmacy orders and text within clinical documents, including surgical pathology reports and radiology reports. In response to a query, i2b2 returns the approximate number of patients matching the search criteria. No patient identifiers or clinical data is revealed. Researchers can store their cohort searches online and later discuss after IRB approval how they might review and extract detailed clinical data on patient cohorts for research purposes.

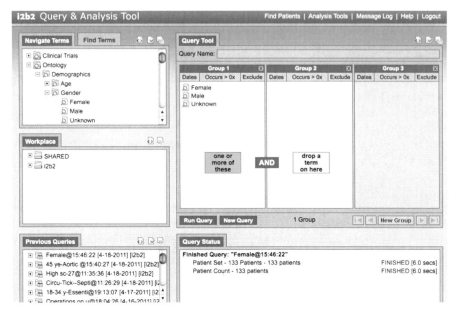

Fig. 17.3 I2b2 workbench

The Shared Health Research Information Network (SHRINE) is a federated query tool for clinical data repositories that can simultaneously send queries to each hospital and display aggregate counts of the number of matching patients. SHRINE had been successfully deployed in a prototype form across four partner hospitals within the Partners HealthCare environment and was designed to be compliant with ethical guidelines, and protective of the interests of the participating hospitals (Weber et al. 2009).

The information can be shared across systems by authorized individuals using peer-to-peer networks. The patient data gathered from two different institutions within the Boston-area CTSA revealed associations between heart attacks and the drugs Vioxx (Brownstein et al. 2007) and Avandia (Brownstein 2010) – links that had previously only emerged after years of expensive clinical studies.

- FARSITE (UK)

  FARSITE (Feasibility Assessment and Recruitment System for Improving Trial Efficiency), developed in UK, aims at querying distributed electronic health records in order to identify patients who match to eligibility criteria and to speed up enrollment of research subjects (Fig. 17.4).

- TRANSFoRm (Europe)

  The FP7 project EU-ADR (http://www.transformproject.eu) consists of 22 public and private partners coordinated by the King's College of London.

  TRANSFoRm will develop rigorous, generic methods for the integration of Primary Care clinical and research activities, to support patient safety and clinical research via:

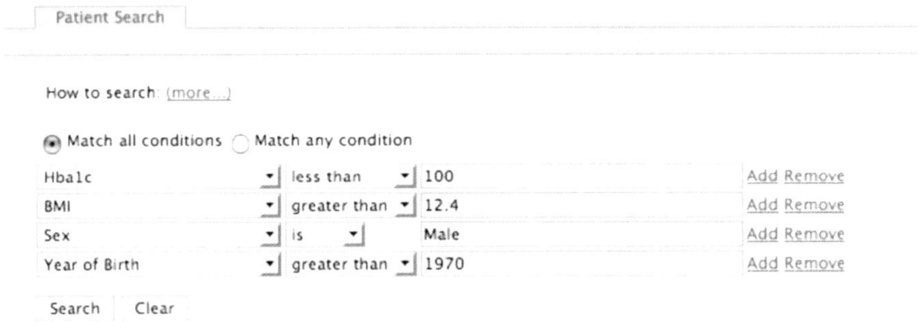

Fig. 17.4 FARSITE workbench

- Rich capture of clinical data, including symptoms and signs rather than just a single diagnosis. A generic, dynamic interface, integrated with electronic health records (eHR), will facilitate both diagnostic decision support and identification of patients eligible for research, thus enhancing patient safety.
- Distributed interoperability of eHR data and other data sources that maintain provenance, confidentiality and security. This will enable large-scale phenotype-genotype association studies and follow up of trials.
- Software tools and services to enable use of controlled vocabulary and standardised data elements in clinical research.
- The concept is demonstrated through three use cases covering the requirements of phenotype-genotype studies, randomized controlled trials and knowledge translation in the form of decision support for diagnosis.

- EHR4CR (Europe)
  The IMI project EHR4CR (http://www.ehr4cr.eu) consists of 32 public and private partners coordinated by Astra Zeneca. EHR4CR aims at providing adaptable, reusable and scalable solutions (tools and services) for reusing data from hospital Electronic Health Record systems for Clinical Research. The project is building a platform (systems, organizational structure, data interoperability, governance model, etc.) to demonstrate the viability and scalability of the EHR4CR business model through pilots. The EHR4CR platform aims to implement four use cases from clinical research lifecycle – clinical protocol feasibility, patient identification and recruitment, clinical trial execution and adverse event reporting – to be demonstrated by 10 pilots throughout five European countries. Implementing the four use cases requires the presence of protocols encoded with protocol-specific metadata that sufficiently describe the protocol's I/E criteria and eCRF content; the existence of patient repositories with sufficient amounts of appropriately encoded demographic, phenotypic and/or genotypic data and the development of a semantic mediation between clinical constraints expressed in protocols and patient data routinely collected in heterogeneous clinical systems to enable automated comparisons of patient data with protocol metadata.

- EMIF (Europe)

  The IMI project EMIF consists of more than 50 public and private partners. EMIF plans to address the logistical challenges of developing a sustainable and scalable information framework which has the potential to access data on a scale and at a level of detail not currently available. The current project focuses on two such research questions to provide focus and some guidance for the framework development: The determination of protective and precipitating factors for conversion from pre-dementia cognitive dysfunction to dementia syndromes as well as conversion from prodromal Alzheimer's disease to typical or atypical Alzheimer's disease.

  The identification and validation of markers that predict such conversion in order to facilitate the development of novel disease modification therapies and lead the way towards stratified selection of individuals for clinical trials.

  The discovery of predictors of the metabolic complications of adult and paediatric obesity, which shall lead to innovative diagnostic tests, pave the way to novel therapeutics targeted to high-risk individuals, and provide the infrastructure to select individuals for such targeted pharmacological interventions.

- EURECA (Europe)

  The FP7 project EURECA (http://eurecaproject.eu) consists of 18 public and private partners coordinated by Philips. EURECA aims at providing semantic interoperability among EHR and clinical trial systems in order to enable secondary use of care data for research, efficient recruitment by matching relevant patient data with eligibility criteria from clinical trials and long term follow-up of patients beyond the end of a clinical trial, for better research and improved safety. The EURECA platform will expose a uniform representation of clinical trial information, validated clinical trial results and other relevant external knowledge and data resources and allow extracting relevant clinical information from the EHR and contextualizing it to the patient case.

- ResearchMatch (US)

  ResearchMatch (https://www.researchmatch.org/), developed by Paul Harris and colleagues at Vanderbilt University, is a nationwide registry of volunteers willing to participate in research studies. Many volunteers have already been directly connected through the system with researchers from CTSA institutions. The registry promises to speed up enrollment of research subjects, a lengthy process that has gummed up clinical trials nationwide.

17.5.2.2 Patient Safety

Clinical trials cannot possibly ensure that a drug will not have unforeseen side effects once it arrives on the market. Post-marketing drug safety monitoring aims to address this problem, but current post market safety studies largely depend on the submission of spontaneous case reports where underreporting. Current ongoing projects propose proactive approaches where patient populations are actively monitored for adverse drug events (ADE). Effective integration and analysis of EHRs/PHRs can help to improve post-market safety activities. Multiple EHRs/PHRs or clinical dataware houses are screened for signal detection. In practice, computable ADEs are compared to patient demographic, phenotypic and/or genotypic characteristics of patients in order to identify ADEs. In addition, mechanisms are implemented in order to simplify ADE reporting. In practice, targeted EHR data are re-purposed and used to prepopulate ADE reporting forms.

- PROTECT (Europe)

  The FP7 project PROTECT (http://www.imi-protect.eu) consists of 33 public and private partners coordinated by the European Medicines Agency.

  The goal of PROTECT is to strengthen the monitoring of the benefit-risk of medicines in Europe. This is achieved by developing a set of innovative tools and methods that will enhance the early detection and assessment of adverse drug reactions from different data sources, and enable the integration and presentation of data on benefits and risks. These methods are tested in real-life situations in order to provide all stakeholders (patients, prescribers, public health authorities, regulators and pharmaceutical companies) with accurate and useful information supporting risk management and continuous benefit-risk assessment.

  A methodological framework for pharmacoepidemiological studies is being developed and tested to enable data mining, signal detection and evaluation in various types of datasets, including data of spontaneous reports, registries and other electronic databases. Means of combining results from clinical trials, spontaneous reporting and observational data will be developed, comparing Bayesian modelling, multi-criteria decision analysis and other analytical methods. Methods for graphical expression of benefit-risk will be tested with different stakeholders. PROTECT trials direct patient data collection in natural languages using web-based, telephone and text messaging systems. It will test the transferability of the data into a common language and explore linkages to data from electronic health records and registries.

- EU-ADR (Europe)

  The FP7 project EU-ADR (http://www.alert-project.org) consists of 18 public and private partners coordinated by the Erasmus University Medical Center. The EU-ADR project aims to develop an innovative computerized system to detect adverse drug reactions (ADRs), supplementing spontaneous reporting systems. To achieve this objective, EU-ADR exploits clinical data from electronic healthcare records (EHRs) of over 30 million patients from several European countries (The Netherlands, Denmark, United Kingdom and Italy). In this project a variety of text mining, epidemiological and other computational techniques are used to analyze the EHRs in order to detect 'signals' (combinations of drugs and suspected adverse events that warrant further investigation). EU-ADR aims to demonstrate that an earlier detection of adverse side effects of drugs is possible by using modern biomedical informatics technologies to efficiently exploit both the massive amounts of available EHRs, and the ever-increasing biological and molecular knowledge.

- SALUS (Europe)

  The FP7 project SALUS (www.salusproject.eu) consists of nine public and private partners. SALUS aims at providing a standard-based interoperability framework that will enable execution of a variety of different intelligent data analysis methods for mining and analyzing real-time patient data in communication with disparate heterogeneous EHR systems. SALUS will provide functional interoperability enabling exchange of EHRs, semantic interoperability solutions enabling meaningful interpretation of the exchanged EHR data and security and privacy mechanisms ensuring EHRs elements are shared in an ethical and safe way.

  Functional interoperability profiles are based on existing IHE profiles (by extending them when necessary) and support open source toolsets enabling EHR systems and clinical research systems to communicate and exchange EHR data for post market safety studies through a common transport protocol; Semantic interoperability solutions are based on the definition of a core common data element set represented as an ontology (based on the existing and evolving standards including CDISC CDASH/ODM, CDISC SHARE, BRIDG Domain model, available HL7 CDA templates, CEN EN 13606 archetypes and OMOP CDM). This core ontology consisting of common data element (CDE) set will be established through a systematic approach by examining the data requirements of the selected SALUS use cases and harmonization of the selected CDEs will be supported through a toolset in conformance to ISO/IEC 11179 standard for metadata registries.

- DiXa (Europe)

  The FP7 project DiXa (http://www.dixa-fp7.eu) consists of seven partners. The general objective of the diXa project is to further develop and adopt a robust and sustainable service infrastructure (e.g. data infrastructure and e-science environment) for harbouring multiplexed data sets as produced by past, current and future EU research projects on developing non-animal tests for predicting chemical safety, in linkage with other globally available chemical/toxicological data bases and data bases on molecular data of human disease.

- Linked2Safety (Europe)

  The FP7 project Linked2Safety (http://www.linked2safety-project.eu/) consists of 12 public and private partners coordinated by INTRASOFT International S.A. The Linked2Safety project will develop a reference architecture, data protection framework, common EHR schema, lightweight semantic model and integrated platform to provide healthcare professionals, clinical researchers and pharmaceutical companies experts with a user-friendly, sophisticated, collaborative decision-making environment for:

  - analyzing all the available data including the genetic, environmental and medical history of subjects that exhibit adverse events occurring in the

frame of clinical trials, based on the clinical care information existing in the specific patients' EHR, leading to the identification of the phenotype and genotype factors that are associated with specific adverse events and thus having direct impact on the patient safety through the early detection of potential patient safety issues.
– wide identification and selection of patients for clinical trials, through the seamless and standardized linking with heterogeneous EHR repositories, providing advice on the best design of clinical studies.

17.6 Conclusion

Translational research aims to facilitate personalized medicine – that is to say, the discovery and the use of the right treatment for the right person (appropriate treatment based on genomic analysis, family history, consideration of environment and eco system).

A major challenge is to promote interdisciplinarity and consequently to associate closely activities of care and research.

The Biomedical Informatics provides solutions for translational research by facilitating the acquisition and the structuring of data, information or knowledge

The analysis of data quality, their integration, their representation and storage in an appropriate format and modalities suitable for treatment is a major challenge for the future.

The adoption of IT platforms for translational research increased over past years. The availability of more robust and available vendor-based and "open source" solutions, coupled with new research initiatives (e.g., CTSA in US, FP7 and IMI programmes in Europe) and regulatory requirements, appear to be contributing to these advances.

In the context of biomedical research, EHRs and PHRs provide a rich repository of medical that has the potential to improve patient recruitment, conduct feasibility studies, refine inclusion/exclusion criteria, enhance safety data and, in general, to inform basic and clinical research. Key challenges in mining EHR revolve around a few key issues, including data privacy and security, and the wide variety of data structures, both standard and non-standard, which can create inconsistencies. There are now many tools being developed and important industry initiatives are under way.

This chapter described the challenges to be addressed in the context of information-intensive, multidisciplinary team-based clinical and translational research projects. There is still a need for a framework that can assist in: (1) identifying major categories of information to be collected, managed, and disseminated throughout the clinical or translational research process and the ways in which they relate to one another, thus enabling the development of integrative platforms capable of addressing such needs in a systematic manner; (2) providing individual researchers with the ability to understand how their activities contribute to a broader goal of generating new knowledge in exchanging or disseminating information in an easily and readily consumable manner; and (3) supporting the modeling and development of technology and socio-technical approaches across patient care and research "silos".

17.7 For More Information

We recommend the reader to consult systematically the websites mentioned in this chapter. The reader should update his/her knowledge about translational bioinformatics and clinical research informatics using the following expressions to query Medline.

- **Translational bioinformatics**: ("medical informatics"[MAJR] OR "medical informatics"[TIAB] OR "clinical informatics"[TIAB] OR "medical computer science"[TIAB] OR "medical computer sciences"[TIAB] OR "medical information science"[TIAB] OR "medical information sciences"[TIAB] OR "Informatics"[MAJR] OR "Informatics"[TIAB]) AND "Genetic Research"[MAJR] OR "Genomics"[MAJR]
- **Clinical research informatics**: ("medical informatics"[MAJR] OR "medical informatics"[TIAB] OR "clinical informatics"[TIAB] OR "medical computer science"[TIAB] OR "medical computer sciences"[TIAB] OR "medical information science"[TIAB] OR "medical information sciences"[TIAB] OR "Informatics"[MAJR] OR "Informatics"[TIAB]) AND ("Biomedical Research"[MAJR] OR "Epidemiologic Study Characteristics as Topic"[MAJR] OR "Evaluation Studies as Topic"[MAJR] OR "Patient Selection"[MAJR] OR "Patient Selection"[TIAB] OR "clinical research"[TIAB] OR "biomedical research"[TIAB] OR "medical research"[TIAB]) NOT "Genetic Research"[MAJR] NOT "Genomics"[MAJR]

Exercises

Q1 A clinical data warehouse can:
What are the correct proposals:

(A). Prescribe medication
(B). Make decision support
(C). Carry out datamining
(D). Conduct feasibility studies for clinical trials
(E). Recruit patients for clinical trials

Exercises (continued)

Q2 Possible uses of Clinical Information Systems in the context of biomedical research:
What are the correct proposals:

(A). Formulating the research hypothesis
(B). Reporting of adverse events
(C). Integration of study results to clinical practice
(D). Recruitment and inclusion
(E). Submission of study findings

Q3 Biomedical informatics provides solutions for translational research:
What are the correct proposals:

(A). Allowing the integration of heterogeneous data
(B). Analyzing the data quality
(C). Solution for storing data
(D). Offering analysis tools
(E). No proposal exact

Q4 The challenges of biomedical informatics research and public health:
What are the correct proposals:

(A). Ethical
(B). Policies
(C). Regulatory
(D). Organizational
(E). Technical

R1 : B C D E
R2 : A B C D E
R3 : A B C D
R4 : A B C D E

References

Aryanto KYE, Broekema A, Oudkerk M, Van Ooijen PMA (2012) Implementation of an anonymisation tool for clinical trials using a clinical trial processor integrated with an existing trial patient data information system. Eur Radiol 22:144–151

Ash JS, Anderson NR, Tarczy-Hornoch P (2008) People and organizational issues in research systems implementation. J Am Med Inform Assoc 15:283–289

Ashburner M, Ball CA, Blake JA et al (2000) Gene ontology: tool for the unification of biology. The Gene Ontology Consortium. Nat Genet 25:25–29

Bernstam EV, Hersh WR, Johnson SB et al (2009) Synergies and distinctions between computational disciplines in biomedical research: perspective from the Clinical and Translational Science Award programs. Acad Med 84:964–970

Brown PJ, Sönksen P (2000) Evaluation of the quality of information retrieval of clinical findings from a computerized patient database using a semantic terminological model. J Am Med Inform Assoc 7:392–403

Brownstein JS, Sordo M, Kohane IS, Mandl KD (2007) The tell-tale heart: population-based surveillance reveals an association of rofecoxib and celecoxib with myocardial infarction. PLoS One 2(9):e840

Burgun A, Bodenreider O (2008) Accessing and integrating data and knowledge for biomedical research. Yearb Med Inform 91–101

Butte AJ (2008a) Medicine. The ultimate model organism. Science 320:325–327

Butte AJ (2008b) Translational bioinformatics: coming of age. J Am Med Inform Assoc 15:709–714

Canim M, Kantarcioglu M, Malin B (2012) Secure management of biomedical data with cryptographic hardware. IEEE Trans Inf Technol Biomed 16:166–175

CDISC (2013) http://www.cdisc.org/. Accessed 5 Oct 2013

Chen R, Mias GI, Li-Pook-Than J et al (2012) Personal omics profiling reveals dynamic molecular and medical phenotypes. Cell 148:1293–1307

Covitz PA, Hartel F, Schaefer C et al (2003) caCORE: a common infrastructure for cancer informatics. Bioinformatics 19:2404–2412

Cuggia M, Besana P, Glasspool D (2011) Comparing semi-automatic systems for recruitment of patients to clinical trials. Int J Med Inform 80:371–388

Dugas M, Lange M, Müller-Tidow C et al (2010) Routine data from hospital information systems can support patient recruitment for clinical studies. Clin Trials 7:183–189

El Fadly A, Rance B, Lucas N et al (2011) Integrating clinical research with the Healthcare Enterprise: from the RE-USE project to the EHR4CR platform. J Biomed Inform 44(Suppl 1): S94–S102

Embi PJ, Payne PRO (2009) Clinical research informatics: challenges, opportunities and definition for an emerging domain. J Am Med Inform Assoc 16:316–327

FDA (2013) Computerized Systems Used in Clinical Investigations. http://www.fda.gov/OHRMS/ DOCKETS/98fr/04d-0440-gdl0002.pdf. Accessed 5 Oct 2013

Fridsma DB, Evans J, Hastak S, Mead CN (2008) The BRIDG project: a technical report. J Am Med Inform Assoc 15:130–137

Gaughan A (2006) Bridging the divide: the need for translational informatics. Pharmacogenomics 7:117–122

IHE QRPH (2013) IHE Quality Research and Public Health. http://www.ihe.net/qrph/index.cfm. Accessed 5 Oct 2013

Kohane IS, Churchill SE, Murphy SN (2012) A translational engine at the national scale: informatics for integrating biology and the bedside. J Am Med Inform Assoc 19:181–185

Kush R, Alschuler L, Ruggeri R et al (2007) Implementing single source: the STARBRITE proof-of-concept study. J Am Med Inform Assoc 14:662–673

Kush RD, Helton E, Rockhold FW, Hardison CD (2008) Electronic health records, medical research, and the Tower of Babel. N Engl J Med 358:1738–1740

Masys DR, Jarvik GP, Abernethy NF et al (2012) Technical desiderata for the integration of genomic data into electronic health records. J Biomed Inform 45:419–422

Murphy SN, Mendis M, Hackett K, et al (2007) Architecture of the open-source clinical research chart from informatics for integrating biology and the bedside. AMIA annual symposium proceedings, pp 548–552

Murphy SN, Gainer V, Mendis M et al (2011) Strategies for maintaining patient privacy in i2b2. J Am Med Inform Assoc 18(Suppl 1):i103–i108

Murphy SN, Dubey A, Embi PJ et al (2012) Current state of information technologies for the clinical research enterprise across academic medical centers. Clin Transl Sci 5:281–284

Nadkarni PM, Brandt CA (2006) The common data elements for cancer research: remarks on functions and structure. Methods Inf Med 45:594–601

Ohmann C, Kuchinke W (2009) Future developments of medical informatics from the viewpoint of networked clinical research. Interoperability and integration. Methods Inf Med 48:45–54

Payne PRO, Embi PJ, Sen CK (2009) Translational informatics: enabling high-throughput research paradigms. Physiol Genomics 39(3):131–140

Prokosch HU, Ganslandt T (2009) Perspectives for medical informatics. Reusing the electronic medical record for clinical research. Methods Inf Med 48:38–44

Ramsey SD, Scoggins JF, Blough DK et al (2009) Sensitivity of administrative claims to identify incident cases of lung cancer: a comparison of 3 health plans. J Manag Care Pharm 15:659–668

Sarkar IN (2010) Biomedical informatics and translational medicine. J Transl Med 8:22

Tan TW, Tong JC, Khan AM et al (2010) Advancing standards for bioinformatics activities: persistence, reproducibility, disambiguation and Minimum Information About a Bioinformatics investigation (MIABi). BMC Genomics 11(Suppl 4):S27

Taylor CF, Field D, Sansone S-A et al (2008) Promoting coherent minimum reporting guidelines for biological and biomedical investigations: the MIBBI project. Nat Biotechnol 26:889–896

Te FC, Chunyan W, Zhiyuan S (2012) The impact of development of population-based study in the biomedical field on laws and regulations: a cross-strait experience on biobank development. J Int Bioethique 23(143–163):181–183

Wang RY (1998) A product perspective on total data quality management. Commun ACM 41:58–65

Warzel DB, Andonaydis C, McCurry B, et al (2003) Common data element (CDE) management and deployment in clinical trials. AMIA annual symposium proceedings, p 1048

Weber GM, Murphy SN, McMurry AJ et al (2009) The Shared Health Research Information Network (SHRINE): a prototype federated query tool for clinical data repositories. J Am Med Inform Assoc 16:624–630

Weng C, Tu SW, Sim I, Richesson R (2010) Formal representation of eligibility criteria: a literature review. J Biomed Inform 43:451–467

Zhan C, Elixhauser A, Richards CL Jr et al (2009) Identification of hospital-acquired catheter-associated urinary tract infections from Medicare claims: sensitivity and positive predictive value. Med Care 47:364–369

Chapter 18
Human Factors and Ergonomics in Medical Informatics

H. Chaudet, F. Anceaux, M.C. Beuscart, S. Pelayo, and L. Pellegrin

Abstract Digital tools cannot be separated from their users, the activities for which they are built, and their utilisation context. They require from professionals a learning of how to work with systems that are more or less easy to use, and a modification, sometimes complete, of their practices and organisations. It is why this chapter introduces a human factor approach of medical informatics, and specifically a work analysis approach, which is a methodological approach aiming to collect essential data in order to describe the psychological, physical, social, technical and economic conditions within which an operator performs a set of tasks or activities that constituted his work.

After presenting the overall context of computerisation pressure associated with the evolution of health work systems and the possible risks and stakes it brings, this chapter describes the concepts of ergonomics and human factors for medical informatics projects. It specifically focuses on work situation analysis, user-centered design with a practical example concerning the implementation of a CPO system, and concludes with a broader approach opening the concept of medical socio-technical system.

Keywords Human factors • Ergonomics • Health work system • Work situation analysis • Usability • User-centered design • Complex socio-technical systems

H. Chaudet (✉)
Faculte de Medecine, UMR 912 – SESSTIM – INSERM/IRD/Aix-Marseille Université, 27, Bd Jean Moulin, Marseille, cedex 5 13385, France
e-mail: herve.chaudet@univ-amu.fr

F. Anceaux
LAMIH-PERCOTEC, UMR CNRS 8201 & UVHC

M.C. Beuscart • S. Pelayo
Evalab/CIC IT, CHRU de Lille & Université de Lille 2

L. Pellegrin
IRBA (Institut de Recherche Biomédicale des Armées)

A. Venot et al. (eds.), *Medical Informatics, e-Health*, Health Informatics,
DOI 10.1007/978-2-8178-0478-1_18, © Springer-Verlag France 2014

After reading this chapter, you should have:

- An understanding of what human factors and ergonomics cover, and their relationships with medical informatics.
- A primary knowledge of the ergonomic and human factors approach during the conception of a health-oriented man–machine system.
- A primary knowledge of the analysis of work situations.
- Knowledge of what is a user-centered design and how to evaluate the usability of a computer system.
- An understanding of what is a complex socio-technical system.

18.1 Place of Human Factors and Ergonomics in Medical Informatics Projects

18.1.1 Computerisation Pressure and Evolution of Health Work System

Medical practice relies on the creation, exchange and use of patient related information. It makes sense that, in a trend to increase the technical level of patient management, informatics has found a prime position. This position has become so preponderant that it is possible to consider health computer systems as omnipresent actors of the patient management process. These systems are manifest in the case of health record systems, or hidden in embedded system controlling technical devices.

In all cases, they act within a man–machine society, where their primary mission is to help man during his information processing needs. As a consequence, human frequently offloads to a large extend his duty onto these systems once the trust in the collaboration is gained. It is the reason why, beyond their technical capabilities, digital tools cannot be separated from their users, the activities for which they are built, and their utilisation context.

We will begin by defining what is a work system.

A work system involves a combination of persons and devices within a given place and environment, associated to interactions between these components within a work organisation. In the field of healthcare, work systems have various conformances, from a minimal structure as a general practitioner office associating the physician and his secretary, until more complex structures as a department, a medical facility, or a regional medical centre gathering several hospitals.

The introduction of digital tools in work systems as new components has *de facto* an impact on usual professional activities. It requires from professionals a learning of how to work with systems that are more or less easy to use, and a modification, sometimes complete, of their practices and organisations. Resulting changes will focus not only on the individuals and their daily tasks but also on the task sharing within the team, sometimes reshaping everybody's roles and functions. At a wider level, organisations, as hospitals, may also be affected by the introduction of an integrated information system.

Following the current observations, the development of information and communication technologies had a major impact in all health sectors, including the patient management. Computerisation reaches step-by-step the core of the medical activity and the health care with, as consequences, a deep transformation of medical practices linked to this technological evolution/revolution.

For example, the introduction of healthcare professional software packages that integrate all patient-related information (patient and family medical history, testing results, treatments, appointments. . .) has deeply modified the traditional physician office workplace.

This software has consequences on the consultation ecology and organisation, including the physician's relationships with his patient and the individual or institutional professional collaborations due to the introduction of emailing tools. The building of the patient's clinical picture is helped by the software functionalities: the physician will gain more quickly access to medical history, encounter and testing history, advices. . . The electronic patient record completeness allows also a focusing on the questions to ask to the patient, improving the diagnosis and the case management during the consultation.

On the other hand, learning difficulties with some software programs, failures during the use of "non intuitive" interfaces or the implementation of inadequate procedures, breakdowns and network access problems, or any other unfortunate use scenario, could be an hindrance to the effective consultation conduct and interfere with the interactions between the physician and his patient. In the same way, the availability of health consumer dedicated web sites on the Internet has modified the traditional relationships between the physician and his patient, allowing this last one to be more involved in the interaction by challenging his physician's knowledge with the information he fetches.

Many examples, coming from organisations such as public or private hospitals and published in medical informatics, show the practice or organisational changes following a department computerisation. All these systems (electronic patient record or health prescription, clinical decision support systems. . .) are related to diverse medical activities, carried out by health professionals with different areas of expertise and skills, and working together.

18.1.2   Caregivers' and Patients' Risks and Stakes Bring by This Evolution

Patient health may be seen as a process or, more generally, as a complex and dynamical environment. As defined by searchers in Cognitive Ergonomics or in Naturalistic Decision Making, actions of caregivers in such dynamical environment must bring on those environments' own dynamic/evolution. Medical activities can then be classified in a broad class of situations in which the operator control on process evolution is incomplete and widely uncertain.

An Example: The Implementation of an Emergency Department Information System in a UK Hospital

An Emergency Department Information System was implemented with the aim of ensuring the complete virtualisation of patient registration and tracking in one of the busiest emergency department in the UK (Vezyridis et al. 2011). This system, which is also interfaced with the hospital main systems, supports most of the daily activities of emergency physicians and nurses for admission and discharge, triage details, clinical notes, prescriptions, consultations and patient tracking in the department or in the hospital. It also helps the medical team to share patient data and records.

After 4 years of current use by the staff members, studies were conducted mainly on nurses for examining the impact of this ED information system on some aspects of their practice, as managing information and coordinating activities with classical supports as whiteboards, paper records or clinical notes. The purpose was to identify if the system had fully abolished or not the traditional paper supports in nursing practice.

The results of the qualitative interviews indicate that the system had effectively a major impact on the organisational and individual practices.

Some of the functionalities proposed by the system have been particularly well accepted by nurses, as the real-time tracking of patients and the digital display of calculated waiting times in place of the traditional whiteboard. In particular, the system gives them a clear view of the patient flows in the department, improving the resources management. In the same way, the system provides a sharing of patients' data to nurses and physicians, as medical records, current location or treatment progress, facilitating the organisational readjustments of activities and redeployment of staff depending on the situation requirements.

By contrast, bed patient-centered activities remained paper-based. Nurses and physicians still used paper forms for documenting observations, prescriptions for medication, and requests for radiological or biological examinations. The justifications they gave were first to preserve a link with the patient by spending more time with him and not to record data in a classical workstation that is located far away the patients' bed.

In this example, the ED information systems had a major impact on organisational practices particularly providing new services to a medical team. It improves the collective patients tracking and management. At the

> **An Example: The Implementation of an Emergency Department Information System in a UK Hospital** (continued)
> same time, some of its physical aspects, particularly those that attempt to replace paper records during nurses' encounters with patients, have not been welcomed and were gradually abandoned.

Since about 40 years, numerous works have studied this class of situation, called "dynamical environments", and some of them have particularly studied the consequences of computerisation and automation of part of the processes on operators' activities.

> **Health Patient Seen as a Process or a Dynamical Environment**
>
> For type 2 diabetic patients, blood sugar level has a proper evolution regulated by a complex process that can go wrong. As breakdowns can be different for each patient, it is necessary to evaluate the interactions between treatments and the proper evolution of the blood sugar level in order to determine the adequate dosage.

Bainbridge (1983), in "Ironies of Automation", suggests that "the increased interest in human factors among engineers reflects the irony that the more advanced a control system is, so the more crucial may be the contribution of the human operator" (pp. 775). Two important changes appeared from this report during systems design: on one hand the taking in account of workers' activities and, on the other hand, their integration in the regulation/control loops which allow them to keep up supervisory function.

The main stake is to build up systems that can be used in natural settings, without major perturbations in the course of the health care activities. Therefore, when such computerized systems are implemented, their power and complexity implicitly lead to systemic risks, which can have important impacts on patient safety. The relinquishment of the "manual" approach during the computerization of paper support, the automation of the pharmaceutical dispensation or with telesurgery, can easily introduce a lack of "good sense" and an over confidence in the systems. Those effects are especially important because the use of such systems is generally linked with the impossibility to verify the enormous amount of data processed and the "invisibility" of the data exchanges.

Moreover, because of the complexity of such systems, an elementary modification of the patients' healthcare process could have significant consequences due to the possible ensuing changes in the system and to the difficulty to predict and anticipate the algorithms' behaviours. IT implementation within complex activities systematically leads to well documented risks' exposure:

- Because some activities are delegated to systems a loss of skills/expertise may appear. A human operator can develop new strategies, for example in order to cope with unusual situations, only by his practice and the efficiency feedback. But he can do that only if he already possesses an adequate knowledge about the functioning of the system under control that is frequently used in order to be accessible when it is required.
- Effects of inadequate trust and complacency in systems have been often reported. Complacency appears when professionals, even experts, adopt the propositions of the system, even when they know its limitations and would not themselves adopt them. In addition to a common over confidence phenomenon, that could lead to dramatic errors. On the opposite, under confidence could urge operators to refuse the use of a system that could help them.

Healthcare activities have also an increasing cooperative component, particularly because of the professionals' specialization. Besides the knowledge distribution between the diverse operators, information is spread out within the different used artefacts, which are paper documents, medical devices or computers. It results that healthcare process may be considered as a distributed cognitive situation between humans and artefacts. It is the reason why many works have studied the impact of computerized or automated systems implementation on the communications that underlie the cooperatives activities of healthcare workers.

The systemic risks described above, in association with system implementations carried out without taking into account the professionals' usage characteristics, may lead to different levels of hazards:

- Hazards in practice, because interfaces could be unadapted to users, and may not supply the relevant information at the expected moment and mislead. When IT systems are designed from a prescribed/theoretical task (see farther) and not from real activity, it may result a loss of adaptability or a lack of resilience (see insert below) as far as they do not allow human to implement adaptive capacities. An increase of mental load may also occur: firstly when interfaces show too much information at the same time, overloading the human work memory; secondly when the operator needs to understand the system functioning in addition to his "normal" activity. Last and most importantly in medical informatics, where most of medical devices are designed to help information gathering, interfaces may lead to the building of an insufficient situation awareness (see insert below). Situation awareness comes from a permanent diagnose of a situation in constant evolution. If the gathered information is too poor (or too important), not integrated in a global understanding or inadequate for anticipating and predicting the process evolution (here, patient health process), it may lead to dramatic consequences.

Some Definitions

Situation awareness (SA) is an internal model of the environment, which allows action selection and implementation. Its construction is a key stage in the decision process, because even experts' decisions can be wrong if the SA is inadequate or incomplete. Endsley's model (1995) illustrates three stages or steps during SA construction: perception, comprehension, and projection.

The resilience of a system is the system's ability to recover from a catastrophic failure or a stress. From a cognitive ergonomics point of view, it can be considered using three temporal spans: before, during and after the event occurrence. In complex situations, as healthcare, human factors are the main source of resilience of the system.

- Organisational hazards, because the provided system can dismantle the services organisation, as the distribution of the various tasks and activities between healthcare professionals is generally dependent on these systems. For example, the case of the prescription-dispensation-administration circuit (detailed farther) shows that, if CPOE implementation improves the prevention of medication errors, it may entail a decrease of communications quality, hinder the cooperation between nurses and doctors or block the synchronization and feedback mechanisms.

However, the introduction of such systems may bring no negative effects if they are well designed, i.e. if they are thought up within the framework of an organisation allowing cooperative activities implementation. On the other hand, if they are designed to substitute the functional communications elaborated during the professionals' practice, they can bring the blocking described above. These organisational hazards can lead to bad perceptions and representations of the systems, generally seen as the causes of work conditions degradation and the sources of psychosocial conflicts.

- Hazards for the patient himself, because IT medical devices are not harmless and have to be particularly reliable as they ensure safety during patient health care. When the system is not well adapted to users and usages, real accidents can be caused by the consecutive negative consequences in the healthcare process. A study proposed in *Pediatrics* (Han et al. 2005) reports the increase of the children's mortality in a department, following the implementation of a patient record's integrated management software. This software was previously designed for adult patients and was unmodified to fit the pediatric unit.

Fundamental qualities are then required for these systems. Among them, their *utility* and *usability* (see definitions in § 19.2.2) will constrain their *acceptability*, and consequently the collaboration level that professionals can reach with them and their long-term use. These three concepts are very important in that they define the scope in which needs of health care professionals will be met by adequate support of these systems to their usual activities.

To determine the best place and role of these computerized « actors » needs to identify the end users "real" needs. This upstream stage relies on a detailed understanding of tasks and activity course. Work ergonomic analysis is a methodological approach aiming to collect essential data in order to describe the psychological, physical, social, technical and economic conditions within which an operator performs a set of tasks or activities that constituted his work. Thus, the approach adopted in this chapter is the Ergonomics' one. The International Ergonomics Association (IEA) gives the following definition of Ergonomics " ... *is the scientific discipline concerned with the understanding of the interactions among humans and other elements of a system, and the profession that applies theoretical principles, data and methods to design in order to optimize human well being and overall system"*. More precisely, we will focus in this chapter on user's point of view and on their usage of those devices.

18.2 Ergonomics and Human Factors Concepts and Methods for Medical Informatics Projects. Example of the Medication Circuit

18.2.1 Work Situation Analysis

Broadly speaking, work analysis is an approach used in ergonomics (and in occupational psychology) for the study of all activities effectively carried out by professionals in their work situation (Fig. 18.1), with the aim of improving the work conditions in adapting them to professionals. In addition, work analysis is also used in other fields, as engineering or computer sciences, for the design of artefacts or computer systems that allow or assist the execution of specific tasks.

Ergonomics pays particular attention to a set of key elements involved during work:

- The operator, i.e. the professional carrying out his duty.
- The task, i.e. what must be done, an aim to reach within given conditions.
- The activity, i.e. what the operator effectively does for fulfilling his task.
- The work context of the operator, his activity and his task.

All these elements are closely linked: the analyse of a work situation is also the description of the relations existing between the operators, their tasks and the activities they carry out for their achievement, the work context related to the situation's scope and particular conditions. A work situation analyse is made up of the identification and the description of these different elements, with a particular

Fig. 18.1 General structure of the ergonomics procedure (work analysis) (From Guérin et al. 2007)

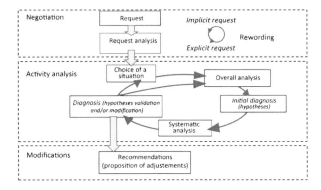

focus on the components that determine the operator's activities and their possible consequences. The activity is contingent on both the task ("external" determinants) and the characteristics of the operators ("internal" determinants), which operate not as simple "automata" that mechanically carry up what they have to do, but make the tasks and the context their own in adapting them. They rely on their knowledge, experience and judgement for setting up the strategies allowing the accomplishment of these tasks.

Ergonomists make a clear distinction between the prescribed task (the definition by the organisation of what the operator must do) and the effective task (what is really done). More precisely, the prescribed task (or prescribed work) covers all that defines in the work organisation everybody's role within his work environment, and then exposes the official way for doing the work. Its description is set up by the orderer, which is the professional organisation hiring the operator.

This description may consider:

- The goals to be met in return of the wages.
- The way of reaching them, the instructions and procedures to follow.
- The technical resources provided.
- The task distribution among the operators.
- The relevant temporal conditions (timetable, time length).
- The social conditions (skills, wages).
- The work physical environment.

The prescribed task description plays a key role upon which the quality and the easiness of the task achievement strongly rely. This description is more or less precise, following an employer's specific format. Job (or post) descriptions are the most usual examples of prescribed task statements for given operator profiles. The nurses' prescribed task that is usually stated in the Countries' Health Regulations and Laws is another example. It is observed that a same task may have several possible descriptions depending on its writer (an engineer has a different view from a human resources manager) or the chosen description format. The precision of this description has a strong influence on the personnel's future activity. Too normative, some instructions may be too heavy to be implemented, making them difficult to be carried out. In opposition, some instructions may be broad or fragmented, allowing the operator the freedom of defining the performance conditions of his task (imprecise prescriptions). Moreover, this does not include the part of implicit matter in the drafting, because all performance conditions are inevitably not specified. As a result, the activity carried out by the operator will attempt to respond to the prescribed task requirements in adapting it to the constraints brought by the work reality. We should then make the distinction between the prescribed task (what is expected) and the effective task (the task the operator effectively achieves), which is the part of the prescribed task's goals and performance conditions that is effectively taken in account and carried out (Fig. 18.2).

Analysing the operator activity must allow understanding how he reaches his goal (prescribed task), how he imagines it, transforms it, adapts it for building his effective task, and, if he fails, what are the reasons. *The activity must be delineated by the work context, the prescribed task, the effective task, and the operator's characteristics.* For building a realistic view of the diverse solutions chosen for achieving the prescribed tasks, it is important during the analysis of the work situation to describe all its components, specifically including the joint analysis of the effective tasks with the operators's activities. The analysis must provide answers to the following questions:

- Who? Who are the operators? Their characteristics? If the objective is to introduce a computer system, who is it intended for?
- Why? What are the goals of the tasks to achieve? What are the objectives of the work situation the new system must cater for?
- What? What is carried out, what tasks and subtask decompositions must be performed? What constraints? In this context, where is the application located?
- How? How the operators really accomplish their activities for achieving their tasks and reaching their goals? How the system backs up its users' work?

The methods that can be implemented for this analysis are varied and complementary (Stanton et al. 2005). They aim at collecting the best possible data for building a model for the activities and tasks. For example, we present three usual methods:

- Verbalisation recordings of discussions, individual or collective free interviews, conducted by an ergonomist during or after the targeted activity, allow the collecting of information about representations, interpretations and expectations, ways of handling the operators' tasks, and encountered difficulties.
- Observations, especially using recording, give access to very detailed data about operators interacting with the system. The operators can be observed during their real use of the system if already in place, or during their work with the traditional support (or another system) before the introduction of the planned new computer system. Whenever possible, these observations are performed in situation, allowing the acquisition of "ecological" data about the activities. Another possibility is to put the operators in simulated situations with laboratories specialised in ergonomic studies each time a real situation is not feasible,

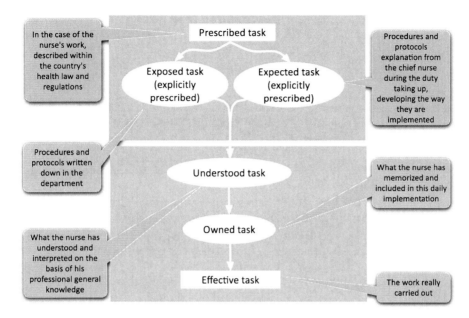

Fig. 18.2 Simplified and annotated schema describing the different kind of tasks (Adapted from Leplat 1993a, b, http://fr.wikipedia.org/wiki/Jacques_Leplat)

or for more extensive studies about tasks that have been specifically identified during previous field observations.

- Questionnaires, usually standardised and validated, are more directive and allow the obtention of structured data about specific subjects or targeted tasks.

These "ergonomic" (or human factors) approaches are in a process of growing presence in medical informatics publications, like the JAMIA (Journal of the American Medical Informatics Association) or the IJMI (International Journal of Medical Informatics), disseminating studies on medical activities assisted by complex computer systems with the objective of building or improving medical numeric tools. For instance, the OLD@HOME project (Hägglund et al. 2010), which offers the sharing of a virtual health record (VHR) between professionals in a healthcare network, has been evaluated on its ability to provide the expected services, especially a differentiated access to relevant information depending on the user profile (general practitioneers, nurses). This study aimed at identifying different patient management scenarios according to each professional's specific activities, with the objective of a better system adaptation to the users' needs. For this purpose, it relied upon the study of the diverse work situations for each user profile, using the tools described above (workgroups, questionnaires, interviews and observations).

18.2.2 User-Centered Design, Usability of Software and Graphical User Interfaces

Bad usability can lead to use errors reducing the expected benefits of a system, even causing harm to the patient. Nielsen (1993), a well-known usability expert, describes a model stressing the importance of usability for the acceptability, and then, the success of a system. He puts the usability within an acceptability concept integrating the practical and the social dimensions. Then, only a good consideration of usability aspects may guarantee the users support. That means usability should be optimized to reduce fatal medical errors related to the bad use of a system, but also to avoid a product that will not be used (or little).

Usability is defined by ISO as "*the extent to which a product can be used by specified users to achieve specified goals with effectiveness, efficiency, and satisfaction in a specified context of use*" (2010). When considering usability, it is therefore important to also consider its usefulness, i.e. its relevance for the intended use. A system may not be used because it is useless, even if it meets all the usability features. A system will be used if it corresponds to the user's needs, their work habits and if users can easily integrate it in their various work organisations and existing procedures. The usability of a system depends also on the system itself but also on its context of use.

The most effective way to ensure a good usability of a system is to adopt for its design the ISO 9241–210:2010 "*Ergonomics of human-system interaction – Part 210: Human-centred design for interactive systems*". This standard describes a design supported by the consideration of the users' needs and characteristics, this at each step of the design and development process. The main goal is to adapt the system to the users rather than impose them modalities of use chosen by designers. The design is therefore users' needs driven rather than technology driven. This user-centered design supposes rigorous methods to collect data about users' tasks and needs in order to transform them in design choices, i.e. ask the users what they want is not sufficient. The integration of the usability methods should be done early in the design process (i.e. at the beginning of the project) and should be iterative (i.e. perform throughout the key steps of the project).

The methods of work analysis support the intertwinement between the task requirements, the environment constraints and the users' resources. The use cases from which the usability evaluation could be performed are elaborated from these work analysis data. As soon as mock-ups or prototypes are available, or pilot sites installed, an iterative step of usability evaluation could begin. Usability flaws are identified and documented, with suggestions to fix them. Only when the new system fits the users' needs, it may go to the market. The three recommended approaches for the usability evaluation of the Graphical User Interfaces are inspection-based, user-based and model-based evaluations. The two first are the most used by usability experts and are largely documented. The model-based approaches are considered limited or immature, expensive to apply and their use is largely restricted to research teams (Cockton et al. 2003).

Usability evaluations can be performed in more or less realistic environments from lab studies, ensuring a good internal validity, to field studies, in a real work environment with real users giving also a better ecological validity.

- The inspection-based evaluation (i.e. heuristic evaluation) is performed in a lab without real users. Evaluators assess the different elements of the dialogue with the interface relying on a set of usability criteria based on standards or established guidelines. The most famous set of criteria is the list of Nielsen (1993) composed of ten criteria (visibility of system status; match between system and the real world; user control and freedom; consistency and standards; error prevention; recognition rather than recall; flexibility and efficiency of use; aesthetic and minimalist design; help users recognize, diagnose, and recover from errors; help and documentation). In France, the most commonly used are those of Scapin and Bastien (1997), a couple of computer scientist and ergonomist researchers. From a synthesis of 900 usability recommendations in the domain of computer, they elaborated a list of 18 criteria structured in 8 dimensions (See Table 18.1). From a practical point of view, several Human Factors/usability specialists independently assess the users' interfaces relying on the criteria. The results describe usability problems in their context, with the consequences on the use of the system and, eventually, the usability suggestions/recommendations for the re-engineering. A severity rate is given to each detected usability problem to prioritize the modifications that need to be performed.
- The user-based evaluation (i.e. the usability testing) is generally performed with real users in a simulated controlled situation, the closest to the actual use. The usability specialist gives to the user orders to perform his or her representative tasks. Within a usability lab, all the users' behaviours, verbalizations and interactions with the system are recorded. These tests suppose the definition of the observation boundary: which functions should be tested? In which context of use? The usability specialist notes the durations, the errors, the misunderstanding, and all events illustrating a difficulty of use. At the end of the test, a debriefing is performed with the user to understand the causes of the problems and a satisfaction questionnaire is given to the user. Usability testing allows the detection of problems of use not detectable with others methods. This method is therefore very informative. But the application of usability testing raises ecological validity issues, especially with complex applications such as Electronic Healthcare Record (EHR) or Computerized Physician Order Entry (CPOE). For instance, the parameterization of a product is very important for its usability. An application installed for a lab is always parameterized by default, also not adapted to the real work situations. Furthermore an application evaluated in a lab cannot be connected to the Clinical Information System (CIS), which hampers the patient data retrieval.
- Another option consists in installing in work situations already functioning with the system a portable lab for the record of the real work (verbalizations and interactions with the system). During this usability testing parameters are

Table 18.1 Ergonomic criteria of evaluation from Scapin and Bastien (1997)

1. Guidance	2. Workload	5. Error management
1.1. Prompting	2.1. Brevity	5.1. Error protection
1.2. Grouping/Distinction of items	2.2.1. Concision	5.2. Quality of error messages
1.2.1. By location	2.1.2. Minimal actions	5.3. Error correction
1.2.2. By format	2.2. Information density	
1.3. Immediate feedback		
1.4. Legibility	**3. Explicit control**	**6. Consistency**
	3.1. Explicit user action	
	3.2. User control	
	4. Adaptability	**7. Significance of codes**
	4.1. Flexibility	**8. Compatibility**
	4.2. User experience	

controlled *a posteriori*. Observations and records should represent all the targeted functions. The duration between the implementation of the system and the usability testing defines the type of usability problems likely to appear. With a new system, usability testing should reveal intuitiveness problems ("ease of learning"). With a system implemented for a long time, usability testing reveals powerful problems (ease of use). The best solution is to sample the test periods according to this variable or to perform a longitudinal study.

18.2.3   Example

18.2.3.1   Risks Related to the Implementation of a CPOE

Applications such as CPOE systems are supposed to ensure patient safety and are considered a progress as far as quality and safety are concerned. However, they are still few hospitals functioning with a medication use process completely computerized, this whatever the country. Moreover, when a part of the work procedures are computerized, problems of use of the systems are observed. For instance, some studies stress doctors' resistance to document their orders, while the others actors (nurses and pharmacists) seem to be more enthusiastic. Doctors doubt the effectiveness of this type of system and perceive an increase of workload. A CPOE can change the distribution of tasks, especially between doctors and nurses. Doctors are constrained to enter their orders themselves instead of dictate them to the nurses, this at a very detailed level. Nurses are no longer in charge of ordering the drugs to the pharmacy, the system taking in charge this function.

18.2.3.2   Context of the Study and Methods

The example of a project of the usability evaluation of a CPOE will illustrate the purpose (Beuscart-Zéphir et al. 2005). The application DxC@re® from the Medasys Company was evaluated. DxC@re®, at the time of the study, combined the functions of a CPOE and of an EHR.

Standards methods of the user-centered design have been performed:

- A work situation analysis of the medication use process, i.e. ordering-administration of drugs, in three hospitals (one functioning with DxC@re® for some months and the two others running a paper-based support) corresponding to seven departments. Semi-structured interviews and observations (more than 500 h) were performed. The observations were focused on (i) the interactions between the healthcare professionals and the supports and (ii) between professionals.
- A usability evaluation of DxC@re® was performed: a usability inspection by three usability specialists relying on the Scapin and Bastien criteria (1997) and usability testing on site in the computerized hospital (16 h of records corresponding to five medication process cycles) with portable labs. All the interactions between the professionals and the system were recorded with the concomitant verbalizations.

18.2.3.3   Work Situation Analysis

The typical organisation observed in paper-documented situations may be described as follows. Usually, the nurse accompanies the physician on his medical rounds. The nurse is the one who handles the patient's record. For each patient, she summarizes, mostly from memory, the patient's case (current treatments, orders or complementary exams pending, physiological and behavioural reactions of the patient). She checks the patient's record or hands the proper document to the physician if he asks for details. The physician may ask the patient some questions, and he eventually performs a clinical exam. He then makes a decision about each item of the patient's therapy (continue, modify, stop, start a new drug) and dictates it to the nurse. The nurse may suggest some modifications, ask some questions about drug interactions or raise some points of difficulty concerning the drug administration (e.g. the patient has difficulties swallowing pills, the patient complained about stomach pains following administration of the drug…). Most often, the nurse writes the order on the order sheet, sometimes including some specific note or reminder for other nurses (mostly about drug administration). Sometimes, the physician signs his orders; even more rarely, he writes the orders himself. After the medical round is completed, the nurse copies or transcribes all the orders in her medication administration record (MAR). Medical students, house officers, and head-nurses may accompany the physician on his round in the academic hospital. In such cases, medical students are the one summarizing the patients' cases.

In the computer-based situation, the organisation is different. The nurse does not accompany the physician during his medical round. Laptops are installed on nursing carts, thus making order entry possible at the point of care. The physician makes his medical rounds usually alone, and wheels the computer to each patient's room. He gets the necessary information about the patient's current status from the PCIS, and documents the entire medication order including time for drug administration. Most of the drugs commonly used in the department are included in order sets, thus facilitating medication order entry. As soon as the physician validates an order, the nurse gets the data on her care plan and medication administration record. Interviews suggested that the doctors are not interested in, and feel no commitment to the exact planning of timing of drug administration, which they ordinarily leave to the nurses. When the order seems unclear or difficult to administer, she may ask the physician, if he is available.

18.2.3.4 Usability Assessment

The heuristic evaluation has identified several problems. For illustration, DxCare® requires two main screens to enter medication orders. The physician selects the proper drug using the first one, and specifies the dosage with the second one. The Fig. 18.3 shows some examples of usability problems of the second screen, mainly guiding problems. The physician may enter a global schedule (e.g. morning – noon – evening) using a list accessible in the upper box. In this case, the system automatically updates the planning table below and sets the specific timing for the administration depending on the organisation of the ward (i.e. 7, 12 am, 6 pm). The physician can also fill in the table directly by clicking in the desired time slot and documenting the dose to be delivered at that time.

The usability testing has confirmed the usability inspection statement. The physicians hardly use the list of pre-set drug administration schedules. They use mainly the timetable to document the planning of medication administration. The physicians generally resent the obligation to enter the exact drug administration timing for each and every order. Moreover, they usually have no idea of the nurses' workflow and the usual times of drug distribution rounds on the ward. When asked to explain the significance of the grey time-slots in the timetable, not a single physician could come up with the correct explanation.

During one of the tests, one physician systematically chose the "wrong" time-slots for the drug administration, setting for example all the "morning" orders at 9 am, while the medication administration morning round in the department usually took place between 6.30 and 7.30 am. On her end of the application, the nurse reading her automatically updated MAR had no way of tracking the physician's intentions with the time schedule. When faced with this "bizarre" planning, the nurse in charge assumed that the physician's intention was a standard morning schedule, which fortunately was the case here. The assumption was that the data in

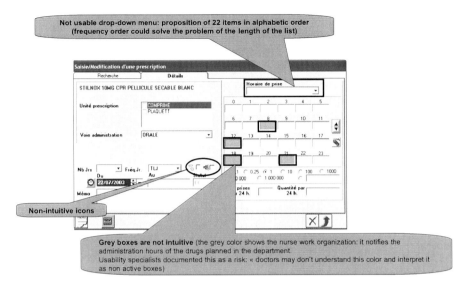

Fig. 18.3 Examples of usability problems detected for the screen of specification of dosage (DxC@re®, Medasys)

her MAR were the results of a "wrong" entry from the physician. She did not ask him any questions, but simply changed all the timings on her own planning in order to fit the usual organisation of the ward, i.e. 8 am.

18.2.3.5   Interpretation of the Results

Except the usability problems for which a reengineering was undertaken, the application did not support effectively the exchanges between physicians and nurses. This led to risks of medical errors. The application transformed the professionals' tasks, the distribution of roles and the organisation of the collective work. But these changes were not only provoked by the application itself. Of course, the implementation of a new system has consequences on the work situations. But the observed impacts are mainly explained by the work organisation.

In case of implementing a system "taylorizing" the tasks, i.e. regarding only the prescribed task, in a work situation inherently collaborative, the tasks are often individualized. For instance, to be compliant with the regulation (traceability for the ordering task for which the doctors are in charge, not the nurses), when a CPOE is introduced, the physicians are required to enter their orders electronically. Progressively the nurse finds him or herself less involved and withdraws from the common rounds, arguing there is no more time for this activity with the new system. A few months after the installation a shift from common rounds organisation to a medical round with doctors alone is completed. Consequently nurses have no more access to doctors ordering task. New errors may also appear because nurses may lack information. For instance, nurses may bad interpret an order as presented in the example of the time of administration. If nurses leave the ordering loop, they may become only "executors" without a global view on the medication process, while they were with the common rounds at each step of the process. In such an organisation of work, the control is made by the CPOE system and the different professionals have only a truncated vision of the process. This cooperation break may lead to risks of errors.

A study addresses the question of the respective impact of CPOE and of organisational factors on nurse-doctor tasks surrounding the medication prescription and administration process (Pelayo et al. 2012). Results show that the organisational factors have a greater impact than a CPOE system on doctor-nurse tasks. If proper attention is paid to organisational determinants of cooperation to keep direct exchanges between doctors and nurses, for instance during common rounds, a mutual understanding is observed which ensure the reliability of the work situation. The study demonstrates that the technical system has no significant impact on the cooperative activities within a given organisational type.

In conclusion, if proper attention is paid to organisational determinants of cooperative activities, it is possible to design future computer-based workplace systems so that institutionalized exchanges are preserved. For example, we have participated in the implementation of a CPOE in a hospital and we could give recommendations to take into account the particular features of the doctors and nurses cooperative activities. Especially, we stressed the importance of keeping the nurse in the ordering loop and to preserve, for instance, common rounds when they already existed. To do so, enough laptops were installed in the departments so that doctors and nurses can each have their own screen with the relevant information in front of them during the medical rounds. After several months, common rounds were maintained.

18.3 Towards Conception and Evaluation of Complex Socio-Technical Systems in Medicine

The different topics that have been introduced in this chapter illustrate the great importance of human-technology interactions within work organisations, healthcare organisations in our case. In others terms, we are faced to a socio-technical system. A socio-technical system refers to interrelationship between human agents and technical systems (and artefacts) belonging to a social structure and which are engaged in common goals. A socio-technical system has some main components: people, prescribed tasks, tools and technics, organisations, work environment. Socio-technical systems are also described as "complex" ones when the interactions and interrelations imply a socially complex organisation that arrange human agents in elaborated work units (production unit, experts team...) and advanced technological systems. These human organisations and technological tools conjointly manage and supervise processes inside global complex structures such as hospitals, healthcare networks, transportation, industrial production, nuclear power plants...). A complex socio-technical system is then based upon strong interactions between human, social, technical and organisational components, which involve adequate resources, high skilled competences, adapted procedures and decision protocols.

Many factors take part in the complexity of the socio-technical systems such as the social organisation and heterogeneity of the agents, and even their geographical distribution. In the case of healthcare systems, they include agents with various roles and competences, coming from different social origins (patients and their families, administrative and technical staff, healthcare providers and medical staff), but having common goals and activities to ensure that the full system correctly runs. These differences, between persons or groups, could lead to conflicts of values and interests which have to be managed in many current healthcare work situation and sometimes in more critical ones. For example, a team in a cardiology service could build specific activities related to patient management, resulting from the specialty requirements, the organisational constraints and the individual expertise, which could be different from those developed in the geriatric service of the same structure or from another hospital cardiology service. In this context, an efficient communication between the agents or social groups is basically necessary to coordinate actions, and it gets to be very difficult in large organisations such as hospitals or healthcare networks. In these social structures, the necessary coordination of agents' activities is further complicated by the fact that people are located in different spaces of work and are supported by sophisticated information and technological systems. We can also consider a medical team as a socio-technical system because its members have to reach common goals, shared the available resources including systems and software, and interact during their common activities. These interactions could be observable as:

- Events joining together healthcare practitioners for collaborative activities such as mutual control and adjustments, interferences between individual goals and actions, delegation and allocation of tasks.
- Actions resulting from individual cognitive activities of team members, such as information management, expert medical diagnosis, situation diagnosis, problem solving and decision-making, in the goal of building a shared representation of the situation by team members.

From the technological "point of view", the growing automation of socio-technical systems has increased the number and the complexity of tasks allocated to computerized agents. As a consequence, human-machine cooperation became as important as the cooperation between the humans in work. In this context, artefacts and technological systems can be seen as "windows" through which human agents access to components of the full complex socio-technical system, providing a mediated representation of the work domain. For example during the supervision of nuclear power plants in control room, the operators cannot directly observe information and data. They have to use elements of the technological system (monitors, scopes, computer interfaces, alarms), to understand and supervise the situation, increasing levels of interpretation. Moreover, complex socio-technical systems are also dealing with situations that could be changing, uncertain and hazardous ones. So the human-system component has to be robust, adaptive to various kinds of conditions, as for example managing rare or emergent events.

For human factors specialists, the actual challenges are to analyse and improve work situations, taking into account the current evolution of society inducing the growing complexity of work tools and social organisations. Such analysis was initiated 20 years ago and lead to the emergence of the "cognitive systems engineering" approach. Cognitive systems engineering (Rasmussen et al. 1994) covers the study, the design and the evaluation of work situations implied in complex socio-technical systems. The aim of this approach is to fully describe different layers of these systems (environment context, organisational structure, workers, technical and engineering systems) and most significantly, their mutual interactions. Because, even if they appear as independent ones, these layers are strongly coupled. The full composition of the layers results in the ultimate performance of the system. The cognitive system engineering proposes models, methods, technics and tools for a "cognitive work analysis" to guide the conception and evaluation of computer-aided systems interacting with humans at work.

Cognitive Work Analysis

The Cognitive Work Analysis (CWA) is a human factors approach proposing a framework for analysing complex socio-technical systems (Vicente 1999). It has its sources from field studies and researches in supervision and control of complex process in hazardous industries. The core goal of CWA approach is to underline what agents could do for achieving the work purposes and their essential goal-directed actions. It is then necessary to analyse and explicitly describe both specific tasks, procedures and existing surrounding constraints, which determine also the agents' behaviours and performances (Fig. 18.4)

This is an ecological approach of work because it places the human as the major actor of an effective management of the system, thanks to its capacities to build autonomous, adaptive and flexible behaviours to control current (and sometimes critical) situations.

It has to ensure (for systems conception) and to verify (for the systems evaluation), that the human agents have built an accurate representation of the situation so that it fits the best as possible to the current evolution of their environment (technical, social, physical and environmental).

The CWA approach proposes a global framework with five main components of the complex socio-technical systems, which have to be explicitly identified: (a) work domain), (b) control tasks and goals to reach, (c) strategies, (d) social organisation and cooperation, skills and expertise of the operator.

Fig. 18.4 Global architecture of various levels describing a complex socio-technical system, as is described by K. Vincente (1999)

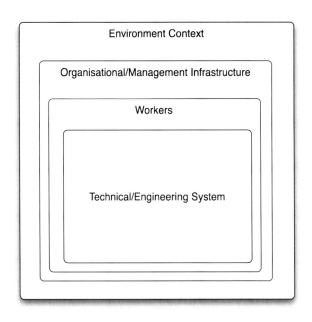

Concerning human-machine system performances, CSE looks beyond the classical human-computer interaction, which conceives agents and computerised systems as separate entities, and argues for as a single entity, people and technology acting together, a "joint cognitive system". Joint Cognitive Systems (JCS) (Hollnagel and Woods 2005) refer to joint human-technology in complex and dynamic work process. Semantically equivalent to the term complex socio-technical systems, it emphasises the nature of coordination, adaptation, control and resilience of goal-oriented human-technology performance in constantly changing contexts. The goal of a JCS is to maintain a control of a situation thanks to its capacity to modify its behaviours on the basis of experience, to resist to disturbances and to sustain task integrity by adjusting its goals and actions. This approach stresses the importance of a joint control and supervision by human and systems in the management of dynamic situations, and it clearly could be applied to healthcare and medical work situations. Controlling a situation is the ability to produce or to obtain specific results during the situation management. Because of the complexity of the situations, control is most of the time partial and, consequently, requires cooperatives actions between agents (humans and systems) and even others JCS. Moreover, controlling a situation is also contextual because it is dependent on changing situation parameters and on skills and knowledge of JCS agents.

For example, a system of real time epidemiological surveillance has been developed for French deployed forces (ASTER) (Pellegrin et al. 2010) in this perspective. ASTER is a complex socio-technical system, which includes a computerised system recording and analysing epidemiological data and a joint social organisation, a surveillance network, which includes field military physicians epidemiologists. Military physicians' goals are locally to identify and to declare patients and epidemiologists have to evaluate and manage outbreaks. These elements of ASTER constitute a co-jointed human-technology–work system that performs collaborative activities related to outbreak alert management. Each network implies specific task allocations between actors and system services, which collectively contribute to common tasks and goals, which are the control of epidemic situations.

In the context of computerised medical systems, the main interest of the cognitive engineering approach is to propose a framework for taking into account all partners (human agents and technical systems. . .) and contextual elements (hospital organisation, patients and families, material constraints. . .) seen as co-actors for the patient management.

The patient management is an expert complex activity, which takes place in a healthcare socio-technical system and includes the evaluation of the patient's current situation, medical diagnosis, decisions of healthcare actions and follow-up which are performed by professionals teams jointed to computerised medical systems.

18.4 For More Information

For deepening the content of this chapter, you may consult the following resources:

- Google Scholar research using the keywords: "medical informatics human factors"
- Beuscart-Zéphir M-C, Elkin P, Pelayo S, Beuscart R (2007) The human factor engineering approach to biomedical informatics projects: state of the art, results, benefits and challenges. Methods Inf Med 46 (Suppl 1):109–27.
- Patel VL, Currie LM (2005) Clinical cognition and biomedical informatics: issues of patient safety. Int J Med Inf 74:869–885.
- Kushniruk AW (2001) Analysis of complex decision-making processes in health care: cognitive approaches to health informatics. J Biomed Inf 34:365–376.

Exercise

Q1

You have to evaluate the usability of an anaesthesiology electronic patient record and to give back re-engineering recommendations before its on-site installation. How do you proceed?

Exercise (continued)
R1

1. You proceed to a work situation analysis within some anaesthesiology departments with the aim of identifying work procedures and users' real needs (both physicians and nurses). With this information you will be able to build the usage scenarios that will allow you to setup the usability evaluations.
2. On the basis of these usage scenarios, you proceed to an ergonomic investigation for detecting the system's ergonomic faults and proposing some re-engineering suggestions.
3. You organise user tests in simulated situation with some system's final users (physician and nurses), and you will then identify usage problems that you have missed with the ergonomic investigation.

References

Bainbridge L (1983) Ironies of automation. Automatica 19:775–779

Beuscart-Zéphir MC, Pelayo S, Anceaux F et al (2005) Impact of CPOE on doctor-nurse cooperation for the medication ordering and administration process. Int J Med Inform 74:629–641

Cockton G, Lavery D, Woolrych A (2003) Inspection-based evaluations. In: Jacko JA, Sears A (eds) The human–computer interaction handbook. Fundamentals, evolving technologies, and emerging applications. Lawrence Earlbaum Associates, New York, pp 1118–1138

Endsley M (1995) Toward a theory of situation awareness in dynamic systems. Hum Factors 37:32–64

Guerin F, Laville A, Daniellou F et al (2007) Understanding and transforming work – the practice of ergonomics. ANACT, Lyon

Hägglund M, Scandurra I, Koch S (2010) Scenarios to capture work processes in shared homecare. From analysis to application. Int J Med Inform 79:e126–e134

Han YY, Carcillo JA, Venkataraman ST, Clark RS et al (2005) Unexpected increased mortality after implementation of a commercially sold Computerized Physician Order Entry system. Paediatrics 116:1506–1512

Hollnagel E, Woods D (2005) Joint Cognitive systems: foundations of cognitive systems engineering. ACM Press Book, Taylor & Francis Group, Boca Raton

International Standards Organisation (2010) Ergonomics of human-system interaction. Part 210: Human-centred design for interactive systems, report no. ISO 9241–210, International Standards Organisation, Geneva

Leplat J (1993a) L'analyse du travail en psychologie ergonomique: Tome 2. Octarès, Toulouse

Leplat J (1993b) L'analyse du travail en psychologie ergonomique: Tome 1. Octarès, Toulouse

Nielsen J (1993) Usability engineering. Morgan Kaufmann, San Franscico

Pelayo S, Anceaux F, Rogalski J, Elkin P, Beuscart-Zephir M-C (2012) A comparison of the impact of CPOE implementation and organisational determinants on doctor-nurse communications and cooperation. Int J Med Inf. pii: S1386-5056(12)00175-X (Epub ahead of print)

Pellegrin L, Gaudin C, Bonnardel N, Chaudet H (2010) Collaborative activities during an outbreak early warning assisted by a decision-supported system (ASTER). Int J Hum Comput Interact 26 (2–3):262–267

Rasmussen J, Pejtersen AM, Goodstein LP (1994) Cognitive systems engineering. Wiley, New York

Scapin DL, Bastien JMC (1997) Ergonomic criteria for evaluating the ergonomic quality of interactive systems. Behav Inf Technol 16:220–231

Stanton NA, Salmon PM, Walker GH et al (2005) Human factors methods: a practical guide for engineering and design. Ashgate, Aldershot

Vezyridis P, Timmons S, Wharrad H (2011) Going paperless at the emergency department: a socio-technical study of an information system for patient tracking. Int J Med Inform 80:455–465

Vicente KJ (1999) Cognitive work analysis: toward safe, productive, and healthy computer-based work. Lawrence Erlbaum Associates, Mahwah

Index

A. Venot et al. (eds.), *Medical Informatics, e-Health*, Health Informatics,
DOI 10.1007/978-2-8178-0478-1, © Springer-Verlag France 2014

Printed by Publishers' Graphics LLC
LMO131204.15.18.90